Engineering Mechanics

Engineering Mechanics

G. Lakshmi Narasaiah
Formerly Professor
Dept. of Aeronautical Engineering
Institute of Aeronautical Engineering
Dundigal, Hyderabad.

A unit of **BSP Books Pvt. Ltd.**
4-4-309/316, Giriraj Lane,
Sultan Bazar, Hyderabad - 500 095

Engineering Mechanics
by G. Lakshmi Narasaiah

Published by:

BSP BS Publications
A unit of **BSP Books Pvt. Ltd.**
4-4-309/316, Giriraj Lane, Sultan Bazar,
Hyderabad – 500 095.
Phone : 040 – 23445688
e-mail:info@bspbooks.net
www.bspbooks.net

ISBN : 978-93-91910-89-1

To my grand children

Vivaan & Sreshta

for enriching my retired life

Preface

Mechanics is a subject dealing with our many experiences in daily life. It is introduced in high school and intermediate education as a part of physics. It is expanded as Engineering Mechanics for meeting the needs of engineering graduates, particularly design engineers in civil, mechanical and aeronautical disciplines.

'See-saw' for kids in play grounds, 'Tug-of-war' in school days, simple balances used in grocery and vegetable shops etc are our daily examples of **coplanar forces** and **equilibrium**. Influence of **centroid** location is seen in every house while suspending a calendar or photo frame vertically. Loss of power due to frictional resistance and need for lubrication to minimise the same in a sewing machine as well as use of friction for stopping a moving vehicle are our daily examples of negative and positive applications of **friction**. Estimates of time and distance while overtaking a vehicle or crossing a road are taught with scientific principles in **kinematics**. Calculation of force required for a particular projectile motion is covered in **kinetics**. Oscillating pendulum of a wall clock is one example of a **dynamic problem**. The list is endless.

All such situations in our daily life are explained through scientific principles and solved by different mathematical approaches. Some commonly used formulae in trigonometry, differential calculus and integral calculus are provided in chapter-1 for a quick review and reference.

Because of the possibility of a large number of problems, which no single book can cover, students often find it tough to do well in this subject. Besides going through the solved problems, they should solve as many problems as possible from different sources. In addition, the following three tips will help many students score well in the examinations –

- Since there is no universal notation for forces, angles etc., it is always essential to draw free body diagrams with these details and form the right link with the examiner
- Most problems are solved from a set of 'n' simultaneous equations in 'n' unknowns. But, the number of equations may not be obvious. Forming equations from specified conditions in problem is essential, with correct interpretation of problem description. Finding adequate number of equations is half the job done
- It is very common in question papers, to find different quantities in different units. They need to be converted into consistent units (km, m, cm; hr, min, sec etc.) before using equations in this subject. Special care should be taken to use angles in degrees or radians in different problems, by setting the calculator in appropriate mode

Syllabus in most universities also includes topics like centroid, area moment of inertia and mass moment of inertia, which are necessary for calculating effects of forces in different situations. Statics portion is covered in this book in five chapters – coplanar concurrent forces, coplanar non-concurrent forces, spatial (3-D) forces, friction and Trusses. Other chapters include kinematics, kinetics, dynamics and work-energy relations. The book covers syllabus of most universities, divided into a number of chapters. All these chapters deal with rigid bodies, without considering deformations and hence this subject is also called ***Rigid body Mechanics***.

Deformations are significant in beams as well as shafts and are considered as deformable structures. They are covered in detail in another subject as '***Mechanics of solids***' or **'Strength of materials'**. Mechanisms are covered in 'Kinematics of machinery'. However, the syllabus of 'Engineering Mechanics' of AICTE mentions beams and mechanisms. Therefore, a brief coverage of beams and a glimpse of mechanisms are also included in this book. Detailed presentation is outside the scope of this subject.

A design engineer needs to understand materials available for a particular application and their limitations. A brief outline is presented in introduction. Detailed presentation is outside the scope of this subject.

With colleges competing for better results, this subject is taught to students with emphasis on scoring marks in examinations and without proper understanding of the basic concepts. With the possibility of a large variety of questions, many such students find it a very difficult subject in engineering course.

With adequate preparation and precautions, students can do well. It is a highly scoring subject, compared to many other subjects. To help students score well, many different solved problems from old question papers and tips for avoiding common mistakes are included in this book.

G. Lakshmi Narasaiah

Ph : 91-9989860244

e-mail: gogineni_ln@yahoo.co.in

Acknowledgements

The problems in this book were collected from many University question papers and solved for students of all branches of Engineering in the first year of Engineering course, over a period of 10 years. There is no unique approach to solving problems in this subject. Some students came out with alternative solutions, which are included in this book.

I am grateful to the managements of the engineering colleges, which chose me for teaching this subject. I am also grateful to the mix of students in each class and in each college – for coming out with alternative solutions and for necessitating different explanations, examples and figures. All of them helped in bringing this book to this shape.

-Author

Contents

Chapter 9

Trusses 225-240

Chapter 10

Beams and Mechanics 241-264

CHAPTER 1

INTRODUCTION

Mechanics is a subject dealing with our many experiences in daily life. It is introduced in high school and intermediate education as a part of physics. Engineering Mechanics is the introductory course for engineers associated with design of any mechanical, civil or aerospace component/structure against failure when subjected to operational loads. It deals with forces acting on bodies and their effects. It is taught in many universities in two parts as –

- Engineering Mechanics or **Mechanics of rigid bodies**, limited to the effects of forces in keeping the body in equilibrium or in producing desired motion and
- Strength of Materials or Solid Mechanics or **Mechanics of deformable solids** considering the deformations of the body such as
  - ➢ axial compression or buckling of columns
  - ➢ bending deformation of beams
  - ➢ angular twist of shafts etc..

The subject is broadly classified as

- Statics – involving calculation of forces & reactions in bodies at rest, even after forces are applied
- Dynamics – involving study of motion of bodies due to forces, covering
  Kinematics – dealing with motion of bodies
  Kinetics – dealing with forces which result in motion of bodies and
  Vibrations – dealing with repetitive or cyclic motion of bodies over an equilibrium position

The subject doesn't use empirical relations and the equations can be checked for matching dimensions on either side. A large variety of problems can be created in this subject. Solution to any problem in Engineering Mechanics involves proper understanding of the physical problem and identifying mathematical equations associated with the specific conditions/situations. Most problems can be solved from a set of 'n' simultaneous equations in 'n' unknowns. They can be solved by many different approaches, using different mathematical (trigonometric) theorems. Hence, if the number of unknown variables and number of equations from the specified conditions match, the problem is half solved.

It is very common in question papers, to find different quantities in different units. The student is expected to use consistent units for all quantities. A brief coverage of units of measurement, arithmetics, trigonometry, differential and integral calculus are added, for a quick reference while solving problems in this subject.

The subject appears to be very well understood, when a student goes through solved problems. **Practice with unsolved problems is essential** for good performance in the examinations. For a better exposure to a large variety of problems, every student is advised to refer to many books and old question papers.

1.1 FORCES

A Force is defined by **Newton's 1st law of motion** as one which changes state of rest or of uniform motion of any physical body. A force differs from other physical properties of a body, by the identifying parameters associated with it.

Every body has properties such as mass, temperature etc., distributed at various points. These properties have magnitudes but are not associated with any specific direction. These are called ***scalars***. They can be added together algebraically.

$$\text{i.e., } \sum m = m_1 + m_2 + m_3 + m_4 + \ldots..$$

A Force such as weight, reaction, centrifugal force etc.. has magnitude as well as direction associated with it. Such a directional quantity is called a ***vector***. Magnitude of a vector **P** is denoted by $|P|$ or p.

A single scalar property such as mass (m) at a point may produce many vectors in different directions such as weight 'mg' (oriented towards center of earth), momentum 'mv' of a moving mass (in the direction of motion), centrifugal force 'mv^2/r' of a rotating mass, acting radially outwards etc.

Effect of a force on a body is not dependent on the point of application, but only on the line of action or direction. A force (F) can be considered as applied from any point on the body (A, B, C or D in the figure) along its line of action. This is called **Principle of transmissibility of forces.**

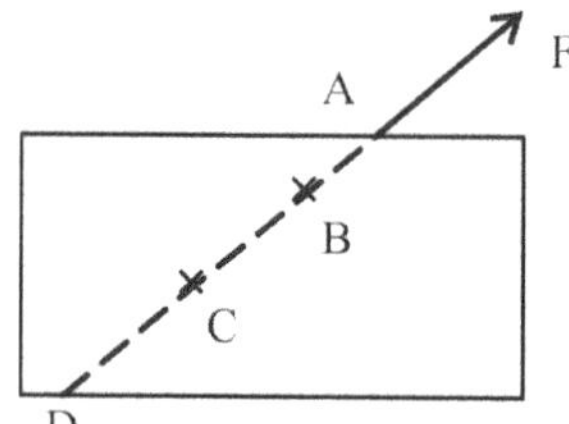

1.1.1 CLASSIFICATION OF FORCES

Forces are broadly categorised, based on their orientation, as ***co-planar (2-D) forces*** (if all the forces lie in a single plane – horizontal, vertical or inclined) and ***spatial (3-D) forces***

They are also categorised, based on their lines of action, as ***concurrent forces*** with lines of action starting from a point or meeting at a point (Ref Fig 1.1 a) and ***non-concurrent forces*** (Ref Fig 1.1 b). Concurrency does not mean forces are acting at one point, but their lines of action, even when they act at different points of a body, meet at one point as shown in the fig 1.1 (a).

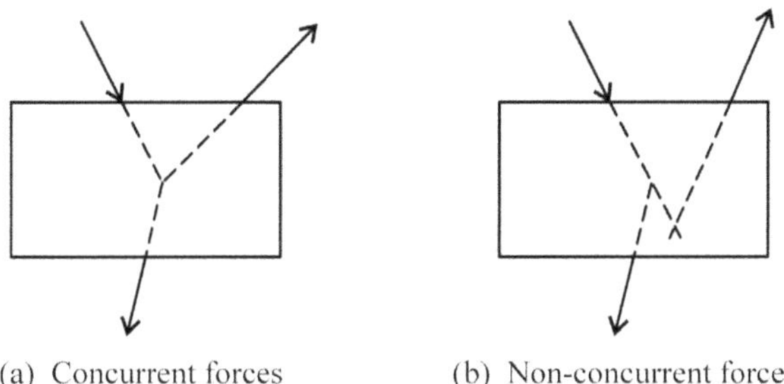

FIGURE 1.1 Concurrent and non-concurrent forces

1.1.2 COMPONENTS OF FORCES

A force oriented in an arbitrary direction is identified by its magnitude and the direction of the force represented by the angles it makes with the coordinate axes. It is more conveniently described by its components along chosen coordinate axes (along X, Y for planar forces and along X, Y and Z for spatial forces). They are explained in more detail with examples in the subsequent chapters.

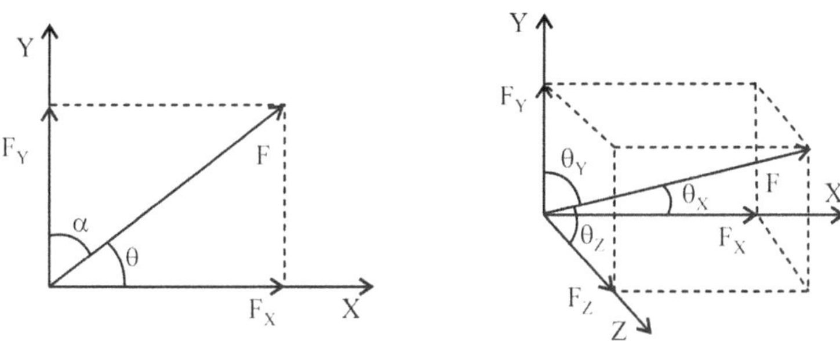

FIGURE 1.2 Components of a force in a 2-D plane & 3-D space

1.2 UNITS

The magnitude of any physical quantity loses its significance, unless associated with the appropriate units. Different countries may use different units. Engineers and scientists in most countries now use SI units – with basic units such as 'meter (m)' for length, 'kilogram (kg)' for mass, 'second (sec)' for time and 'degree Kelvin (K)' for temperature and derived units of 'Newton (N)' for force, 'Pascal (Pa)' for pressure or stress and 'Watt (W)' for power etc. The basic units of m, kg, sec for length, mass and time respectively are together called MKS system.

1.2.1 SOME DERIVED UNITS

(a) According to Newton's 2nd law of motion, magnitude of force is proportional to rate of change of momentum

$$F \;\alpha\; d(m \times v)/dt$$

where

v is the velocity of the body due to the applied forcé

$F = k \times m \times (dv/dt)$ where, k is the proportionality constant

$= k \times m \times a$ if mass of the body (m) does not change with time

and

$a = dv/dt$ is the acceleration of the body

In SI units, $F\,(1\text{ N}) = m\,(1\text{ kg}) \times a\,(1\text{ m/sec}^2)$ is a derived unit from kg, m and sec with $k = 1$ for the given set of units

We refer to weight (forcé with which mass, m is attracted to the ground) in kg_f

It is defined as $F\,(1\text{ kg}_f) = m\,(1\text{ kg}) \times a\,(9.81\text{ m/sec}^2)$

Thus, $1\text{ kg}_f = 9.81\text{ N} \approx 10\text{ N}$

(b) Applied pressure (p) or internal stress (σ) is defined as force per unit area.

Thus, $p \text{ or } \sigma = F/A$ or $1\text{ Pa} = 1\text{ N/m}^2$

Magnitude of 1 Pa is very low and in many problems,

Pressure or stress is expressed as M Pa (Mega Pascal or 10^6 Pa)

Modulus of elasticity or modulus of rigidity (in the same units as stress) is expressed as G Pa (Giga Pascal or 10^9 Pa).

In our day-to-day life, we use MKS system of units. Accordingly, we are familiar with ***weight in 'kg_f', pressure in 'kg_f/sq cm'*** etc. This may pose some difficulty or confusion to the students. However, enginners and scientists would prefer to use unambiguous units like 'kg' for mass and 'N' for force (weight,..) as primary or basic units and 'Pa' or 'N/sq m' for pressure etc as derived units.

(c) Power (P), extensively used in this subject, is defined as

Power = Work done per unit time

where, work done, W = Force × distance

Thus, P (1 W) = Work done (1N × 1m) / time (1 sec)

or 1 W = 1 Nm/sec

1.2.2 METRIC SYSTEM OF UNITS

A brief understanding of metric system (with smaller and larger units as negative and positive powers of 10) will be useful in this subject. Table 1.1 gives standard symbols used in metric units system

Table 1.1 Metric system of units

10^{-9}	10^{-6}	10^{-3}	10^{-2}	10^{-1}	Basic unit	10^1	10^2	10^3	10^6	10^9
Nano (n)	**Micro (μ)**	**Milli (m)**	**Centi (c)**	**Deci (d)**		**Deca (da)**	**Hecta (h)**	**Kilo (k)**	**Mega (M)**	**Giga (G)**
nm	μm	mm	cm	dm	**meter**	da m	hm	km	Mm	Gn
ng	μg	mg	cg	dg	**gram**	da g	hg	kg	Mg	Gg
nW	μW	mW	cW	dW	**Watt**	da W	hW	kW	MW	GW
nPa	μPa	mPa	cPa	dPa	**Pascal**	da Pa	hPa	kPa	MPa	GPa

Conversion factor: A commonly used unit of velocity in Kinematics chapter is 1 m/sec = 1 × 10^{-3} km / (1/3600)hr = 3.6 Km/hr or kmph

1.3 ENGINEERING MATERIALS

A design engineer should have some knowledge of materials to be used in a particular component and their properties as well as limitations. Detailed study of the materials is outside the scope of this book.

1.3.1 MATERIAL TYPES

Many structures may consist of members made up of different materials with distinctly varying material properties, such as Modulus of elasticity (E), Modulus of rigidity (G), coefficient of linear thermal expansion. These properties may be

(a) constant (linear stress-strain relationship) or variable (non-linear stress-strain relationship) over the range of load

(b) same in all directions (isotropic) or vary in different directions (anisotropic or orthotropic)

(c) constant over the temperature range or vary with temperature, particularly when the temperature range over the component is large

These variations may be inherent in the material or induced by the manufacturing processes like rolling, casting, welding etc. Treatment of non-homogeneous material, with varying properties at different locations of a component, is difficult and is also very unusual. In most cases, variation of material properties with the direction and with temperature may not be significant and hence neglected. So, an isotropic, homogeneous material is most often used in the analysis of a component.

Engineering materials used in examples of this book are assumed to be

- **homogeneous** – having same properties at all points of the component (in reality most alloys are non-homogeneous at micro level, but their gross or average properties are considered homogeneous and used in calculations)
- **isotropic** – having same properties in all directions (in reality materials like wood, fiber reinforced composite materials etc are non-isotropic)
- **stress is within the linear elastic range** – when loads are applied, the resulting stress is proportional to strain, satisfying Hooke's law (even though at times, stress may be in the non-linear elastic or plastic range) and
- **properties are independent of type of load** – some materials may behave differently when subjected to tensile, compressive, shear and torsional loads (some materials like bricks, concrete, wood are strong in compression and weak in tension)

1.4 LINEAR ANALYSIS

It is based on linear stress-strain relationship (Hooke's law) and is usually applicable to stress below the elastic limit (or yield stress), at any point in the component. In this analysis, ***linear superposition of results*** obtained for individual loads on a component is valid in order to obtain stresses due to any combination of these loads acting simultaneously. In some designs, it is necessary to check for many combinations such as pressure and thermal loads at different times of a start-up transient of a steam turbine. In such a case, analysing for unit pressure; multiplying the stress results with the pressure corresponding to that particular time of the transient and adding to the stresses due to temperature distribution will be economical.

1.5 A BRIEF REVIEW OF MATHEMATICS

- **Solution of a quadratic polynomial** of the form $a \times x^2 + b \times x + c = 0$ is given by

$$x = -b \pm \sqrt{(b^2 - 4 \times a \times c)} / (2 \times a)$$

- **Sum of an arithmetic progression**

$$n + (n - 1) + (n - 2) + \ldots\ldots. + 1 = n \times (n + 1)/2$$

1.5.1 OVERVIEW OF TRIGONOMETRY

Elementary concepts of trigonometry are used while solving many problems in this subject. A review of the basics is considered desirable, without proof for the same. Books on trigonometry may be referred for additional concepts as well as for proof of the theorems mentioned here.

- If two straight lines are inclined at an angle θ, then **opposite angle** is also θ (Ref Fig 1.3 a)
- If two parallel lines are cut by a straight line, the **corresponding angles** are equal (Ref Fig 1.3 b)
- If two parallel lines are cut by a straight line, **internal opposite angles** are equal, by a combination of the first two rules (Ref Fig 1.3 c)
- If two straight lines AB and AC are inclined at an angle θ, then their **perpendiculars DE and FG are also inclined at the same angle θ** (Ref Fig 1.3 d)

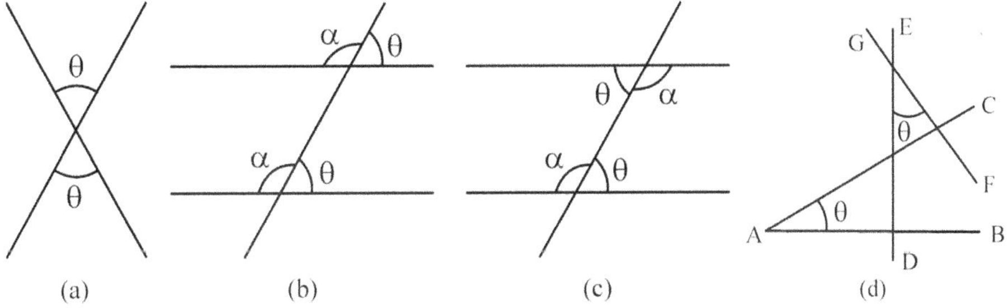

FIGURE 1.3 Some useful theorems of Trigonometry

In a right angled triangle (one angle is 90^0),

Sine θ or Sin θ = Opposite side / Hypotenuse = AB/AC

Cosine θ or Cos θ = Adjacent side / Hypotenuse = BC/AC

Tangent θ or Tan θ = Opposite side / Adjacent side

= Sin θ/Cos θ = AB/BC

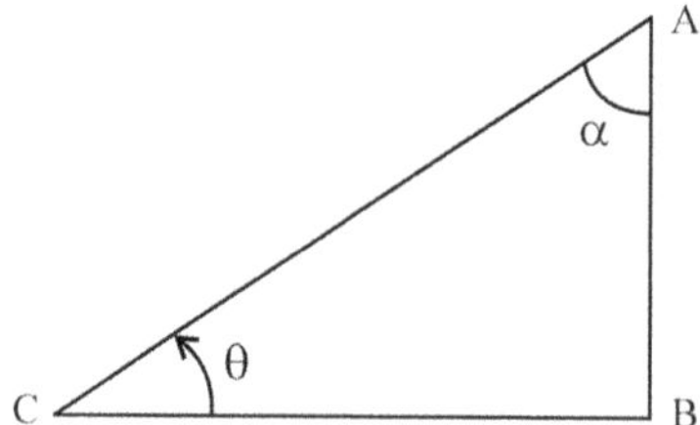

Cosecant θ or Csc θ = 1 / Sin θ = Hypotenuse / Opposite side = AC/AB

Secant θ or Sec θ = 1 / Cos θ = Hypotenuse / Adjacent side = AC/BC

Cotangent θ or Cot θ = 1 / Tan θ = Adjacent side / Opposite side = BC/AB

It is obvious that, Sin θ = Cos α and Cos θ = Sin α

Minimum of (Sin θ or Cos θ) = 0

and Maximum of (Sin θ or Cos θ) = +1 or -1

Angle is usually measured in the counter-clockwise direction from +ve X-axis. For different values of angle θ, nature of values of Sin θ, Cos θ and Tan θ are given in Table 1.2.

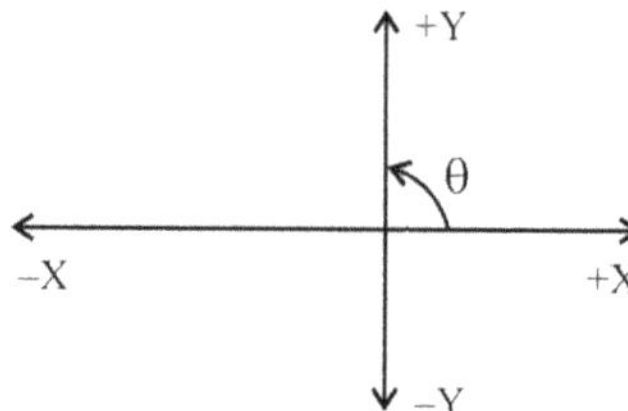

Table 1.2 Values of trigonometric functions

Angle θ (degrees)	Angle θ (radians)	Sin θ	Cos θ	Tan θ
0^0 or 360^0	0 or 2π	0	1	0
$> 0^0 - < 90^0$	> 0 – < π/2	+ ve	+ ve	+ ve
90^0	π/2	1	0	∞
$> 90^0 - < 180^0$	> π/2 – < π	+ ve	– ve	– ve
180^0	π	0	–1	0
$> 180^0 - < 270^0$	> π – < 3π/2	– ve	– ve	+ ve
270^0	3π/2	–1	0	∞
$> 270^0 - < 360^0$	> 3π/2 – < 2π	– ve	+ ve	– ve

Signs of these trigonometric functions are remembered by the notation – All Silver Tea Cups, 1st letter of each word indicating +ve functions (All, Sin θ, Tan θ and Cos θ) in the four quadrants in sequence i.e., ALL in 1st quadrant (+X,+Y), Sin in 2nd quadrant (–X,+Y), Tan in 3rd quadrant (–X,–Y), Cos in 4th quadrant (+X,–Y)

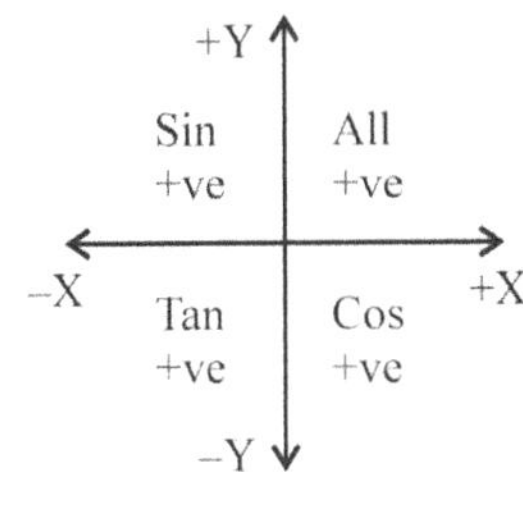

Necessary changes must be made, when angles are measured from –ve X-axis, +ve Y-axis or –ve Y-axis

From Pythagoras theorem, $AB^2 + BC^2 = AC^2$

or $AB^2/AC^2 + BC^2/AC^2 = 1$

$\text{Sin}^2\,\theta + \text{Cos}^2\,\theta = 1$

If α and β are two adjacent angles, trigonometric properties (Sine and Cosine) for the sum of the angles are given by

$\text{Sin}\,(\alpha + \beta) = \text{Sin}\,\alpha \times \text{Cos}\,\beta + \text{Cos}\,\alpha \times \text{Sin}\,\beta$

$\text{Sin}\,2\,\alpha = 2\,\text{Sin}\,\alpha \times \text{Cos}\,\alpha$

$\text{Cos}\,(\alpha + \beta) = \text{Cos}\,\alpha \times \text{Cos}\,\beta - \text{Sin}\,\alpha \times \text{Sin}\,\beta$

$\text{Cos}\,2\,\alpha = \text{Cos}^2\,\alpha - \text{Sin}^2\,\alpha$

$= 2\text{Cos}^2\,\alpha - 1 = 1 - 2\text{Sin}^2\,\alpha$

$\text{Sin}^2\,\alpha = (1 - \text{Cos}\,2\alpha)\,/\,2$

$\text{Cos}^2\,\alpha = (1 + \text{Cos}\,2\alpha)\,/\,2$

$\text{Tan}^2\,\alpha = \text{Sin}^2\,\alpha\,/\,\text{Cos}^2\,\alpha = (1 - \text{Cos}^2\,\alpha)\,/\,\text{Cos}^2\,\alpha = \text{Sec}^2\,\alpha - 1$

$\text{Cot}^2\,\alpha = \text{Cos}^2\,\alpha\,/\,\text{Sin}^2\,\alpha = (1 - \text{Sin}^2\,\alpha)\,/\,\text{Sin}^2\,\alpha = \text{Csc}^2\,\alpha - 1$

1.5.2 DIFFERENTIAL & INTEGRAL CALCULUS

Some formulae of differential and integral calculus are given below, useful for solving problems in centroid, Moment of inertia, kinematics etc. Students interested in a detailed explanation and proof may refer to appropriate books in mathematics.

Differentiation	Integration
$d(x^n)/dx = n \times x^{n-1}$	$\int x^n\, dx = x^{n+1} / (n+1)$
$d[(ax+b)^n]/dx = a\times[n\times(ax+b)^{n-1}]$	$\int (ax+b)^n\, dx = (ax+b)^{n-1} / [(n-1)\times a]$
$d(\sin\theta)/d\theta = \cos\theta$	$\int (\cos\theta)\, d\theta = \sin\theta$
$d(\cos\theta)/d\theta = -\sin\theta$	$\int (\sin\theta)\, d\theta = -\cos\theta$
	$\int (\tan\theta)\, d\theta = -\log_e(\cos\theta)$
	$\int (\cot\theta)\, d\theta = \log_e(\sin\theta)$
$d(\tan\theta)/d\theta = \sec^2\theta = 1+\tan^2\theta$	$\int (\sec^2\theta)\, d\theta = \tan\theta$
$d(\cot\theta)/d\theta = -\text{cosec}^2\theta = -(1+\cot^2\theta)$	$\int (\csc^2\theta)\, d\theta = -\cot\theta$
$d(\sec\theta)/d\theta = \sin\theta / \cos^2\theta$ $= \sec\theta \times \tan\theta$	$\int(\sin^2\theta)\, d\theta = \int [(1-\cos 2\theta)/2]\, d\theta$ $= \theta/2 - (\sin 2\theta)/4$
$d(\csc\theta)/d\theta = -\sin\theta / \cos^2\theta$ $= -\csc\theta \times \cot\theta$	$\int(\cos^2\theta)\, d\theta = \int [(1+\cos 2\theta)/2]\, d\theta$ $= \theta/2 + (\sin 2\theta)/4$
$d(\log_e x)/dx = 1/x$	$\int dx/x = \log_e x$
$d^2(x^m y^n)/(dx\, dy) = d(x^m)/dx \times d(y^n)/dy$ $= (mx^{m-1}) \times (ny^{n-1})$ if x and y are mutually independent	$\int\int x^m y^n\, dx\, dy = \int x^m\, dx \times \int y^n\, dy$ $= [x^{m+1}/(m+1)] \times [y^{n+1}/(n+1)]$ if x and y are mutually independent

1.6 PRECAUTIONS / USEFUL TIPS TO STUDENTS

- Questions may contain different quantities in different units. Care should be taken to ***use all physical quantities in one consistent system of units*** (N or kN for force; km or m or cm for distance; Pa or MPa or GPa for pressure; hr, min or sec for time; ...). Units of quantities in all equations on either side should match. Otherwise results can be erroneous, even after using correct method and correct formulae.
- Mere numbers in answers will not fetch full marks. ***Students should practice using units along with all physical quantities*** like distance, forcé, stress, speed, velocity, acceleration, .. in answers. It will also help them get a physical feel of quantities like distance, force, stress etc in solutions and correct absurd answers.
- ***Angles are in degrees in some chapters / problems, while they are in radians in some other chapters. Students should use calculator in the appropriate mode***. Otherwise, solutions will be incorrect.

- Different students may use different coordinate systems (X, Y, Z axes) and differnt notations (Force components P_1, P_2, P_3,.. or F_1, F_2, F_3,..or P_X, P_Y, P_Z,.. ; angles α, β, γ, θ, φ etc) while solving problems in examinations.
- ***It is a good practice to draw Free body diagram (FBD) in all relevant problems*** which will also indicate notation of forces, angles and coordinate system used by a student thereby avoiding a posible misunderstanding with the examiner while evaluating answer paper

CHAPTER 2

COPLANAR CONCURRENT FORCES

2.1 RESOLVING A FORCE INTO ITS COMPONENTS

Force acting on anybody is a vector, having magnitude and direction. For convenience of representation of direction, a force may be identified by its components along three mutually perpendicular Cartesian coordinates (X, Y and Z) while a force in 2-D plane is resolved along two mutually perpendicular Cartesian coordinates (X and Y) as shown in Fig 2.1.

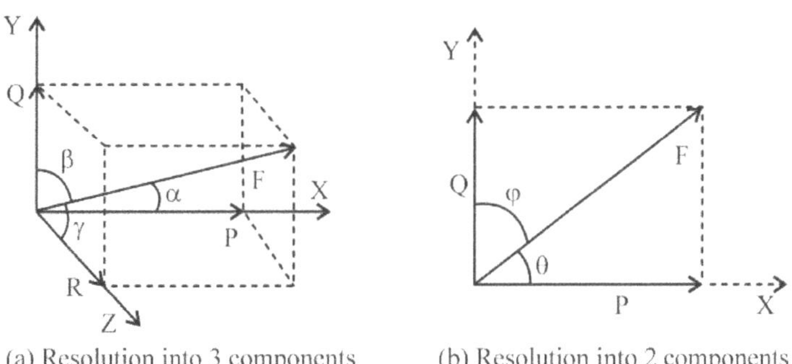

FIGURE 2.1 Resolution of a force vector along coordinate axes

If α, β and γ are the angles made by a force vector F in 3-D space with the three coordinate axes X, Y and Z,

$$P = F \cos \alpha \,; \qquad Q = F \cos \beta \,; \qquad R = F \cos \gamma$$

Since components along these directions are cosine functions of the force, these angles are also called ***direction cosines***.

$$|F| = \sqrt{|P|^2 + |Q|^2 + |R|^2}$$

Similarly, if θ and φ are the angles made by a force vector F in 2-D plane with the coordinate axes, $P = F\cos\theta$; $Q = F\cos\varphi = F\sin\theta$ and

$$|F| = \sqrt{|P|^2 + |Q|^2}$$

If a force 'F' is inclined at an angle 'θ' with the positive X-axis, the two components of the force can also be identified more conveniently by $F_X = F\cos\theta$ and $F_Y = F\sin\theta$. The components are so oriented that they start at the starting point of force vector or end with the end point of the force vector, as shown in Fig 2.2

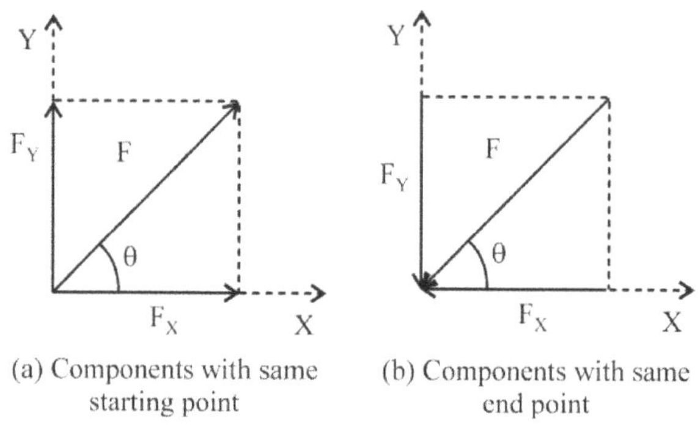

FIGURE 2.2 Force components in 2-D plane

Proper signs are assigned for the two components of a force by the trigonometric functions sin θ and cos θ, if θ is measured from +X axis in counter-clockwise direction, as shown in the following four possible orientations of the force vector. Signs for force components need to be assigned, if acute angle of force vector 'α' is specified from the nearest X-axis.

(a) If F is oriented along N-E direction, or 1st quadrant, F_X is +ve and F_Y is +ve

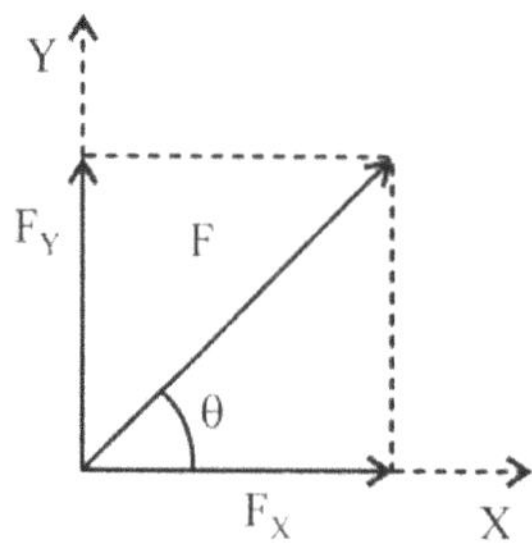

(b) If F is oriented along N-W direction, or 2nd quadrant, F_X is –ve and F_Y is +ve

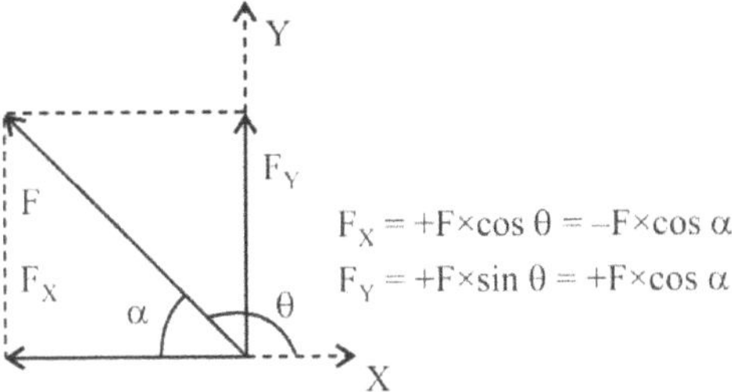

(c) If F is oriented along S-W direction, or 3rd quadrant, F_X is –ve and F_Y is –ve

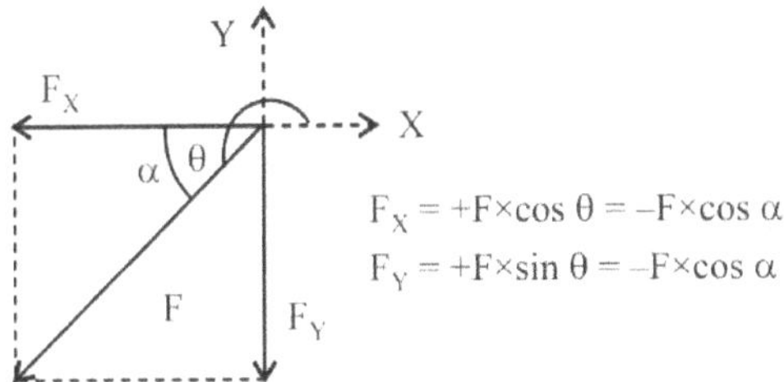

(d) If F is oriented along S-E direction, or 4th quadrant, F_X is +ve and F_Y is –ve

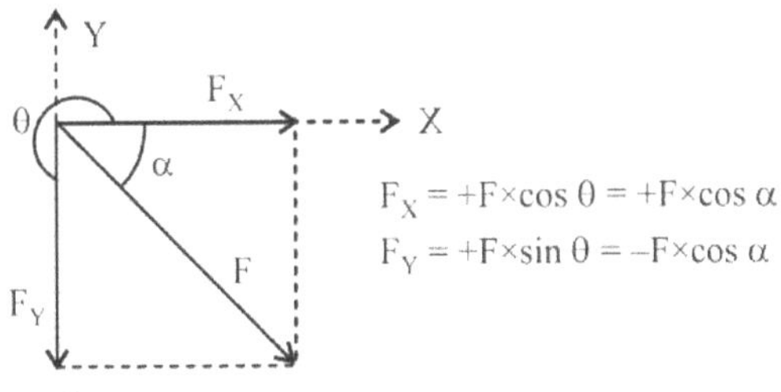

Obviously, if $\theta = 0^0$ or 180^0, $F_X = \pm F$; $F_Y = 0$

and if $\theta = 90^0$ or 270^0, $F_X = 0$; $F_Y = \pm F$

FIGURE 2.2 Force components in the 4 quadrants of 2-D plane

2.1.1 REPRESENTING A FORCE BY VECTOR NOTATION

Any force vector $\boldsymbol{F}$ can be expressed in vector notation. If $\boldsymbol{i}$, $\boldsymbol{j}$ and $\boldsymbol{k}$ represent unit vectors along the three mutually perpendicular axes (X, Y and Z), using its components P, Q and R along the three axes, as $\boldsymbol{F} = \text{P}\,\boldsymbol{i} + \text{Q}\,\boldsymbol{j} + \text{R}\,\boldsymbol{k}$

Its magnitude is given by $|F| = \sqrt{|P|^2 + |Q|^2 + |R|^2}$

For a planar force vector, the vector simplifies as $\boldsymbol{F} = P\,\boldsymbol{i} + Q\,\boldsymbol{j}$

Its magnitude is given by $|F| = \sqrt{|P|^2 + |Q|^2}$

2.2 RESULTANT OF TWO FORCES ACTING ON A BODY

If two forces are acting on a body, their **vector sum** is called ***Resultant***. It is also defined as the single force acting on the body which produces the same effect as that of the two forces acting together. Magnitude of the resultant force is, in general, not equal to the algebraic sum of the magnitudes of the two forces, unless they act in the same direction.

(a) Resultant of two forces by ***parallelogram law***

From any point O, lines OA and OB are drawn representing forces P and Q in magnitude (to some convenient scale) and direction. Lines parallel to OA from B and parallel to OB from A are drawn to intersect at C. Line OC represents the sum or resultant of the two force vectors, both in magnitude and direction. But, the measured values of magnitude and inclination of resultant may not be of sufficient accuracy.

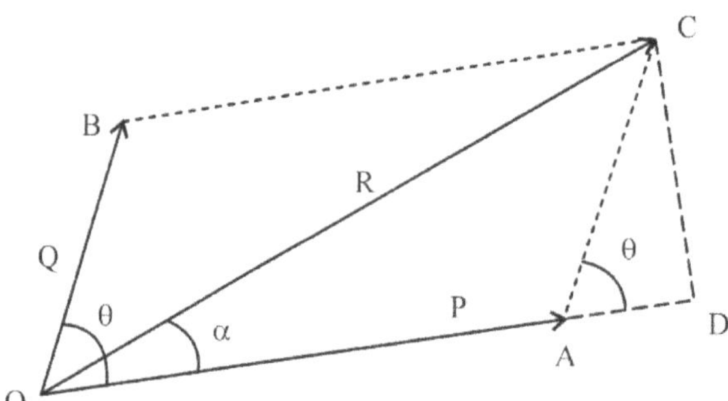

FIGURE 2.3 Parallelogram law of forces

Its magnitude can be derived by dropping a perpendicular line from C on OA extended, to intersect at D. If the two forces are inclined to each other at an angle θ, then, magnitude of the resultant is obtained from the right angled triangle ODC as

$$OC^2 = OD^2 + CD^2 = (OA+AD)^2 + CD^2$$

$$= (OA + AC\cos\theta)^2 + (AC\sin\theta)^2$$

$$= (OA^2 + 2\,OA\,AC\cos\theta + AC^2\cos^2\theta) + AC^2\sin^2\theta$$

$= OA^2 + AC^2 + 2\ OA\ AC \cos\theta$

$= OA^2 + OB^2 + 2\ OA\ OB \cos\theta$

$= P^2 + Q^2 + 2\ P\ Q \cos\theta$

or $R = \sqrt{P^2 + Q^2 + 2\ P\ Q \cos\theta}$

The angle 'α' made by the resultant with the direction of force P, is obtained from

$\tan\alpha = CD / OD = CD / (OA+AD)$

$= AC \sin\theta / (OA + AC \cos\theta)$

$= OB \sin\theta / (OA + OB \cos\theta)$

$= Q \sin\theta / (P + Q \cos\theta)$

(b) Resultant of two forces by ***triangle law***

Parallelogram law is also identical to the triangle law of forces, constructed with the forces P and Q as shown, Q starting from the end point A of force P so that the line joining starting point 'O' of force vector P with the end point 'C' of force vector Q represents the resultant force 'R' in magnitude and direction. The magnitude and direction of resultant can also be calculated by following same construction as in parallelogram law

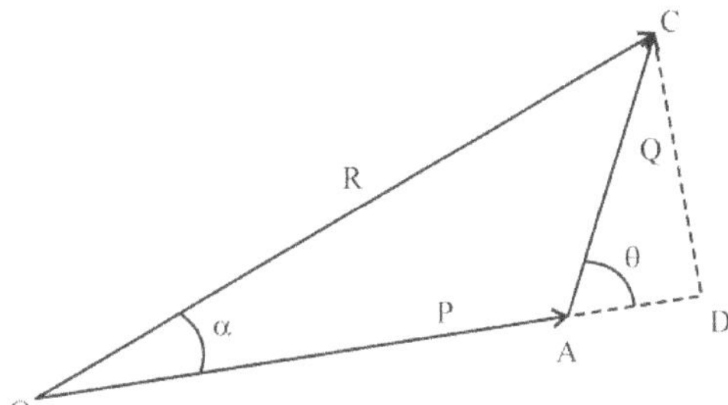

FIGURE 2.4 Triangle law of forces

(c) Special cases

- If P and Q are perpendicular to each other, $\theta = 90^0$, $\sin\theta = 1$ and $\cos\theta = 0$

  Therefore, $R = \sqrt{(P^2 + Q^2)}$ and $\tan\alpha = Q \sin\theta / (P + Q \cos\theta) = Q / P$

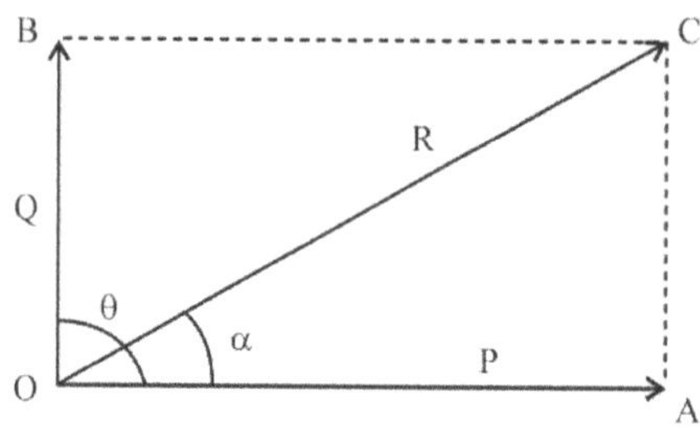

- If P and Q are collinear and in the same direction, $\theta = 0^0$, $\sin\theta = 0$ and $\cos\theta = 1$

  Therefore, $R = \sqrt{P^2 + Q^2 + 2\,P\,Q} = \sqrt{(P+Q)^2} = P + Q$

  and $\tan\alpha = Q\sin\theta / (P + Q\cos\theta) = 0$

  or R acts along P and Q

This is identical to the scalar addition

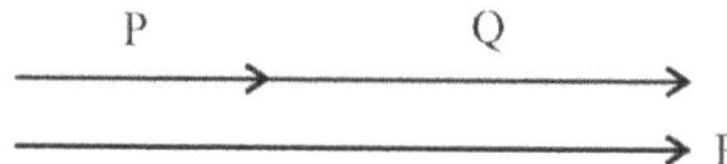

- If P and Q are collinear and opposite in direction, $\theta{=}180^0$, $\sin\theta = 0$ and $\cos\theta = -1$

  Therefore, $R = \sqrt{P^2 + Q^2 - 2\,P\,Q} == \sqrt{(P-Q)^2} = P - Q$

  and $\tan\alpha = Q\sin\theta / (P + Q\cos\theta) = 0$

  or R acts along the larger of P and Q

If $P > Q$, $R = P - Q$

If $P < Q$, $R = Q - P$

(d) Important rules

- In the particular case when $P = Q$ and $\theta = 180^0$, $R = 0$

  i.e., A point O, acted upon by 2 equal, opposite and collinear forces remains in its state of equilibrium. Ex: Tug of war

Q P $\equiv$ Q P

- Two forces acting on a body will always have a non-zero resultant force (leading diagonal of parallelogram or third side of force triangle), unless they are equal and opposite ($\theta=180^0$)

2.3 RESULTANT OF MANY CONCURRENT FORCES ACTING ON A BODY

Resultant of multiple concurrent forces acting on a body can be obtained by repeated application of (a) parallelogram law or (b) triangle law as shown

(a) If a number of forces P_1, P_2, P_3, P_4, ... P_n are acting on a body, sum (or resultant) of all these forces can be obtained by repeatedly applying parallelogram law as shown in Fig 2.4

Resultant R_2 is obtained as the diagonal of the parallelogram with sides P_1 and P_2

Resultant R_3 is obtained as the diagonal of the parallelogram with sides R_2 and P_3

Resultant R_4 is obtained as the diagonal of the parallelogram with sides R_3 and P_4 ......

Resultant R_n is obtained as the diagonal of the parallelogram with sides R_{n-1} and P_n

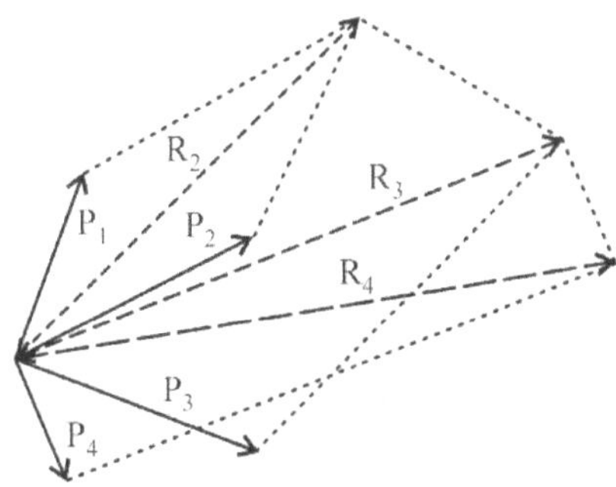

FIGURE 2.4 Resultant of many concurrent forces by parallelogram law

(b) Repeated construction of Parallelograms is a tedious process, for any system of multiple forces. A simpler method is graphical construction of a ***force polygon***, which is constructed by repeated application of triangle law. In this method, forces P_1, P_2, ... P_n are represented to some scale, in sequence, such that 2nd force starts from the end point of the 1st force, 3rd

force starts from the end point of the 2nd force,..., parallel to their lines of action. A line drawn from the starting point of the 1st force vector to the end point of the last force vector indicates resultant force in direction and magnitude (to the same scale as the other forces). This is an extension of the triangle law of forces, with intermediate resultants R_2, R_3,.. Force polygon is just a representation of forces acting on a body in proper directions, without any consideration for the shape or size of the object.

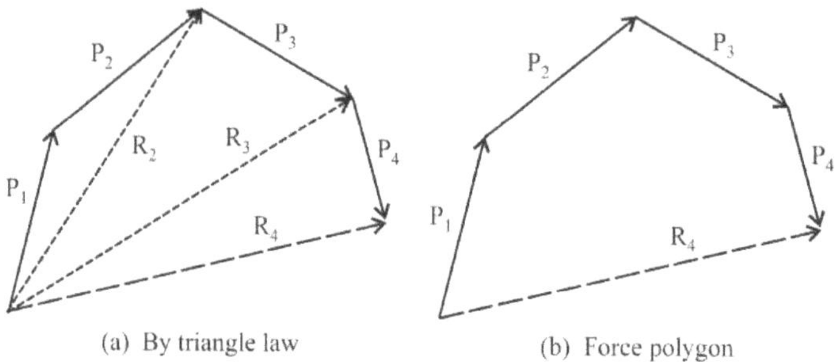

FIGURE 2.5 Resultant of many concurrent forces

(c) Resultant of multiple concurrent forces from their components

A simpler method of obtaining resultant of multiple concurrent forces is to resolve each of the forces into its components along the chosen coordinate directions. Then, considering a system of planar forces,

$P_{1X} = P_1\cos\theta_1$; $P_{1Y} = P_1 \sin \theta_1$; $P_{2X} = P_2\cos\theta_2$; $P_{2Y} = P_2 \sin \theta_2$; .....

All components along any direction, with appropriate sign, can be algebraically added together (similar to scalars). For example, in the following system of three concurrent forces

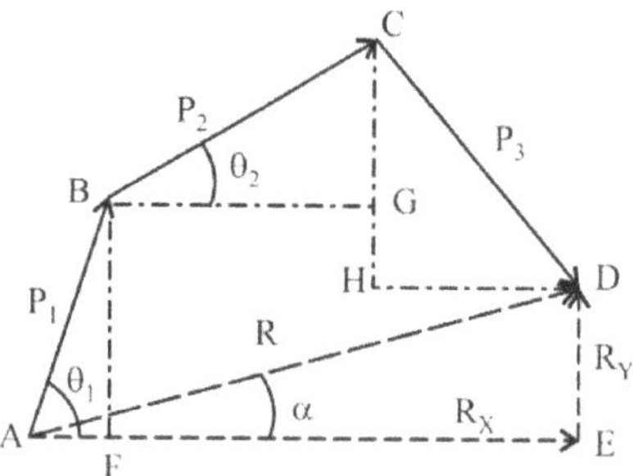

FIGURE 2.6 Resultant of multiple concurrent forces by components method

$$\sum P_X = P_{1X} + P_{2X} + P_{3X} = AF + BG + HD = AE = R_X$$

and $$\sum P_Y = P_{1Y} + P_{2Y} + P_{3Y} = FB + GC - CH = ED = R_Y$$

$$R = \sqrt{R_X^2 + R_Y^2} \quad \text{and} \quad \alpha = \tan^{-1}(R_Y/R_X)$$

In general, for a system of spatial forces,

$$R_X = \sum P_X \ ; \quad R_Y = \sum P_Y \quad \text{and} \quad R_Z = \sum P_Z$$

and $$R = \sqrt{R_X^2 + R_Y^2 + R_Z^2}$$

The components of forces is approach is simpler and more accurate than graphical method.

2.4 EQUILIBRIUM OF BODIES

If all forces lie in a single, 2-D plane (horizontal, vertical or inclined) and their lines of action start from a point or meet at a point, they are called ***Co-planar concurrent forces***. Concurrency does not mean forces are acting at one point, but their lines of action, even when they act at different points of a body, as shown in the figure, meet at one point. We will consider in this chapter, systems of coplanar concurrent forces.

If a body, acted upon by a system of co-planar forces, is in static equilibrium, then all the forces balance each other and the magnitude of their resultant is zero. It, therefore, implies that $R_X = \sum P_X = 0$; $R_Y = \sum P_Y = 0$.

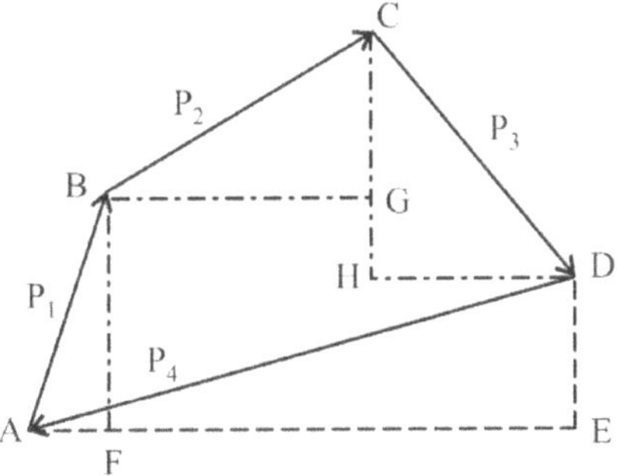

FIGURE 2.7 Concurrent forces on a body in equilibrium

For example, in the system of four concurrent forces in equilibrium,

$$\sum P_X = P_{1X} + P_{2X} + P_{3X} + P_{4X}$$
$$= AF + BG + HD - EA = 0$$

Similarly, $$\sum P_Y = P_{1Y} + P_{2Y} + P_{3Y} + P_{4Y}$$
$$= FB + GC - CH - DE = 0$$

It implies that force polygon constructed for a system of co-planar forces in equilibrium is a ***closed polygon*** i.e., the end point of the last force vector, in a force polygon, coincides with the starting point of the 1st force vector.

Two forces acting on a body can keep it in equilibrium only when they are equal in magnitude, opposite in direction and collinear.

2.4.1 FREE BODY DIAGRAM (FBD)

Free Body Diagram (FBD) is a representation of all concurrent forces with their lines of action on a body as vectors, without referring to the size or shape of the body as well as points of application of forces (Ref Fig 2.8 a &b). Size or shape of the body as well as points of application of forces have no influence on the effect of these forces on the body.

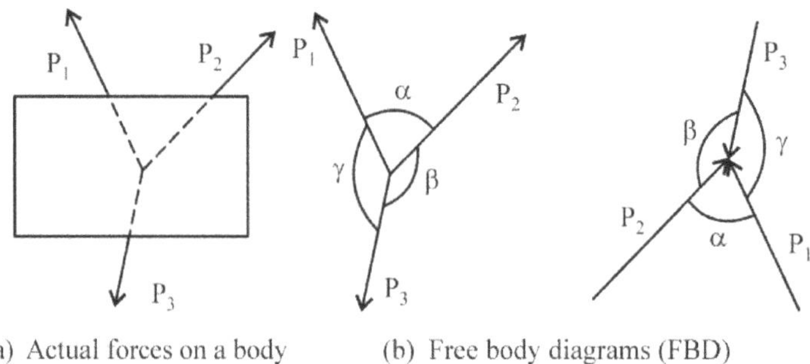

FIGURE 2.8 Actual forces and free body diagram

All forces should be drawn starting at a single point or ending at a single point in the free body diagram as shown in Fig 2.8 (b). Otherwise, angles between the forces can be different and may result in wrong answers.

2.4.2 EQUILIBRANT

A force, necessary to keep a body in equilibrium, is called equilibrant. If a system of 'n' co-planar concurrent forces, acting on a body keep it in equilibrium, then the resultant of any n–1 forces should be equal in magnitude and opposite in direction to the nth force to satisfy $\sum F_X = 0$; $\sum F_Y = 0$ and $\sum F_Z = 0$. Thus, each of the n forces is called the equilibrant of the remaining n-1 forces.

In the following example of three forces P_1, P_2 and P_3 acting on a body and keeping it in equilibrium,

$$\text{Resultant of } P_2 \text{ and } P_3,\ R_{23} = -P_1 = \text{Equilibrant of } P_1 \text{ or } E_1$$

Similarly, Resultant of P_1 and P_2, $R_{12} = -P_3$ = Equilibrant of P_3 or E_3

Resultant of P_1 and P_3, $R_{13} = -P_2$ = Equilibrant of P_2 or E_2

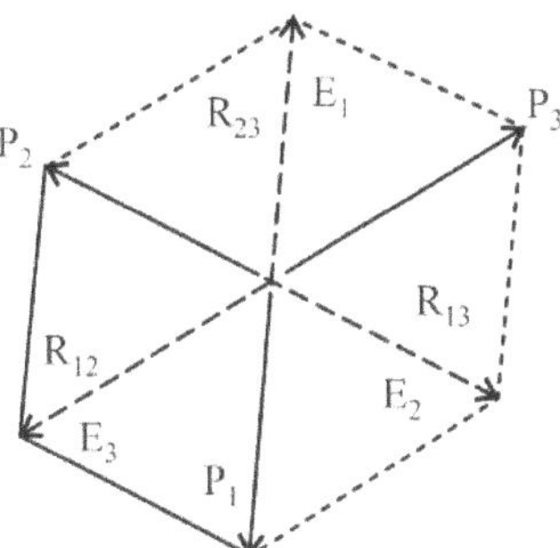

FIGURE 2.9 Equilibrant of a system of coplanar concurrent forces

2.4.3 LAMI'S THEOREM

If three concurrent forces acting on a body are in equilibrium, force polygon as already explained is a closed triangle. All the properties of a triangle are also, therefore, applicable to the triangle of forces. Sine rule of triangle, which states that the ratio of length of any side to the sine of the angle opposite to it is a constant.

$$a / \sin \alpha = b / \sin \beta = c / \sin \gamma$$

This rule is quite often used for solving problems of coplanar concurrent force systems.

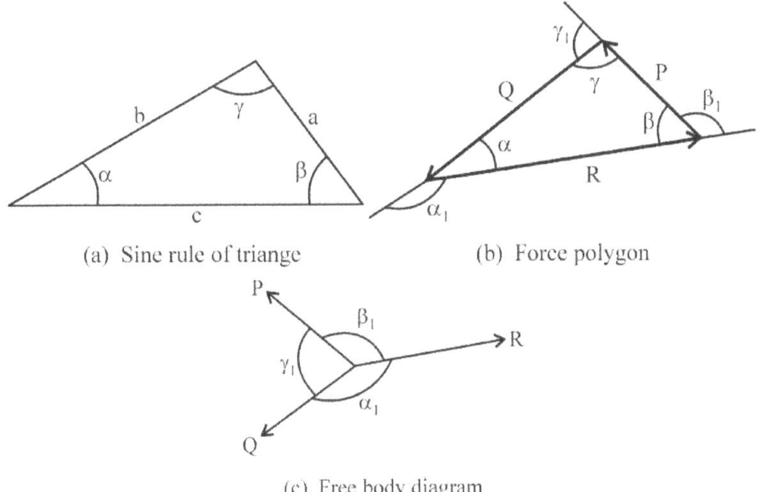

FIGURE 2.10 Lami's theorem for 3 coplanar concurrent forces in equilibrium

If three coplanar concurrent forces P, Q and R acting at a point are in equilibrium, and these forces are represented by a force polygon (Fig 2.10 b) which is a closed triangle, then Lami's theorem states that

$$P / \sin \alpha = Q / \sin \beta = R / \sin \gamma$$

Since $\sin \alpha = \sin (180 - \alpha)$, it can also be written as

$$P / \sin (180 - \alpha) = Q / \sin (180 - \beta) = R / \sin (180 - \gamma)$$

i.e., $$P / \sin (\alpha_1) = Q / \sin (\beta_1) = R / \sin (\gamma_1)$$

Thus, sine rule can be applied on the free body diagram of three coplanar concurrent forces in equilibrium (Fig 2.10 c), without actually drawing a force polygon.

2.4.4 CONDITION FOR EQUILIBRIUM

According to **Newton's 1st law of motion**, every body continues to be in its state of rest or of uniform motion unless it is compelled by an external force.

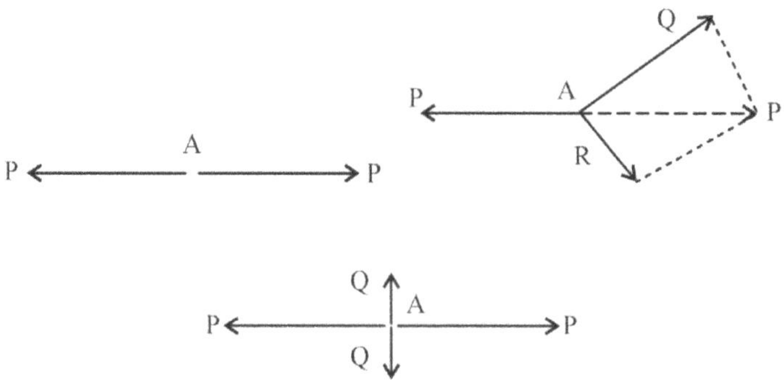

The law does not refer to the individual forces acting on the body, but only to the resultant. Point A, acted upon by multiple forces in the above examples, remains in equilibrium. Thus, for a system of concurrent forces, a body is in equilibrium only if the resultant (or vector sum of the forces) is zero. i.e.,

$$\boldsymbol{R} = \sum \boldsymbol{F} = 0$$

It is difficult to compute vector sum of a number of forces, using parallelogram law. A more convenient approach is to resolve each of the forces along three (two, for co-planar forces) perpendicular axes. Then all force components along an axis (positive and negative, depending on their direction) can be algebraically added just like scalars.

Then, $\boldsymbol{R} = R_X\boldsymbol{i} + R_Y\boldsymbol{j} + R_Z\boldsymbol{k} = 0$ is satisfied only when

$R_X = 0$; $R_Y = 0$ and $R_Z = 0$ are simultaneously satisfied.

Therefore, a body is in equilibrium if the vector sum (also, equal to the algebraic sum) of the components of forces along the three perpendicular axes is zero.

i.e., $R_X = \sum F_X = 0$; $R_Y = \sum F_Y = 0$; $R_Z = \sum F_Z = 0$

Proper understanding of the physical problem is essential while solving problems correctly. Practice with a wide range of problems and a good knowledge of basic trigonometry, to identify the orientation of forces and resolving them into components along two perpendicular axes, will help in scoring well in this subject. A few examples are included here to explain the basic concepts. Some wrong interpretations are also given to explain common mistakes.

Example 1

A force of 1000N must be applied along the mid-stream to tow a boat. For some reason, it is not possible to apply the force along the mid-stream. So, two forces are applied at 30^0 and 45^0 on either side of it in the same plane. Determine the magnitudes of these forces.

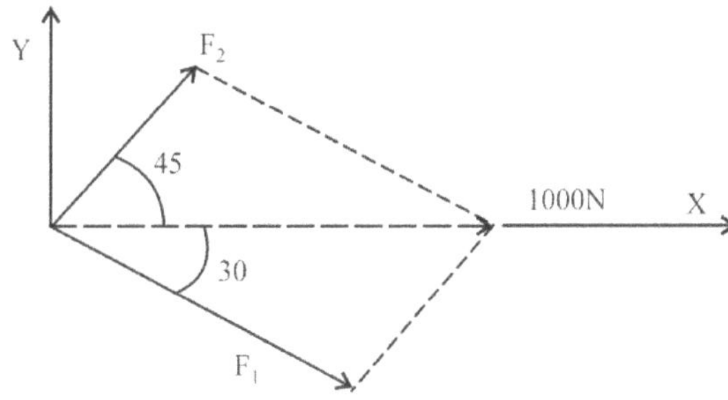

Solution:

Since the boat is ***not*** moving along Y-direction, there can be no resultant force along Y direction. Resolving forces into their components along X and Y directions,

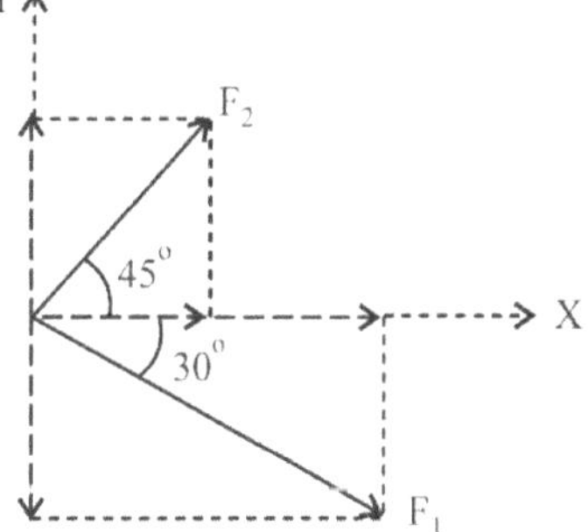

$$\sum F_Y = F_2 \sin 45 - F_1 \sin 30 = 0$$

It implies that $F_1 = F_2 \sin 45 / \sin 30 = F_2\sqrt{2}$

The boat is moving along the stream due to a resultant force of 1000 N along it

or $\sum F_X = F_1 \cos 30 + F_2 \cos 45 = 1000$ N

Substituting $F_1 = F_2\sqrt{2}$,

$$F_2 \times \sqrt{2}\cos 30 + F_2 \times \cos 45 = F_2 \times \sqrt{2} \times (\sqrt{3}/2) + F_2 \times (1/\sqrt{2})$$

$$= 1000 \text{ N}$$

Therefore, $F_2 \times (\sqrt{3}+1)/\sqrt{2} = 1000$ or $F_2 = 517.64$ N

and $F_1 = F_2\sqrt{2} = 732.05$ N

Example 2

Forces 10N, 20N, 30N, 40N and 50N are acting on one of the angular points of a regular hexagon, towards the other five angular points taken in order. Find the direction and magnitude of the resultant.

Different interpretations: Some students consider the problem by a wrong combination of forces, as shown in Fig-a. Here, all forces are not acting at one of the corners and the given 5 forces can not be represented along the sides of a 6-sided polygon. Also, there is no unique solution to the problem. ***Many correct solutions can exist***, each for the particular geometry considered, as shown in Fig-b, Fig-c and Fig-d. Detailed solution is presented for one particular geometry, different from the above.

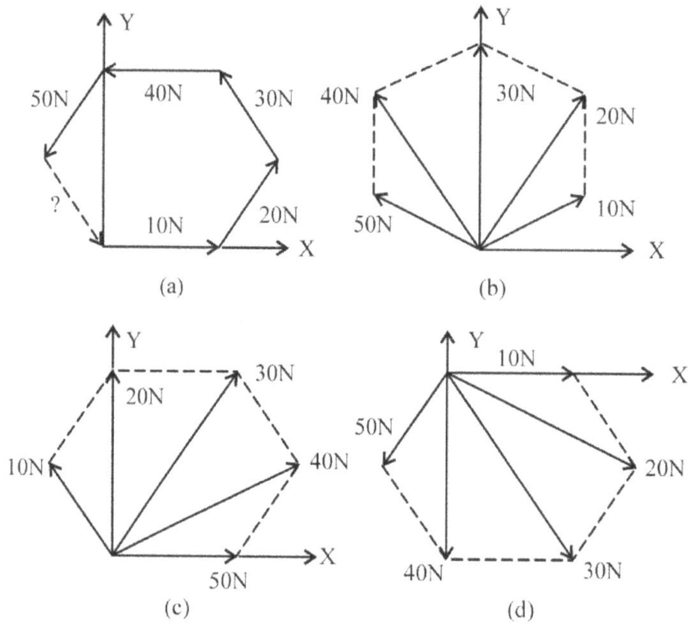

Solution:

Before solving the problem, for the particular geometry shown here, it is essential to find the correct orientation of the forces, from the properties of a regular hexagon, so that each of these forces can be resolved into components along the chosen X and Y axes.

In a regular hexagon, each internal angle = 120^0

Consider Δ ABC. AB = BC and $\angle B = 120^0$

Therefore, $\angle CAB = \angle ACB = (180 - 120)/2$
$= 30^0$

Similarly, from the Δ AFE, $\angle EAF = 30^0$

$\angle BAF = 120^0$ and AD divides the hexagon into 2 equal parts.

Therefore, $\angle DAB = \angle DAF = 120^0/2 = 60^0$

and, hence, $\angle CAD = \angle DAE = 30^0$

Note: It should be understood that the lengths of lines AB, AC, AD, AE and AF along which the forces act, have no relation to the magnitudes of respective forces.

Resolving forces along X and Y,

$$R_X = \sum F_X = 10 \cos 0 + 20 \cos 30 + 30 \cos 60 + 40 \cos 90 + 50 \cos 120 = 17.32 \text{ N}$$

$$R_Y = \sum F_Y = 10 \sin 0 + 20 \sin 30 + 30 \sin 60 + 40 \sin 90 + 50 \sin 120 = 119.28 \text{ N}$$

Resultant, $R = \sqrt{R_X^2 + R_Y^2} = 120.53 \text{ N}$

Angle between R and X, $\theta = \tan^{-1}(R_Y/R_X) = 81^0 44'$

Example 3

A derrick crane carries a load of 10kN as shown. If AB, BC and CA are 4m, 4m and 2m respectively, calculate forces in BC and CA.

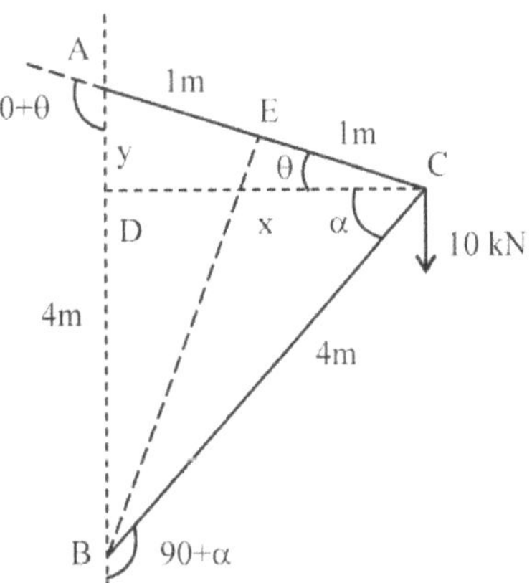

Solution:

Angles of inclination of the members AC and BC need to be evaluated first, in order to find forces in them. There are many methods.

(i) Let CD = x and AD = y

Then, from Δ ACD, $x^2 + y^2 = 2^2$

and from Δ BCD, $x^2 + (4 - y)^2 = 4^2$

$\Rightarrow$ y = 0.5 m and x = 1.9365 m

(ii) Alternatively,

Area of ΔABC = $\sqrt{s \times (s-a) \times (s-b) \times (s-c)}$

where, s = (a + b + c) / 2 = (2 + 4 + 4) / 2 = 5m

Also,

Area of Δ ABC = AB × CD/2 = 4 × x/2 = 2 × x

Then, $\sqrt{[5 \times (5-4) \times (5-4) \times (5-2)]} = \sqrt{15} = 2 \times x$

Therefore, x = 1.9365 m and $y = \sqrt{(2^2 - x^2)} = 0.5$ m

Also, sin θ = y/2 or θ = 14.48^0 and sin α = (4–y)/4 or α = 61.05^0

(iii) Since BA=BC=4m, the bisector of ∠B is the perpendicular bisector of AC.

Thus, AE = EC = 1m and $BE = \sqrt{(BA^2 - AE^2)} = \sqrt{15}$

Then, ∠EAB=∠ECB = $\sin^{-1}$(BE/BA) = $\sin^{-1}\left(\sqrt{15}/4\right) = 75.52^0$

From Δ ACD, ∠ACD= θ = 90 –∠CAD= 90 – 75.52 = 14.48^0

and ∠DCB= α =∠ACB –∠ACD = 75.52 – 14.48^0 = 61.04^0

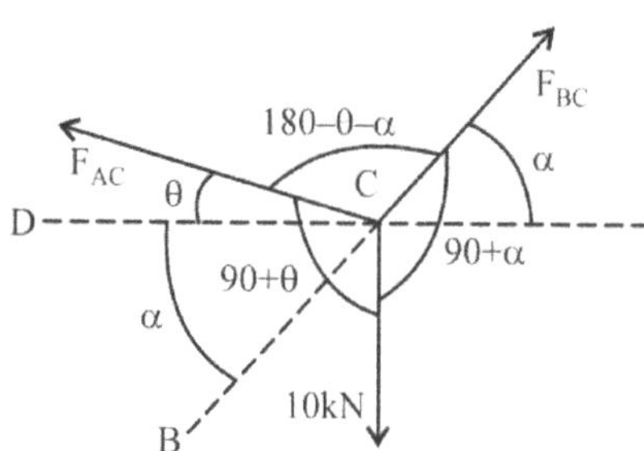

Using the free body diagram, with forces in the members assumed as shown, ***Lami's theorem can be applied with the values of θ and α*** calculated earlier by different methods.

$$10 / \text{Sin}(180 - \theta - \alpha) = F_{BC} / \text{Sin}(90 + \theta) = F_{AC} / \text{Sin}(90 + \alpha)$$

or $F_{BC} = 10 \text{ Sin}(90 + \theta) / \text{Sin}(180 - \theta - \alpha) = 10 \text{ Cos}\theta / \text{Sin}(\theta + \alpha)$

$= 10 \text{ Cos } (14.48) / \text{Sin}(75.52) = 10$ kN (Compressive)

and $F_{AC} = 10 \text{ Sin}(90 + \alpha) / \text{Sin}(180 - \theta - \alpha) = 5 \text{ Cos}\alpha / \text{Sin}(\theta+\alpha)$

$= 5 \text{ Cos } (61.04) / \text{Sin}(75.52) = 5$ kN (tensile)

Some systems of co-planar non-concurrent forces can be solved using the equations of co-planar concurrent forces, by applying them at multiple points on the system.

An example of such a solution is presented below.

Example 4

Two ropes connected to A and B carry a bar CD with a load of W at D and 25kN at C, as shown. Find the value of W so that the bar remains horizontal.

Solution:

Considering equilibrium of point C, forces in members AC and CD can be calculated for the given value (25kN) of vertical load at C.

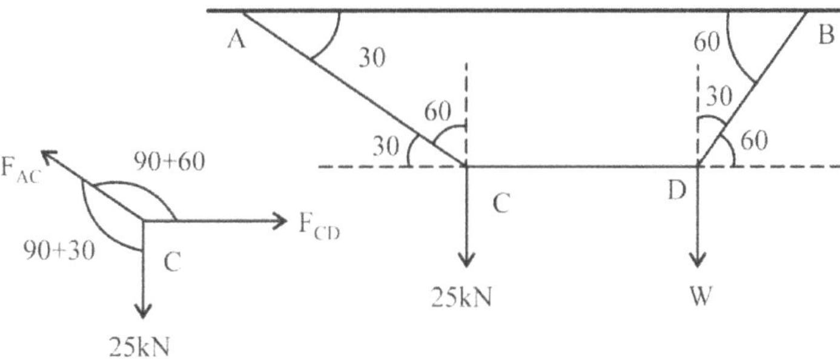

Using Lami's theorem at C,

$25/ \sin 150 = F_{AC} / \sin 90 = F_{CD} / \sin 120$

or $F_{CD} = 43.3$ kN

Next, considering equilibrium of point D, force in member BD and vertical load 'W' can be calculated for the now known value of force in member CD.

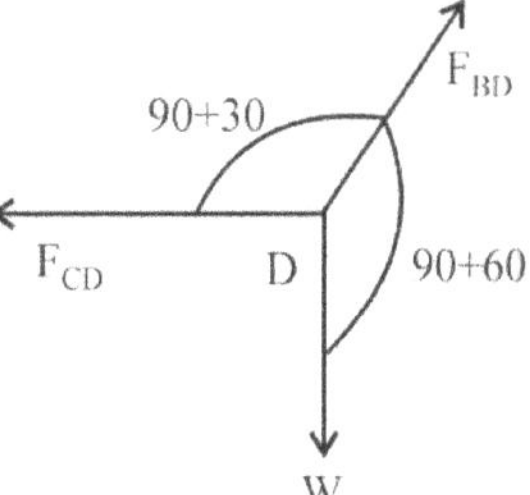

Using Lami's theorem at D,

$W/ \sin 120 = F_{BD} / \sin 90 = F_{CD} / \sin 150$

or $W = 75$ kN

Example 5

The resultant of two forces $F_1$and F_2 acting at a point is R. If F_2 is doubled, R is also doubled. What is the ratio of $F_1:F_2:R$?

Solution:

If two forces F_1 and F_2 are inclined at an angle θ, we know the resultant R is given by

$$R^2 = F_1^2 + F_2^2 + 2\, F_1 \times F_2 \cos\theta \qquad .....(2.1)$$

Since R is doubled when F_2 is doubled,

$$(2R)^2 = F_1^2 + (2F_2)^2 + 2\, F_1 \times (2F_2) \cos\theta$$

or

$$4R^2 = F_1^2 + 4F_2^2 + 4\, F_1\, F_2 \cos\theta \qquad .....(2.2)$$

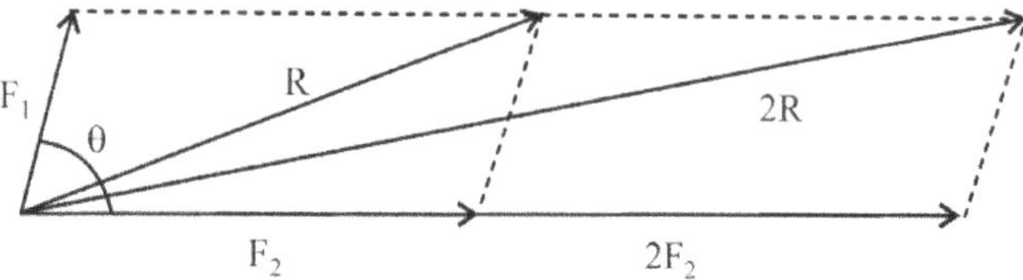

Multiplying eq. (2.1) with 4 and subtracting eq.(2.2) from it,

$$3F_1^2 + 4\, F_1\, F_2 \cos\theta = 0 \Rightarrow 4\, F_1\, F_2 \cos\theta = -\, 3F_1^2 \qquad .....(2.3)$$

From eq (2.2) and (2.3),

$$4R^2 = -2F_1^2 + 4F_2^2 \Rightarrow 2R^2 = -F_1^2 + 2F_2^2 \qquad .....(2.4)$$

This equation is satisfied only when

$$\mathbf{F_1 : F_2 : R = \sqrt{2} : \sqrt{3} : \sqrt{2}}$$

<u>Check</u>: If $F_1 = \sqrt{2}\ P$ and $F_2 = \sqrt{3}\ P$, then from eq. (2.4),

$$-F_1^2 + 2F_2^2 = -2P^2 + 2(3P^2) = 4P^2 = 2R^2 = 2\sqrt{2}\ (P)^2$$

Therefore, $\mathbf{F_1 : F_2 : R = \sqrt{2}\ P : \sqrt{3}\,P : \sqrt{2}\,P = \sqrt{2} : \sqrt{3} : \sqrt{2}}$

Example 6

Two forces equal to 2P and P respectively act on a particle. If first be doubled and the second increased by 12N, direction of the resultant is unaltered. Find the value of 'P' ?

Solution:

Let the two forces P and 2P be inclined at α.

Since direction of the resultant θ is unaltered,

$\tan\theta = P \sin\alpha / (2P + P \cos\alpha)$

$= (P + 12) \sin\alpha / [4P + (P + 12) \cos\alpha]$

Cross multiplying and eliminating α, P = 12 N

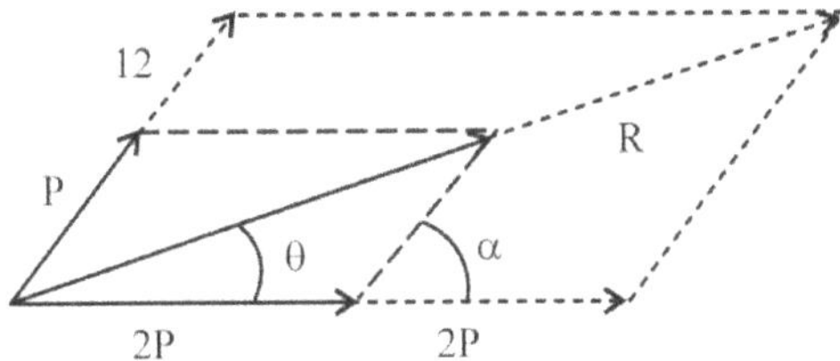

Alternative Method

Since BC and DE are parallel, from similar Δs ABC and ADE,

$BC/DE = AB/AD \Rightarrow P/(P + 12) = 2P/4P \Rightarrow P = 12\text{ N}$

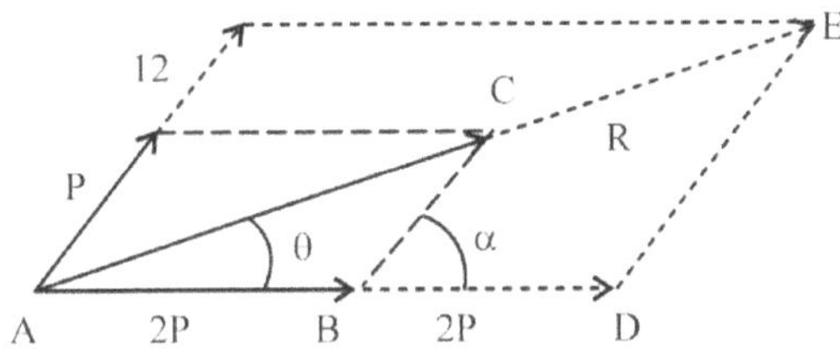

Example 7

The resultant of two forces P and Q is 14kN when they act at 60^0. The same forces produce a resultant of 12kN when they act at 90^0. Determine the magnitudes of the two forces.

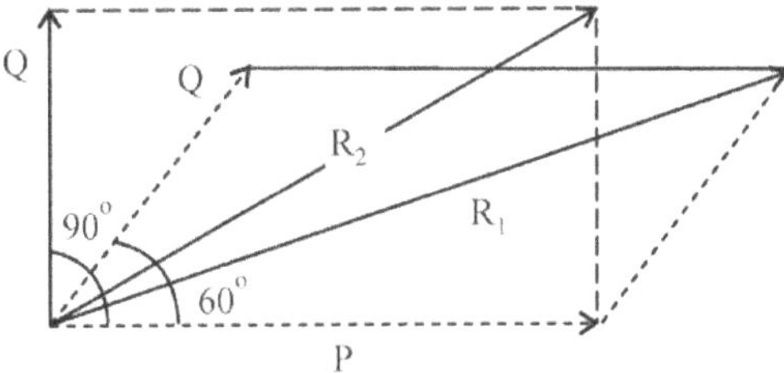

Solution:

When the two forces are inclined at 60^0,

$$P^2 + Q^2 + 2PQ\cos 60 = R_1^2 = 14^2 \quad .....(2.5)$$

When the two forces are inclined at 90^0,

$$P^2 + Q^2 = R_2^2 = 12^2 \quad .....(2.6)$$

Substituting eq.(2.6) in eq.(2.5),

$$12^2 + 2\,P\,Q\,(1/2) = 14^2 \Rightarrow PQ = 14^2 - 12^2 = 52$$

Then, $(P + Q)^2 = (P^2 + Q^2) + 2\,(P \times Q) = 12^2 + 2 \times 52 = 248$

$$\Rightarrow P + Q = 15.748 \qquad .....(2.7)$$

and $(P - Q)^2 = (P^2 + Q^2) - 2\,(P \times Q) = 12^2 - 2 \times 52 = 40$

$$\Rightarrow P - Q = 6.325 \qquad .....(2.8)$$

By adding & subtracting equations (2.7) & (2.8), we get

$$P = (15.748 + 6.325)/2 = 11.0365\text{kN}$$

and $Q = (15.748 - 6.325)/2 = 4.7115\text{kN}$

2.5 NORMAL REACTION

Whenever two bodies are in contact with each other and are in equilibrium, a force exerted by one body will produce a matching reaction from the other body. This is in line with the ***Newton's third law of motion*** – To every action, there is an equal and opposite reaction.

A body of weight 'W' resting on a table, is in static equilibrium, due to an equal and opposite reaction 'N' by the table onto the body. The reaction has no fixed value and varies with increasing load or its component at the point of contact. It always acts along the common normal to the two bodies at the point of contact. Hence, it is also called ***Normal Reaction***.

For a cylinder or sphere resting on a plane, the common normal (at A) passes through the center of the cylinder or sphere.

Mere contact between two surfaces (at B or C) **does not imply presence of reaction force between them**. Reaction comes into play between the two surfaces in contact, only when a force is applied through the point of contact.

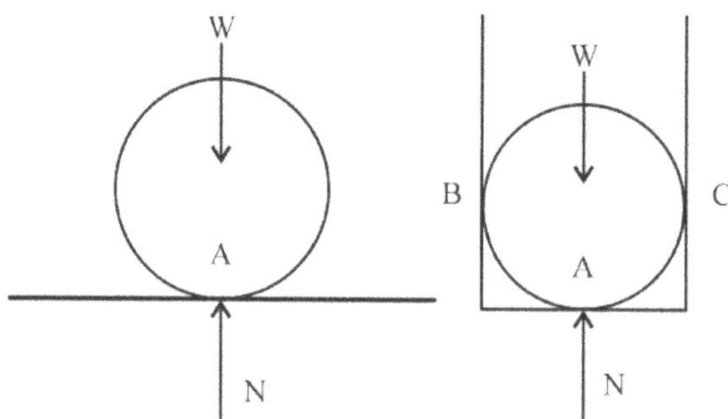

FIGURE 2.11 Contact points and normal reactions

These simple principles apply to every practical situation. A few examples illustrate application of these principles for solving problems.

Example 5

A sphere of weight 120N is tied to a smooth wall by a string. Find the tension in the string and reaction R of the wall.

Solution:

Weight 'W' is to be supported by the vertical component of force in the string. Since the string is inclined to the vertical at an angle 30^0, it will have a horizontal component against the wall. This action produces a normal reaction by the wall against the sphere, passing through its center.

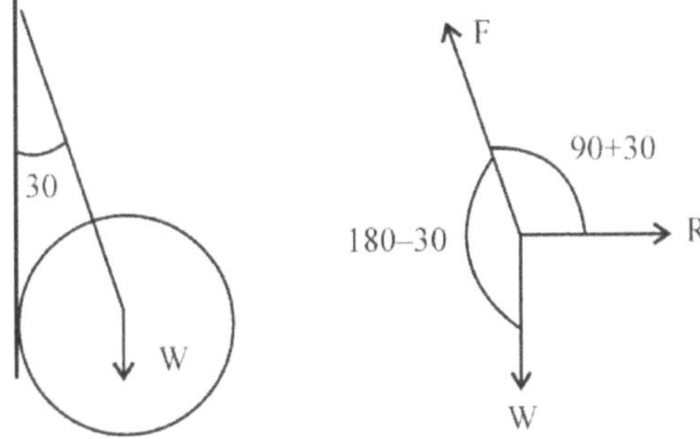

Using Lami's theorem at the center of sphere,

W/ sin 120 = F/ sin 90 = R / sin 150

or F = W/ sin 120 = 136.6 N

and R = F sin 150 = 68.3 N

Example 6

Two smooth spheres A and B, each of radius 10cm and weight 'W' of 100N, rest in a 30cm wide horizontal channel having vertical walls, as shown. Find the reactions on the wall and the floor, at the points of contact.

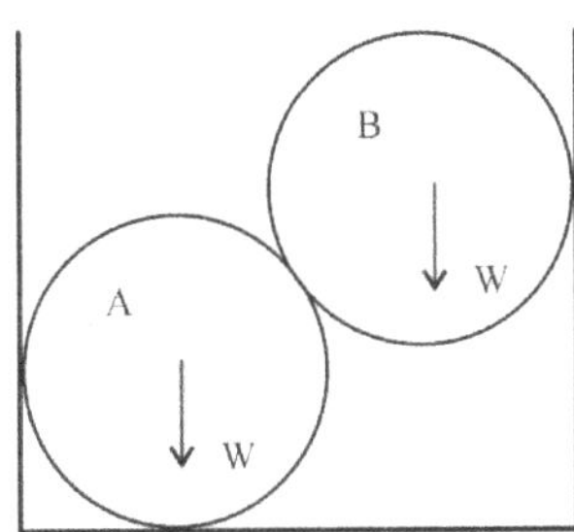

Solution:

Let us consider the equilibrium of sphere B. Its weight has to be supported by forces at the points of contact C and D. The vertically downward force 'W' acts on sphere A through the point of contact C between them. This action produces a reaction 'R_C' along the common normal from A to B, such that its vertical component balances the weight.

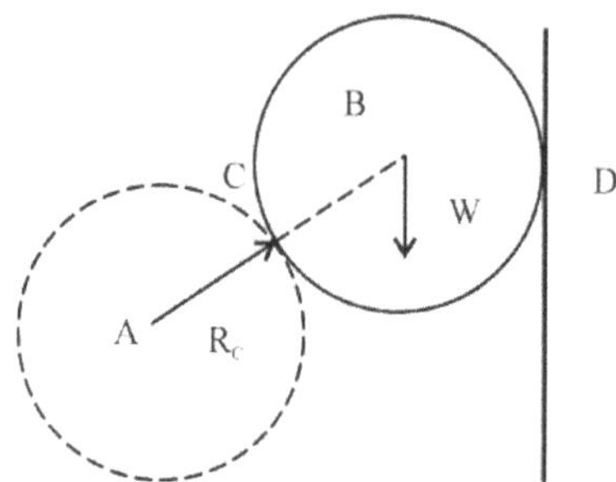

The normal reaction R_C also has a horizontal component which acts on the side wall at D. This action produces a normal reaction 'R_D' by the wall. Normal to a sphere at any point on the boundary passes through the center. Thus, the two normal reactions through C and D, supporting sphere B, pass through the center of sphere B.

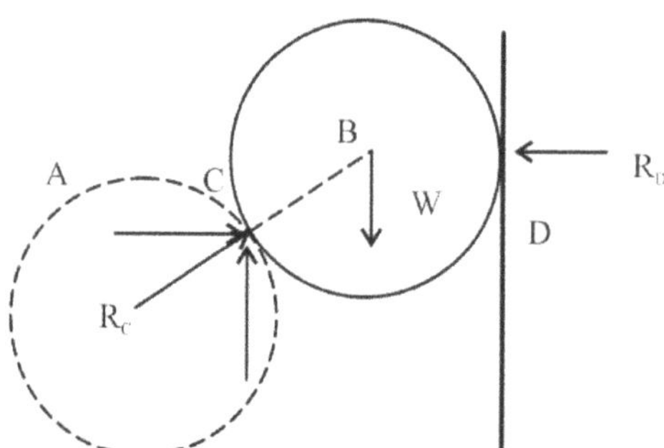

An important step in solving the problem is the calculation of the direction of reaction at C, which is not specified but can be calculated with the given geometrical data. In Δ ABE, the horizontal distance between the centers of the spheres,

$$AE = FG - FA - EG = FG - FA - BD$$
$$= 30 - 10 - 10 = 10 \text{ cm}$$

and $AB = AC + CB = 10 + 10 = 20$ cm

Then, $\cos\theta = AE / AB = 10 / 20$
$$= 0.5 \quad \text{or} \quad \theta = 60^0$$

The free body diagram at the center of the sphere B, with all force vectors drawn away from B, now has three forces – weight 'W', and reactions R_C and R_D at C and D respectively as shown. Since the sphere B is in equilibrium under the influence of these three concurrent forces, Lami's theorem can be applied to solve for the two unknown reactions.

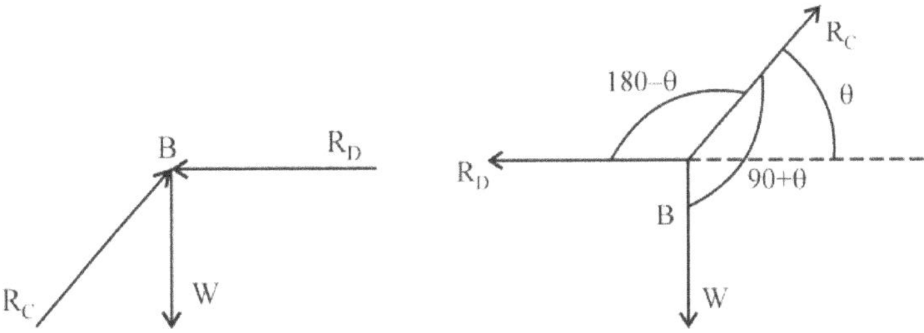

Using Lami's theorem at the center of sphere B,

$$R_D / \sin(90 + \theta) = R_C / \sin 90 = 100 / \sin(180 - \theta)$$

or $\quad R_D = 100 \sin(90 + \theta) / \sin(180 - \theta) = 100 \cos\theta / \sin\theta$

$$= 100 / \sqrt{3} = 57.75 \text{ N}$$

and $\quad R_C = 100 \sin 90 / \sin(180 - \theta) = 100 / \sin\theta = 100 \times 2 / \sqrt{3} = 115.5 \text{ N}$

Sphere B applies force on sphere A in the form of normal reaction through C along BA. The vertical component of this reaction as well as weight of sphere A are balanced by the vertical normal reaction at I while the horizontal component of force at C has to be balanced by the normal reaction at F.

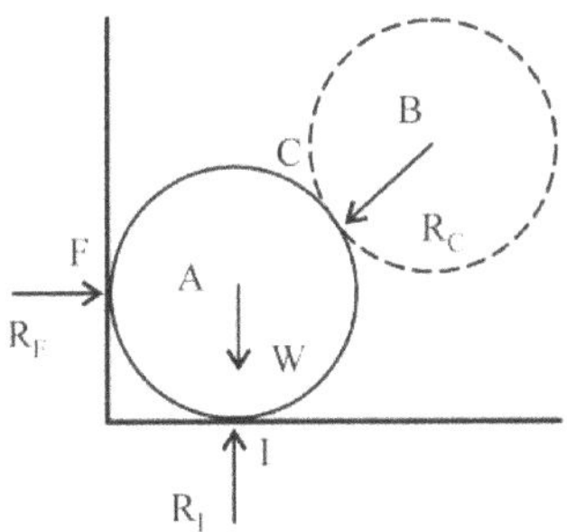

The free body diagram at the center of the sphere A now has three forces – weight 'W', and reactions R_C, R_E and R_I at C, E and I respectively as shown.

Considering equilibrium of sphere A, with X and Y axes as shown,

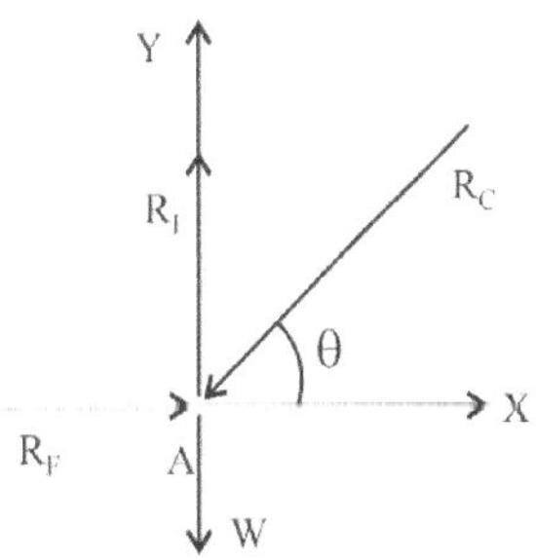

$$\sum F_X = R_F - R_C \cos\theta = 0$$

⇨ $R_F = R_C \cos\theta = 115.5 \times 0.5 = 57.75$ N

and $\sum F_Y = R_I - W - R_C \sin\theta = 0$

⇨ $R_I = W + R_C \sin\theta = 100 + 115.5 \times \left(\sqrt{3}/2\right)$

$= 100 + 100 = 200$ N

Example 7

Three identical cylinders, each weighing 'W', are stacked as shown on smooth surfaces inclined at an angle 'θ' with the horizontal. Determine the smallest angle θ to prevent the stack from collapsing.

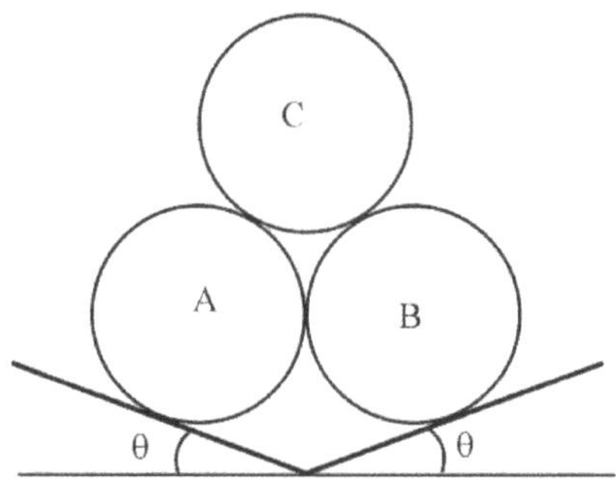

Solution:

An important aspect that has to be understood to solve such problems is - *On the verge of collapse, the cylinders A and B lose their contact and hence, reaction between them is zero*. Normal reactions are, therefore, considered at the four points of contact – D, E, F and G

Weight of cylinder C has to be supported by the reactions R_{AC} and R_{BC} from the cylinders A and B through the points of contact F and G, towards the center of cylinder C.

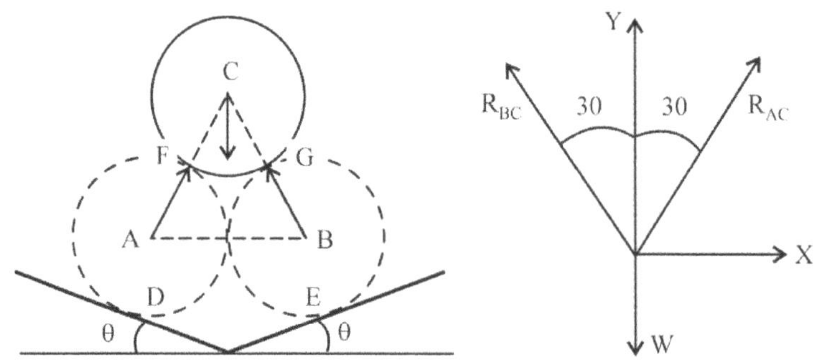

From the equilateral triangle ABC, $\angle C = 60^0$. Since it is symmetrically placed w.r.t. the vertical line through C, orientation of reactions at F and G w.r.t. the vertical is 30^0.

Considering equilibrium of cylinder C,

$$\sum F_Y = R_{AC} \cos 30 + R_{BC} \cos 30 - W = 0$$

and $\sum F_X = R_{AC} \sin 30 - R_{BC} \sin 30 = 0$

Solving them, we get $R_{AC} = R_{BC} = W / (2 \cos 30) = W/\sqrt{3}$

Weight of cylinder C acts on cylinders A and B through reactions R_{AC} and R_{BC} at F and G. These forces as well as weights of cylinders A and B have to be supported by the reactions R_D and R_E from supports at D and E.

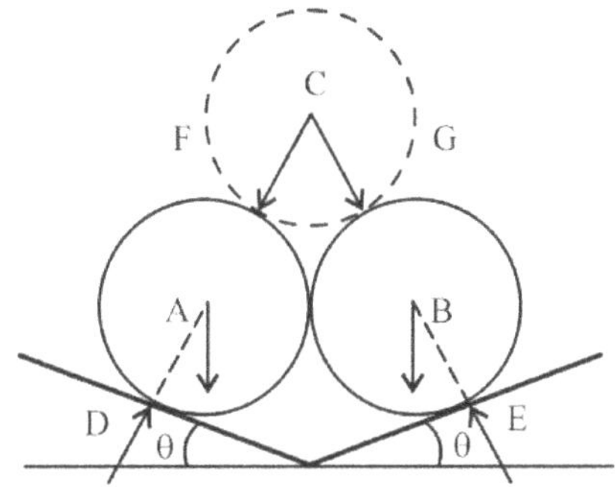

Note : Inclination of DA and EB are related to the inclination of inclined plane while Δ ABC is always an equilateral triangle of side '2 × radius of cylinder'. Hence, DA and AC are not collinear, as also EB and BC.

Since the cylinders are placed symmetrically w.r.t. the vertical line through C, the reactions at D and E will be identical in magnitude.

Thus, considering equilibrium of cylinder A,

$$\sum F_Y = R_D \sin (90-\theta) - R_{AC} \cos 30 - W = 0$$

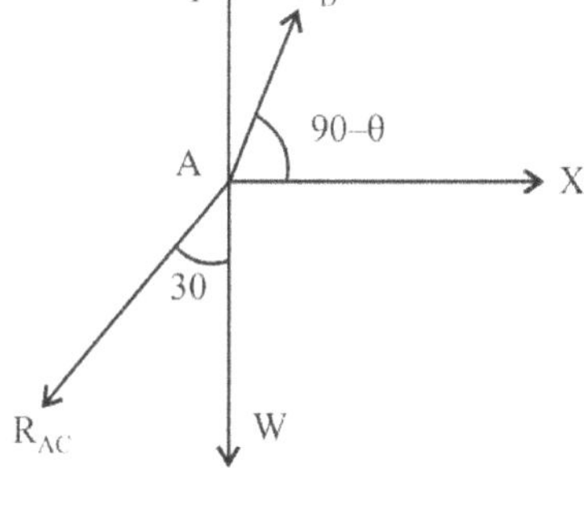

and $\sum F_X = R_D \cos (90-\theta) - R_{AC} \sin 30 = 0$

Solving them, we get

$$\tan \theta = R_{AC} \sin 30 / (R_{AC} \cos 30 + W)$$

$$= (W / 2\sqrt{3}) / (W / 2 + W)$$

$$= 1 / (3\sqrt{3})$$

or $\theta = 10.9^0$

If the angle $\theta < 10.9^0$, the horizontal components of reactions R_D and R_E will not be adequate to balance the horizontal components of R_{AC} and R_{BC} and the cylinders will fall apart.

Example 8

A uniform wheel 600mm diameter and weighing 30kN rests against a rigid block 150mm thick. Find the least pull through the center of the wheel to just turn the wheel over the corner of the block.

Solution

An important aspect that has to be understood while solving such problems is - *On the verge of lifting up, the wheel loses its contact at B with the ground and hence, reaction between them is zero*. The wheel is, therefore, in equilibrium under the influence of weight 'W', applied force 'P' and reaction 'R' from the point of contact A through the center of the wheel.

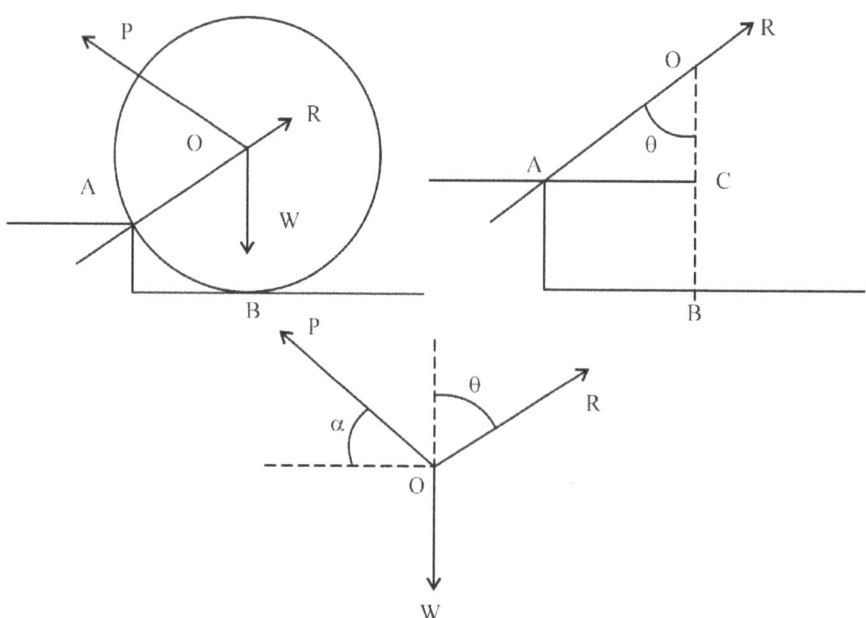

In order to solve the problem, orientation of the reaction has to be determined with the given geometric data.

Here, OA = OB = Radius of cylinder = 300/2 mm

From Δ AOC, $\cos\theta = OC / AO = (OB - BC) / AO$

$= (300 - 150) / 300 = 1/2$ or $\theta = 60$

Method-1: The resulting free body diagram is shown here, with the inclination of force P with the horizontal assumed to be α.

Angle between R & W = $180 - \theta = 120^0$

Angle between P & W = $90 + \alpha$

and Angle between P & R = $90 + \theta - \alpha = 150 - \alpha$

For equilibrium of the wheel, applying Lami's theorem,

$P / \sin 120 = W / \sin (150-\alpha) = R / \sin(90+\alpha)$

or $P = W \sin 120 / \sin (150-\alpha)$

Therefore, P is minimum when the denominator is maximum

i.e., $\sin (150 - \alpha) = 1$ or $\alpha = 60^0$

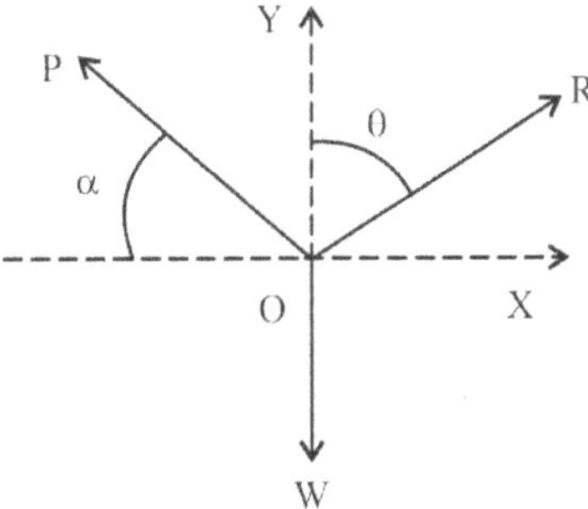

Method-2 : Resolving forces along X and Y axes,

$\Sigma F_X = R \sin \theta - P \cos \alpha = 0 \Rightarrow R = P \cos \alpha / \sin \theta$

$\Sigma F_Y = R \cos\theta + P \sin \alpha - W = 0$

$\Rightarrow$ $(P \cos\alpha/\sin \theta) \cos\theta + P \sin \alpha = W$

$P [\cos\alpha \cos\theta + \sin \alpha \sin \theta] = W \sin \theta$

$P (\cos \alpha + \sqrt{3} \sin \alpha)/2 = (\sqrt{3} / 2) W$

P is minimum when $\cos \alpha + \sqrt{3} \sin \alpha$ is maximum

or $\dfrac{d}{d\alpha} (\cos \alpha + \sqrt{3} \sin \alpha) = 0$

$-\sin \alpha + \sqrt{3} \cos\alpha = 0$ or $\tan \alpha = \sqrt{3}$ $\Rightarrow$ $\alpha = 60^0$

SUMMARY

1. Resultant 'R' of two forces 'P' and 'Q', inclined at an angle θ is $R = \sqrt{P^2 + Q^2 + 2\,P\,Q\cos\theta} = \sqrt{P^2 + Q^2}$ if P and Q are perpendicular. The angle 'α' made by the resultant 'R' with the direction of force P, is obtained from $\tan\alpha = Q\sin\theta / (P + Q\cos\theta) = Q / P$ if P and Q are perpendicular

2. Components of any force 'F' along two perpendicular axes X and Y are

   $F_X = F_X \times \cos\theta$ where θ is the angle between 'F' and +X-axis

   $F_Y = F_X \times \sin\theta$ or $= F_X \times \cos\alpha$ where α is the angle between 'F' and +Y-axis

3. For a system of co-planar concurrent force $P_1, P_2, P_3, \ldots$, components of the resultant are $R_X = \sum P_X = P_{1X} + P_{2X} + P_{3X} + \ldots$

   and $R_Y = \sum P_Y = P_{1Y} + P_{2Y} + P_{3Y} + \ldots$

   Resultant force is given by $R = \sqrt{R_X^2 + R_Y^2}$ and $\alpha = \tan^{-1}(R_Y/R_X)$ since R_X and R_Y are perpendicular

4. According to Newton's 1st law of motion, if a body, acted upon by a system of co-planar concurrent forces, is in static equilibrium, then ***equations of equilibrium*** to be satisfied simultaneously are

   $$R_X = \sum P_X = 0 \quad \text{and} \quad R_Y = \sum P_Y = 0$$

5. Two forces acting on a body can keep it in equilibrium only when they are equal and opposite.

6. Force polygon of three concurrent forces in equilibrium (forces represented to a common scale and drawn one after the other with proper orientation) forms a closed triangle. Sine rule of triangle applied to a force polygon is called ***Lami's theorem***, which states that the ratio of length of any side (representing a force) to the angle opposite to it is a constant.

   $$P / \sin\alpha = Q / \sin\beta = R / \sin\gamma$$

   It is also equal to $P/\sin(180-\alpha) = Q/\sin(180-\beta) = R/\sin(180-\gamma)$

   or $P / \sin\alpha_1 = Q / \sin\beta_1 = R / \sin\gamma_1$

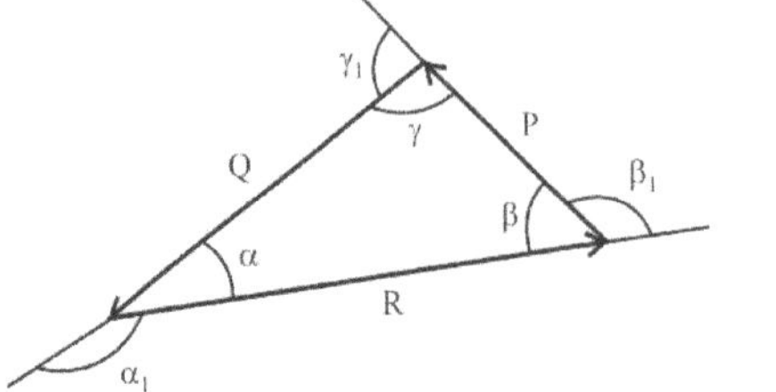

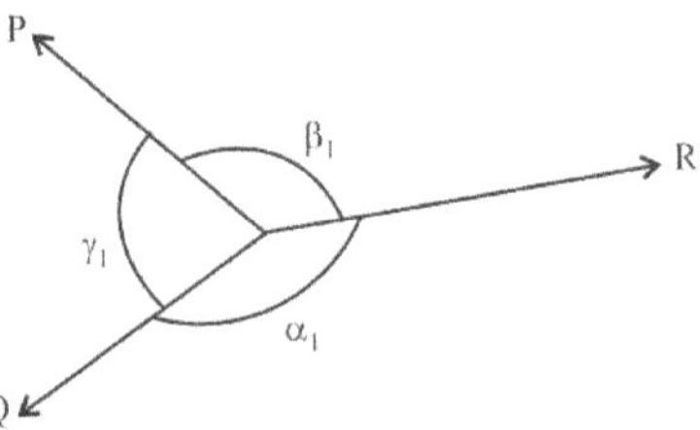

PROBLEMS FOR PRACTICE

Problem 1: An electric light fixture weighing 20N hangs from a point by two strings inclined at 30^0 and 45^0 to the vertical respectively. Determine forces in the strings.

(*Ans:* 14.64N, 10.35N)

Problem 2: Two identical rollers, each of weight 100N, are supported by an inclined plane and a vertical wall as shown in figure. Assuming smooth surfaces, find the reactions induced at the points of support A, B and C.

(*Ans:* 86.6N, 144.34N, 115.47N)

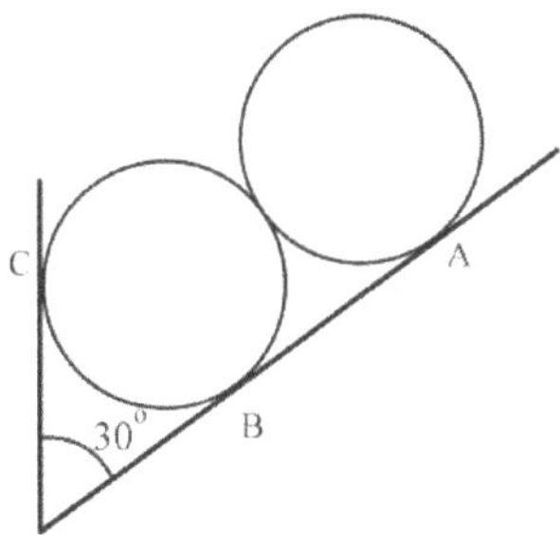

Problem 3: Two forces equal to 2P and P respectively act on a particle. If first be doubled and the second increased by 12N the direction of the resultant is unaltered, find the value of 'P' ?

(*Ans:*12N)

Problem 4: A 4500N load is attached to a pin at C as shown. Determine the forces acting in members AC and BC.

(*Ans:*7.5kN, 6.0kN)

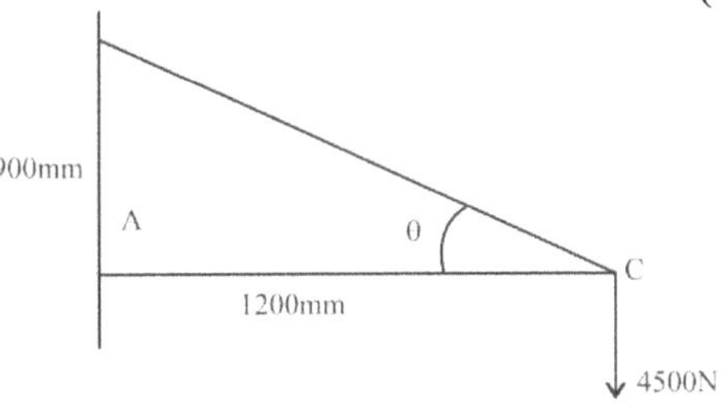

Problem 5: A rubber band has an unstretched length of 20cm. It is fixed at its two ends and pulled in the middle by a force of 6N, until its length is 25 cm. Calculate the tension in the rubber band.

(*Ans:*5N)

Problem 6: Three identical cylinders ofradius 50cm and weight 1000N are kept on a flat plate, as shown, with the centers of the two lower cylinders connected by a 120cm long string. Calculate tension in the string and support reactions.

(***Ans:***375, 1500,1500N)

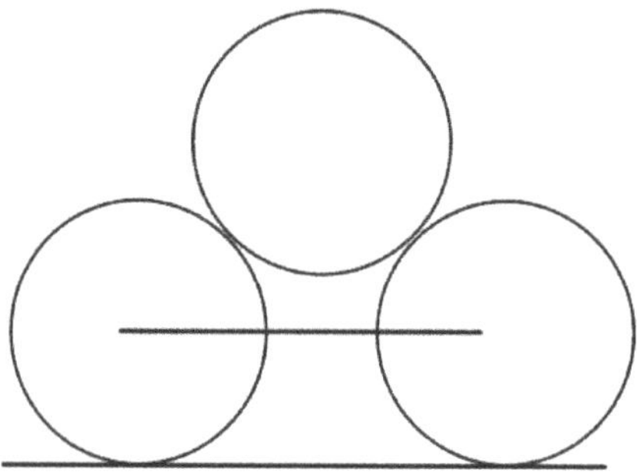

Problem 7: Two equal spheres A and B of radius 'r' and weight 'W' are in equilibrium within a smooth cup of radius '3r'. Calculate reactions at the points of contact.

(***Ans:***1.155W ; 1.155W ; 0.5775W)

Problem 8: Two rollers A and B of weights 35N and 65N rest in a bigger cylinder, with C as its center as shown, and are connected by a rod AB. If the length AB and the radius of the bigger cylinder are such that $\angle ACB = 90^0$, find the force in the rod and the angle θ that it makes with the horizontal, when the system is in equilibrium.

(***Ans:***41.08N, $18^0 26'$)

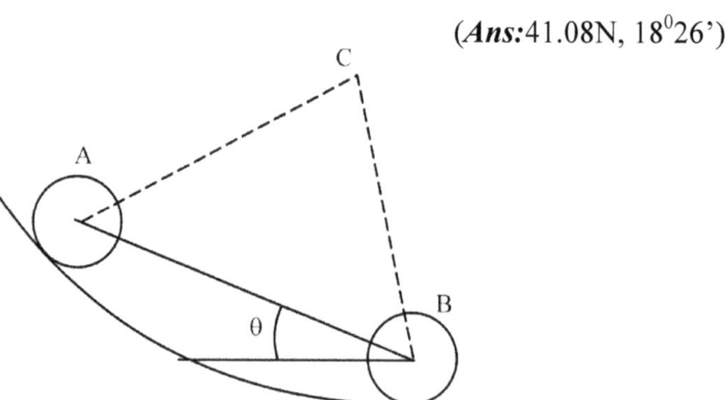

Problem 9: Three equal strings of negligible weight are knotted together to form an equilateral triangle ABC and a weight of 100N is suspended from A. If the triangle and the weight are supported by two strings attached at B and C so that BC is horizontal, as shown, determine the tension in BC.

(*Ans:*19.405 N)

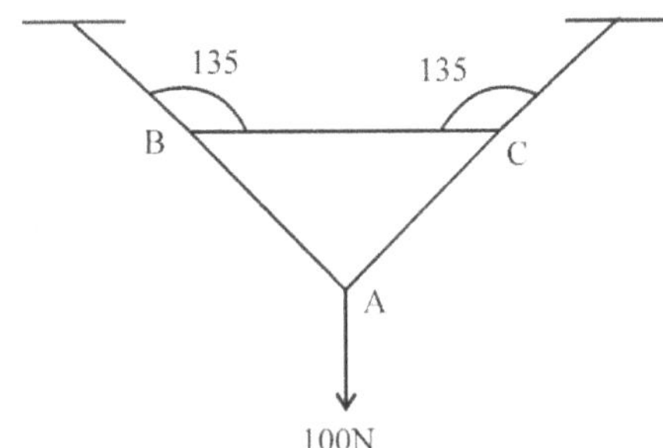

Problem 10: Two unequal forces inclined to one another at an angle of 120^0 have a resultant of 86.6N at an angle of 30^0 with one of the forces. Find the magnitudes of the two forces

(*Ans:*50N, 100N)

Problem 11: The resultant of two forces acting at an angle of 60^0 is 70N. If they act at right angles, their resultant is $\sqrt{3400}$N. Find the magnitude of the forces

(*Ans:*30N, 50N)

Problem 12: Forces of magnitudes 10N, 20N, 30N, 40N, 50N and 60N respectively act from the center of a regular hexagon towards its vertices, in order (starting from –X axis in clockwise direction). Find the magnitude and direction of the resultant force

(*Ans:*60kN at 60^0 from +X axis in 4th quadrant)

Problem 13: Concurrent forces of 100N at 30^0 from +X axis, 400N at 225^0 and 100N at 270^0 are acting at a point. Find the magnitude and direction of force P acting at the same point, so that the resultant of these four forces is 400N along +X axis.

(*Ans:*682.85N at 150.75^0 from +X axis)

Problem 14: The resultant of two forces, one of which is double the other, is 260N. If the direction of the larger force is reversed and the other remains unaltered, the resultant reduces to 180N. Determine the magnitude of the forces and the angle between them

(*Ans:*100N, 200N, 63.89^0)

Problem 15: Four forces of magnitudes 10kN, 15kN, 20kN and 40kN are acting at a point. The orientation of these forces with +X axis are 30^0, 60^0, 90^0 and 120^0 respectively. Find the magnitude and direction of the resultant

(*Ans:*72.74kN, 93.03^0)

Problem 16: A cylinder of weight 200N is held against a smooth inclined plane by means of a weightless rod PQ as shown. Determine the forces exerted on the cylinder by the plane and the rod.

(***Ans:***192.8N by plane and 150.5N by rod)

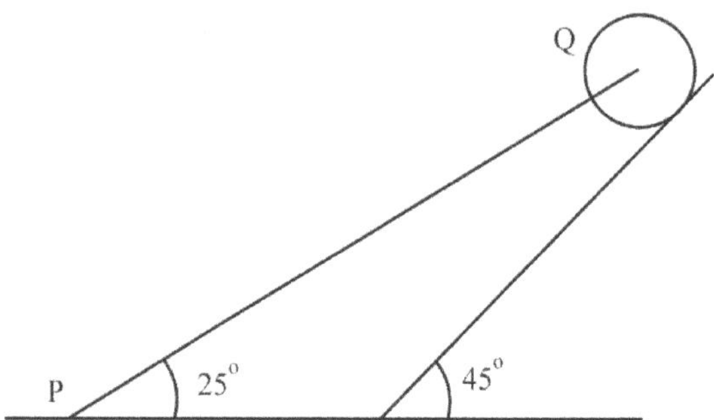

CHAPTER 3

COPLANAR NON-CONCURRENT FORCES

If the lines of action of different forces acting on a body do not meet at a single point, they are called ***non-concurrent forces***. A special case of two parallel (non-collinear) forces P and Q acting on a body can not keep the body in equilibrium. Such forces tend to rotate the body in clockwise (Ref Fig 3.1 a) or counter-clockwise or anti-clockwise (Ref Fig 3.1 b) direction, depending on the way forces P and Q act on the body. This effect of a force to rotate a body is called ***moment of the force***. Therefore, equilibrium of a body depends not only on forces but also on moments produced by these forces.

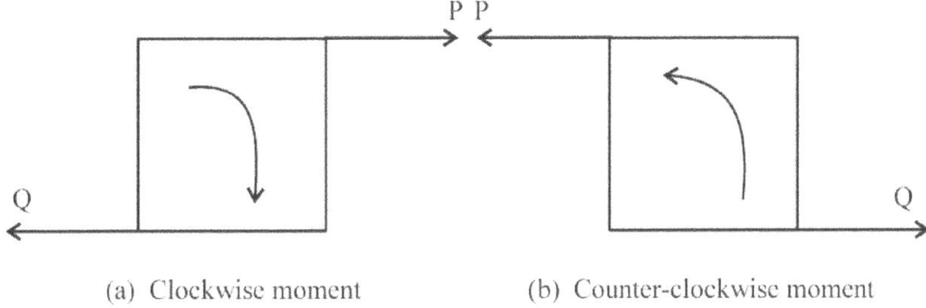

FIGURE 3.1 Two equal parallel and non-collinear forces producing moment

3.1 MOMENT OF A FORCE

It is the rotational effect of any force vector in X-Y plane about any point, whose magnitude is the product of force and its normal distance 'd' from the point (or moment center) 'O'. It is represented by clockwise or counter-clockwise rotation about the point, following right hand screw rule. For the forces in the X-Y plane, moments are identified about Z-axis and, therefore, represented by M_Z. Moment of a force about any point on the line of action of force is zero, since the normal distance 'd' from the point 'O' to the force 'P' is zero. Moment of a force 'P' passing through a point 'A', about a point 'O' at a distance 'd' can be obtained either by determining normal distance 'a' or by resolving the force into two components 'P_X' and 'P_Y' along OA and perpendicular to OA and adding the moments of the force components.

$$M_Z = P_X \times o + P_Y \times d = P_Y \times d = (P \cos \theta) \times d$$
$$= P \times (d \cos \theta) = P \times a$$

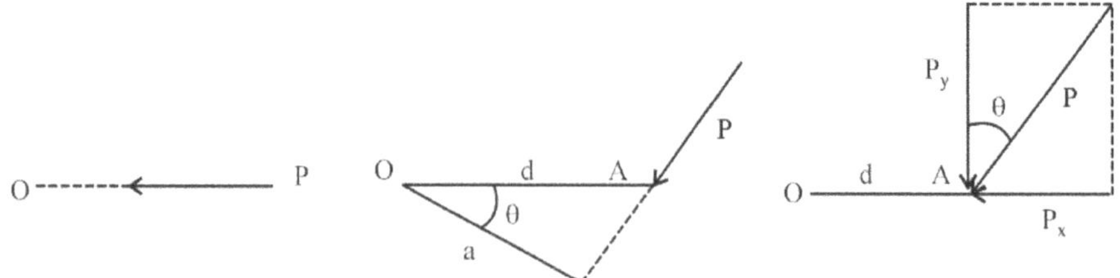

FIGURE 3.2 Moment of a force about a point O

3.1.1 SIGN CONVENTION

Counter-clockwise or anti-clockwise moment is designated as +ve (just as any angle is measured) and clockwise moment as –ve. Moments of forces in the two different cases are given in Fig 3.3.

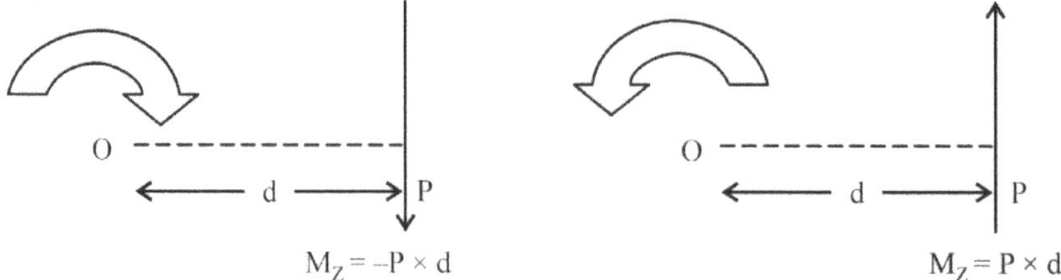

FIGURE 3.3 Sign convention for Moments of a force

3.1.2 COUPLE

It is a pair of equal and parallel (non-collinear) forces in opposite directions, separated by a distance 'a'. Resultant moment of these two forces at a distance of 'd' and 'd+a' from any point 'O' in that plane (Ref Fig 3.4) is

$$M_Z = P \times (d + a) - P \times d = P \times a$$

Thus, magnitude of the moment of a couple depends on the magnitude of the forces 'P' and the normal distance between the forces 'a'. It is independent of the point 'O' about which moments of both the forces are considered.

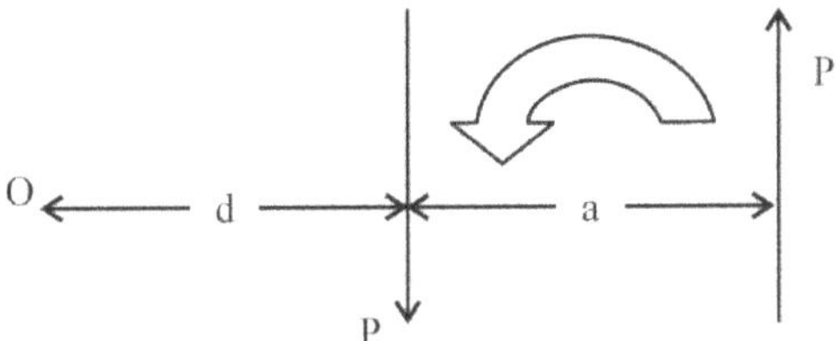

FIGURE 3.4 Moment of a couple

3.1.3 VARIGNON'S THEOREM (OR LAW OF MOMENTS)

Moment of resultant of a number of coplanar forces about any point is equal to the algebraic sum of moments of individual forces about the same point.

For **any** system of coplanar non-concurrent forces, in **equilibrium**,

- The sum of all force components along any direction, acting on the component, is equal to zero. Usually force components are resolved along two perpendicular directions and, therefore,

$$H = \sum F_X = 0 \quad \text{and} \quad V = \sum F_Y = 0$$

- The sum of moments of all the forces acting on the component, about any point, is equal to zero. $M_Z = \sum (M_Z)_i = 0$

3.1.4 SOME PRACTICAL EXAMPLES OF MOMENTS

We come across many situations in our daily life, where moments of forces are used.

(a) See-saw with equal weights: In any play ground, it is one of the most commonly used facilities by children. It uses a system of parallel forces and demonstrates moment balance for equilibrium

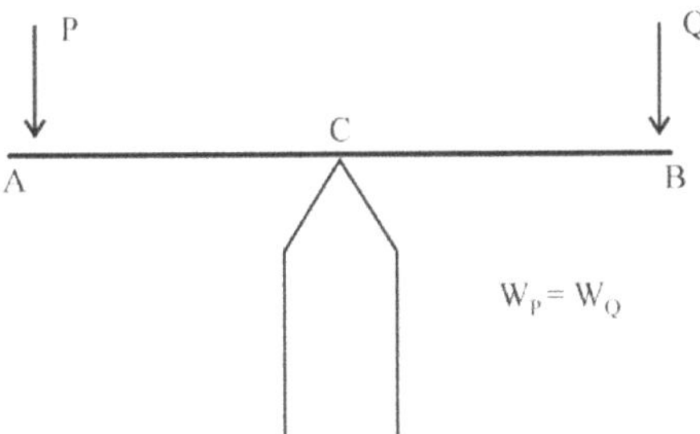

If the two persons P and Q are of equal weight and the fulcrum or pivot C is at the mid point of bar AB such that AC = CB, taking moments about C,

$$M = W_P \times AC - W_Q \times BC = 0$$

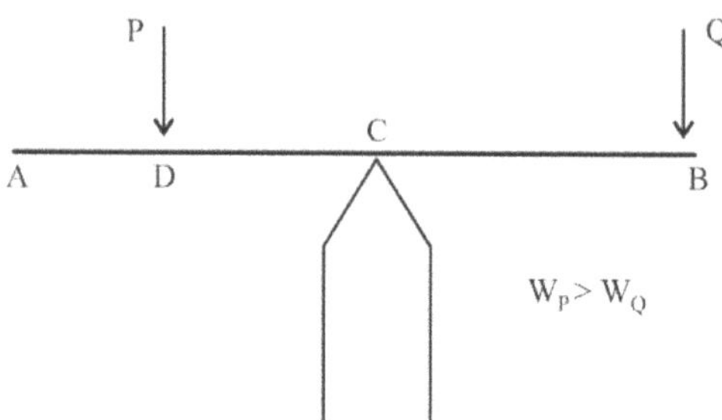

Therefore, the bar AB is in equilibrium and remains horizontal.

(b) See-saw with unequal weights: If the two persons P and Q are of unequal weight ($W_P > W_Q$) and the person P is seated at a point D closer to the fulcrum or pivot C such that

$$M = W_P \times DC - W_Q \times BC = 0$$

Then, the bar AB is in equilibrium and remains horizontal.

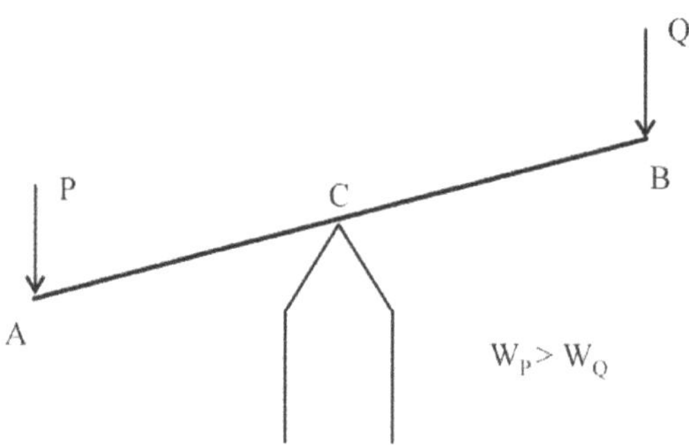

But, if the two persons P and Q, of unequal weight, are seated at equal distance from fulcrum C, then

$$M = W_P \times DC - W_Q \times BC \neq 0$$

Therefore, the bar tilts down until it touches the ground

(c) Simple balance with equal weights: When items like sugar, pulses, .. are purchased in a shop, a balance with known weight P on one side and the material purchased Q on the other side will produce moments about the fulcrum C.

If AC = BC and the bar AB remains horizontal

$$M = W_P \times AC - W_Q \times BC = 0$$

$$\Rightarrow \quad W_Q = W_P$$

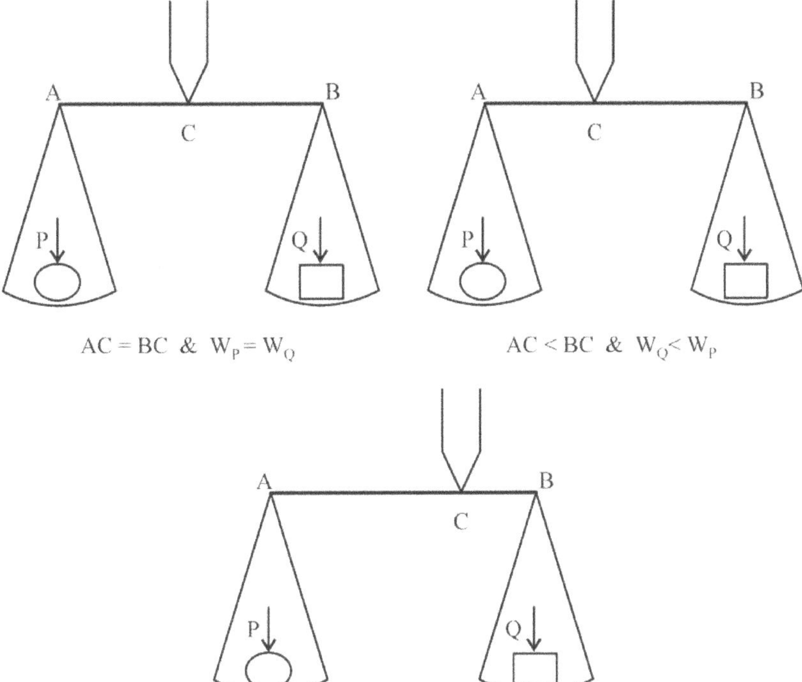

(d) Simple balance with unequal weights: If the seller wants to cheat the customer by using a balance with AC < BC,

$$M = W_P \times AC - W_Q \times BC = 0$$

and the bar AB remains horizontal implies that the customer gets lesser quantity ($W_Q < W_P$) than he is paying for and $W_Q = W_P \times AC / BC$

The same idea is also used correctly in some balances to weigh large items using small weights (ex. Railway stations)

In these balances, $AC >> BC$

so that $M = W_P \times AC - W_Q \times BC = 0$

Actual measuring weight, $W_P << W_Q$, weight of the booked item

3.1.5 IMPORTANT RULE

Three forces acting on a body *can* keep it in equilibrium only if they are parallel (Fig. a) **or concurrent** (Fig. b). Otherwise, resultant 'R' of any two forces, even if equal in magnitude and opposite in direction to the third force, ***can not*** keep the body in equilibrium due to unbalanced moment (Fig c, d).

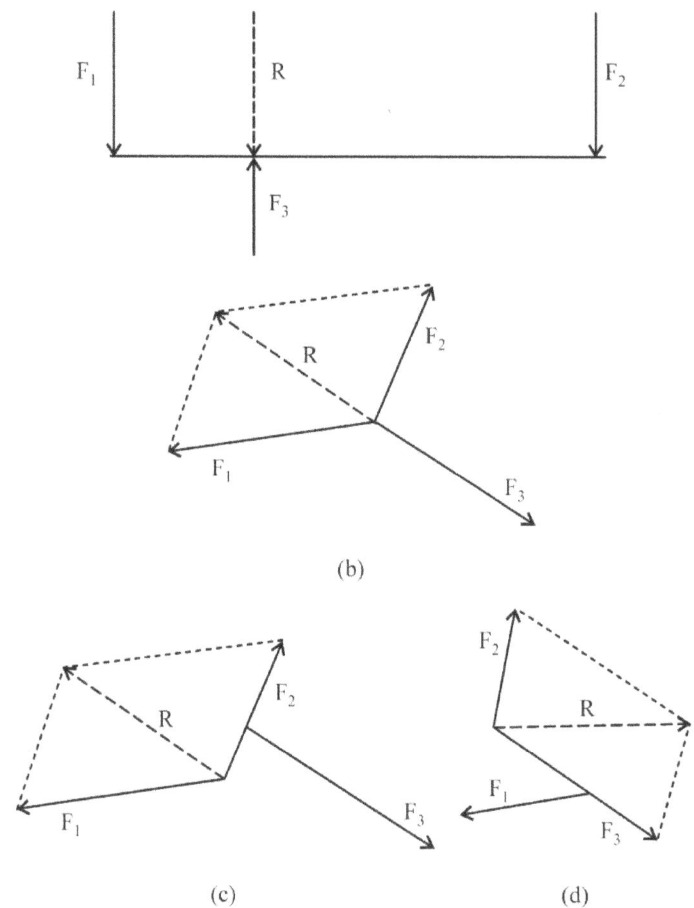

Example 1

A weight of 80N is suspended from a weightless bar AB, which is supported by a cable CB and a pin at A. Determine the tension in the cable and the reactions at A

Solution:

Considering equilibrium of point B, the weight 80N is balanced by the vertical component of force in CB. Horizontal component of force in CB needs to be balanced by the force in AB. Reaction at A will be equal and opposite to the force in member AB.

$$\text{Sin } \alpha = 4\sqrt{(4^2 + 9^2)} = 0.4061$$

Method-1 : This system of three concurrent forces at B, represented by ab, cb and db in the Free body diagram, can be solved by using Lami's theorem

F_{BC} / sin ∠abd = F_{AB} / (sin ∠cbd) = 80 / sin ∠cba

i.e., F_{BC} / sin 90 = F_{AB} / sin (90 + α) = 80 / sin (180 – α)

F_{BC} = 80 × sin 90 / sin (180–α)

= 80 × 1 / sin α = 80 / 0.4061 = 197N

and F_{AB} = 80 × sin (90 + α) / sin (180 – α)

= 80 × cos α / sin α = 80 cot α

= 80 × (9/4) = 180N

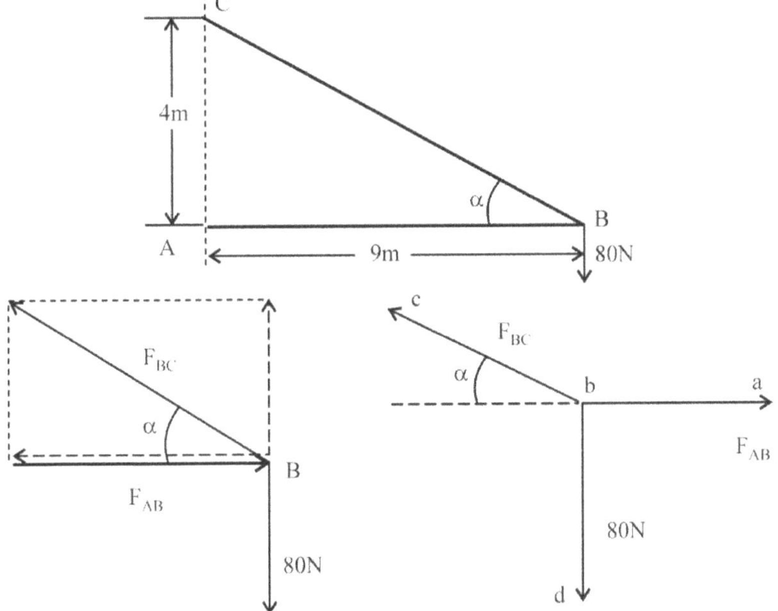

Method-2 : This problem can also be solved by taking moments

Taking moments about C, $\sum M_C = F_{AB} \times AC - 80 \times AB = 0$

i.e., $F_{AB} = 80 \times AB / AC = 80 \times 9 / 4 = 180N$

F_{BC} can now be obtained

(i) by using $\sum F_V = F_{BC} \text{Sin}\, \alpha - 80 = 0$

⇒ $F_{BC} = 80 / \text{Sin}\, \alpha = 80 / 0.4061 = 197\ N$ or

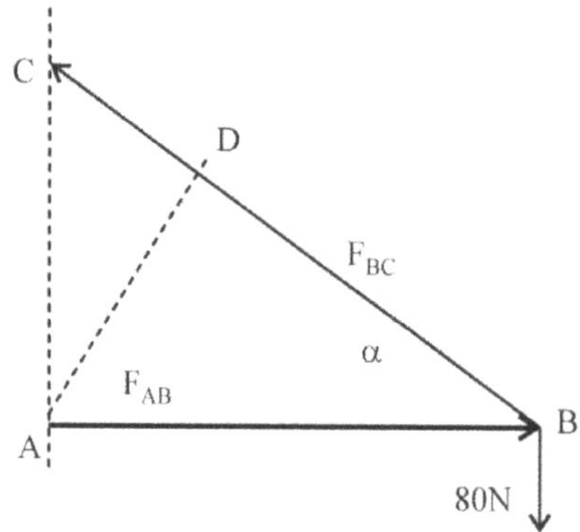

(ii) by taking moments about A, $\sum M_A = AD \times F_{BC} - 80 \times AB = 0$

$\Rightarrow \quad F_{BC} = 80 \times AB / AD = 80 \times 9 / (AB \sin \alpha)$

$= 720 / (9 \times 0.4061) = 180$ N

Example 2

How far out on the plank can a 80 kg mass person walk if the allowable crushing force on the rollers at A and B is 1500 N?

Solution:

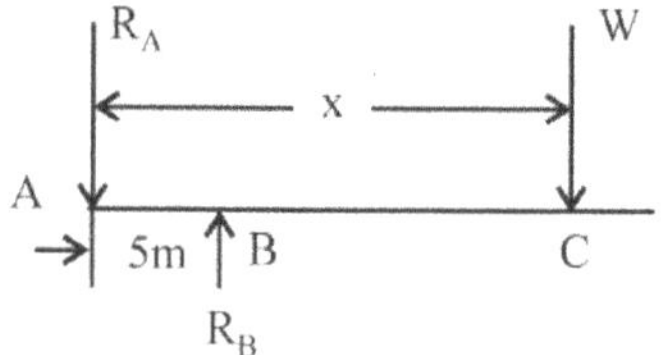

In order to balance the moments, the reactions R_A and R_B have to be in the opposite directions, as shown. Both the reactions should be limited to max 1500N. Let the max distance a person of weight 'W' can walk be 'x' from A.

Taking moments about B,

with $W = \text{mass} \times g = 80 \times 9.81$N,

$\sum M_B = 5 \times R_A - (x - 5) \times (80 \times 9.81)$

$\Rightarrow \quad x = 5 + 5 \times R_A/(80 \times 9.81) = 14.6$m

for $(R_A)_{max} = 1500$N

Magnitude of the other reaction, $R_B = R_A + 80 = 1580\text{N} >$ allowable max value.

Therefore, a solution with $(R_A)_{max} = 1500$N gives a higher than allowable value for R_B.

Let us now find a solution with $(R_B)_{max} = 1500$N.

Taking moments about A,

$\sum M_A = 5 \times R_B - x \times 80 \times 9.81$

$$x = 5 \times R_B / (80 \times 9.81) = 9.56 \text{ m} \quad \text{for } (R_B)_{max} = 1500 \text{ N}$$

Corresponding value of the other reaction,

$$R_A = R_B - 80 = 1420\text{N} < \text{allowable max value}$$

Hence, Maximum distance that the person can walk on the plank,

$$x_{max} = 9.56\text{m}$$

Example 3

A 2 m uniform bar weighs 400 N. One end is resting on smooth floor while the other end is resting against a smooth wall. Its lower end is connected with a rope to the corner, as shown, to prevent the bar from falling down. Determine tension in the rope AC.

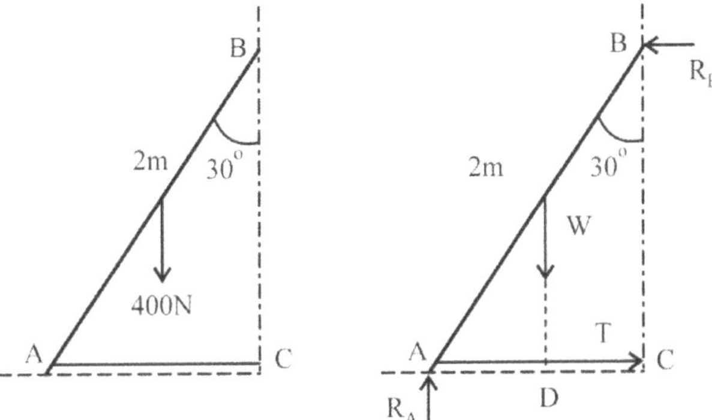

Solution:

The free body diagram, shown here, consists of weight 'W', normal reactions R_A and R_B and tension 'T' in the rope AC. Taking moments about A,

$$\sum M_A = R_B \times BC - W \times AD = 0$$

or

$$R_B = W \times AD / BC$$

$$= W \times (AB/2) \sin 30^0 / AB \cos 30^0$$

$$= W \times (2/2) \times (1/2) / [2 \times (\sqrt{3}/2)] = 115.5 \text{ N}$$

From $\sum F_X = T - R_B = 0, \quad T = R_B = 115.5$ N

Example 4

A uniform door with 18 kg mass is hinged at A and B. Determine hinge reactions at A and B, assuming vertical components of reactions at A and B to be equal.

Solution:

Weight of the door ($W = m \times g$) is equally shared by the vertical components of reactions R_{AV} and R_{BV} at A and B.

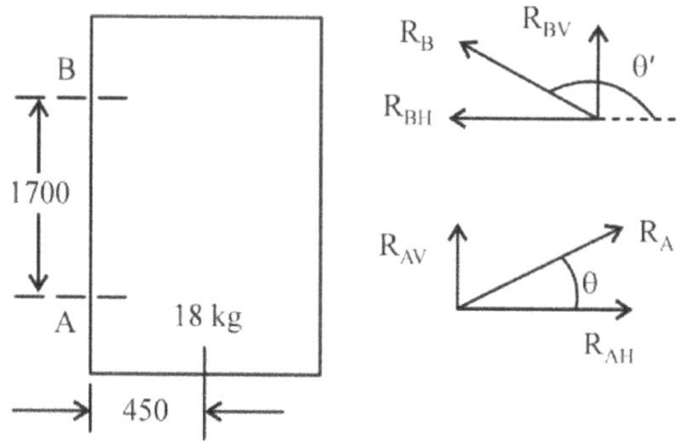

$$R_{AV} = R_{BV} = 18 \times 9.8 / 2 = 88.2 \text{ N}$$

However, since the reactions and the weight are not collinear, there is an unbalanced moment, which is balanced by equal and opposite horizontal components of reaction R_{AH} and R_{BH} at A and B, as shown.

Taking moments about A,

$$\sum M_A = R_{BH} \times 1700 - 18 \times 9.8 \times 450 = 0$$

$\Rightarrow$ $R_{BH} = 46.69\text{N}$ along – X direction

From $\sum F_X = R_{AH} + R_{BH} = 0$, $R_{AH} = -R_{BH} = 46.69$ N along +X direction

$$R_A = \sqrt{R_{AH}^2 + R_{AV}^2} = \sqrt{88.2^2 + 46.69^2} = 99.8 \text{ N}$$

and $R_B = \sqrt{R_{AH}^2 + R_{AV}^2} = 99.8 \text{ N}$

$\tan\theta = R_{AV} / R_{AH} = 88.2 / 46.69 = 1.889 \Rightarrow \theta = 62.1^0$

and $\tan\theta' = R_{BV} / R_{BH} = 88.2 / (-46.69) = -1.889 \Rightarrow \theta' = 117.9^0$

Example 5

Determine the tension in cable BC for lifting a load of 800N as shown, if $\angle CAB = 40^0$; $\angle ABC = 20^0$; AE = 120 cm and EB = 60 cm

Solution:

Tension in cable BC can be obtained by taking moments about any point, not lying along BC. In order to calculate moment about A, the normal distance for moment about A can be calculated by extending the line BC and dropping a perpendicular AD onto it from A. Then, taking moments about A,

$$\sum M_A = T_{BC} \times AD - 800 \times AG = 0$$

From Δ ABD, AD = (AE+EB) sin ∠ABD

From Δ AEF, AG = EF = AE sin ∠EAF

Substituting these values, in the moment equation,

$$T_{BC} \times (AE + EB) \sin \angle ABD = 800 \times AE \sin \angle EAF$$

$$\Rightarrow \quad T_{BC} \times (120 + 60) \sin 20 = 800 \times 120 \sin 40$$

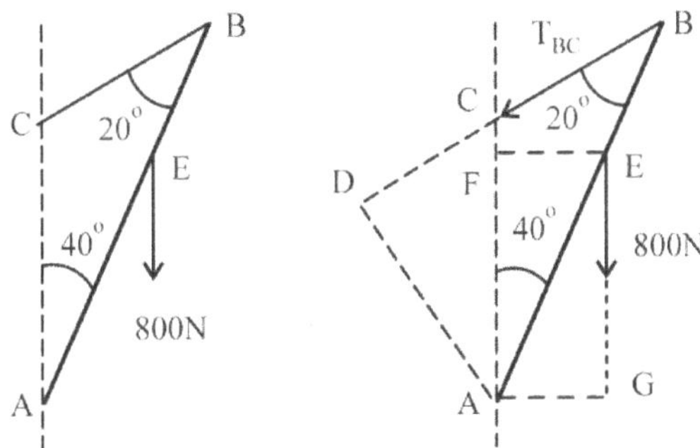

$$\Rightarrow \quad T_{BC} = [800 \times 120 \sin 40] / [(120 + 60) \times \sin 20]$$

$$= 1002.34 \text{ N}$$

Example 6

Three bars pinned together at B and C and supported by hinges at A and D are loaded as shown. Determine the value of P that will prevent motion.

Solution:

This system of coplanar non-concurrent forces can be solved treating it as two systems of coplanar concurrent forces at B and C.

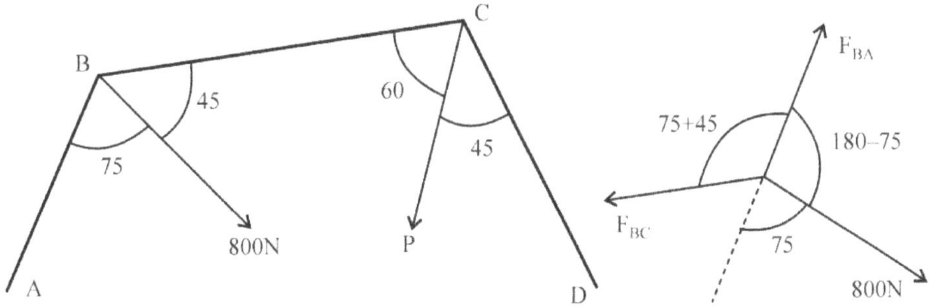

B is in equilibrium under the influence of applied force of 800N and the forces in members BA and BC. Considering the equilibrium of B, and applying Lami's theorem,

$$F_{BC} / \sin (180 - 75) = 800 / \sin (75 + 45)$$

or $\quad F_{BC} = 800 \times \sin(180 - 75) / \sin(75 + 45)$

$= 1017.8$ N

F_{CD}

45+60

F_{CB}

180–45

180–60

P

C is in equilibrium under the influence of applied force P and the forces in members CB and CD.

Since, $F_{CB} = F_{BC}$ (equal and opposite), considering equilibrium of C,

$$P / \sin(45 + 60) = F_{CB} / \sin(180 - 45)$$

or $\quad P = F_{CB} \times \sin(45 + 60) / \sin(180 - 45)$

$= 1390.6$ N

Example 7

A cylinder of weight 'W' and radius 'r' is supported in horizontal position against a vertical wall by a bar AB of negligible weight. The bar is hinged to the wall at A and supported at B by a horizontal rope BC. Find the angle θ that the bar should make with the wall so that tension in the rope is minimum.

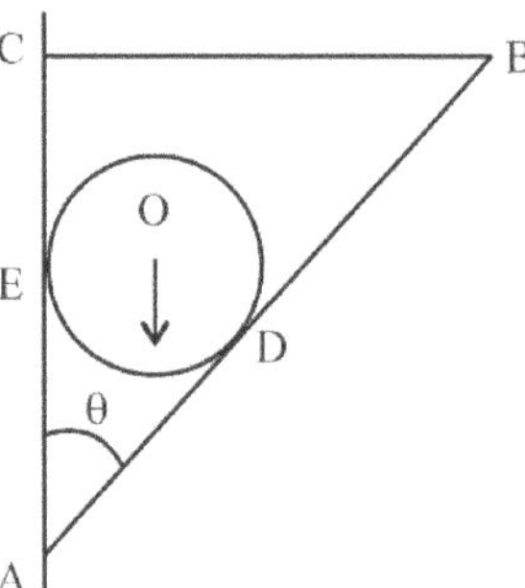

Solution:

The cylinder is supported by the normal reactions from the bar at D and from the wall at E, as shown by the free body diagram. The vertical component of reaction R_D should balance weight of cylinder while the horizontal component is balanced by reaction R_E.

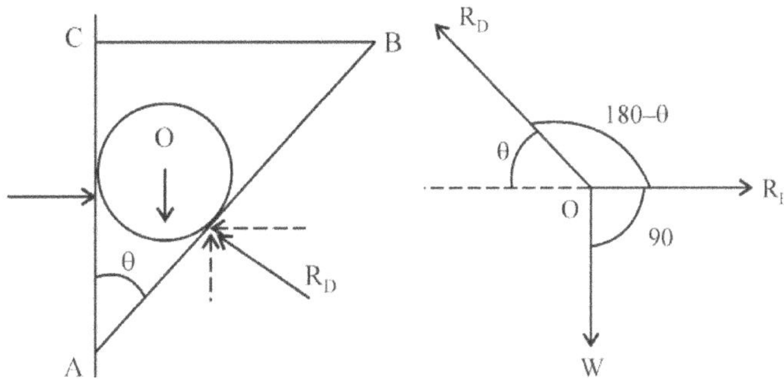

Considering equilibrium of the cylinder and applying Lami's theorem,

$$W / \sin(180 - \theta) = R_D / \sin 90$$

or $$R_D = W \sin 90 / \sin(180 - \theta) = W / \sin\theta$$

Triangles AEO and ADO are similar, with AO as the common side, EO = DO = radius of cylinder and $\angle E = \angle D = 90^0$.

Therefore, $\angle EAO = \angle DAO = \theta / 2$

and $$AD = DO \times \cot(\theta/2) = r \times \cot(\theta/2)$$

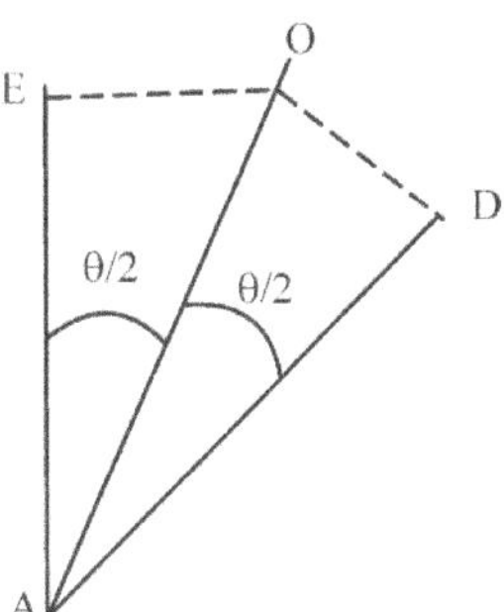

Weight 'W' of the cylinder acts on the bar as R_D, in the direction opposite to that of the reaction at D, supporting the cylinder.

Considering equilibrium of bar AB and taking moments about A,

$$\sum M_A = T \times AC - R_D \times AD = 0$$

where $$AC = AB \cos\theta = L \cos\theta$$

Hence, $$T = R_D \times AD / AC = R_D \times AD / (L \times \cos\theta)$$

$$= (W / \sin\theta) \times [r \times \cot(\theta/2)] / [L \times \cos\theta]$$

$$= [W / \{2 \sin(\theta/2) \times \cos(\theta/2)\}] \times [r \times \cot(\theta/2)] / [L \times \cos\theta]$$

$$= W \times r / [2 \sin^2(\theta/2) \times (L \times \cos\theta)]$$

$$= W \times r / [(1 - \cos\theta) \times (L \times \cos\theta)]$$

Tension in the rope T is minimum,

when $\cos\theta \times (1 - \cos\theta)$ is maximum

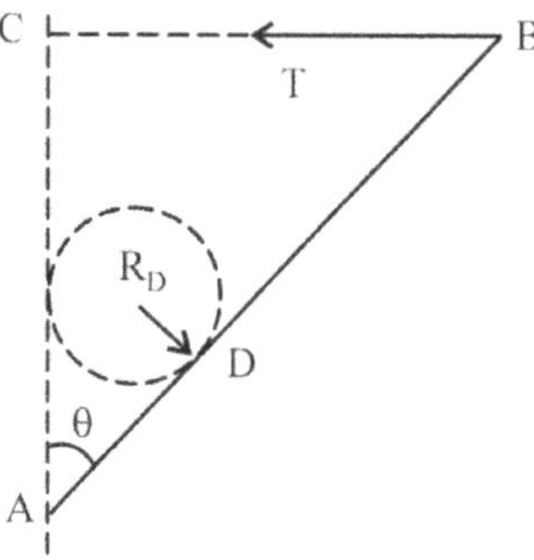

$$\Rightarrow \quad \frac{d}{d\theta}[\cos\theta(1-\cos\theta)] = \frac{d}{d\theta}[\cos\theta - \cos^2\theta)] = 0$$

or $\quad -\sin\theta - 2\cos\theta(-\sin\theta) = \sin\theta(2\cos\theta - 1) = 0$

i.e., $\quad \sin\theta = 0 \quad \Rightarrow \quad \theta = 0$

which is not physically feasible

or $\quad 2\cos\theta - 1 = 0 \Rightarrow \cos\theta = ½ \quad$ or $\quad \theta = 60^0$

Example 8

A pulley of 1m radius, supporting a load of P = 5000N is mounted at B on a horizontal beam. If the beam weighs 2000N and pulley weighs 500N, find the hinge force at C. AB = 3m and BC = 2m

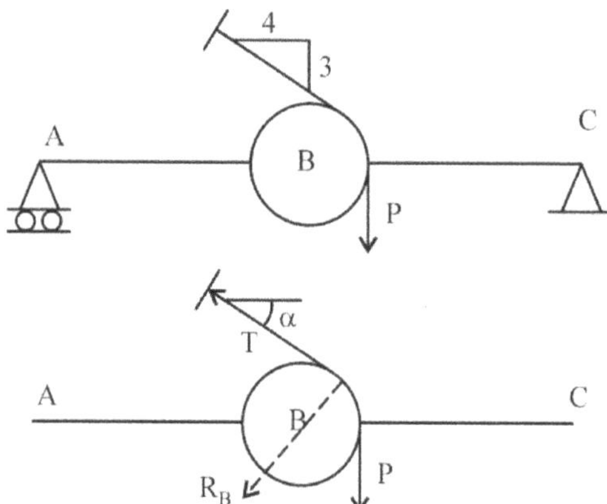

Solution:

Tension in the rope remains constant, if friction over the pulley is neglected.

Therefore, $T = P = 5000$ N

For the given inclination of rope, $\sin \alpha = 3/5$, $\cos \alpha = 4/5$

The resultant R_B of rope tension T and load P acts through the support point B and its horizontal and vertical components are given by

$$(R_X)_B = -T \times \cos \alpha = -5000 \times (4/5) = -4000 \text{ N}$$

and $$(R_Y)_B = -P + T \times \sin \alpha = -5000 + 5000 \times (3/5)$$

$$= -2000\text{N}$$

Weight of the bar 'W_B', weight of pulley 'W_P' acting vertically down through the centroid or mid-point of the bar AC and the resultant of rope tension are balanced by the reactions at A and C. The roller support at A can have only vertical reaction R_A while the reaction at the hinge C can act in any direction, which can be conveniently identified by its horizontal component $(R_X)_C$ and vertical component $(R_Y)_C$

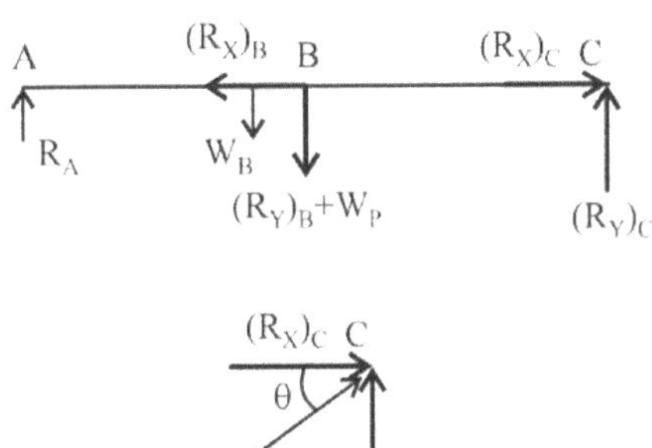

Taking moments about A,

$$\Sigma M_A = -W_B \times (AC/2) - [(R_Y)_B + W_P] \times AB + (R_Y)_C \times AC$$

$$\Rightarrow \quad (R_Y)_C = [\,2000 \times (5/2) + (2000 + 500) \times 3\,] / 5 = 2500 \text{ N}$$

$$\Sigma F_X = (R_X)_C - (R_X)_B = 0 \;\Rightarrow\; (R_X)_C = (R_X)_B = 4000 \text{ N}$$

Resultant reaction at C, $R_C = \sqrt{[(R_X)_C^2 + (R_Y)_C^2]}$

$$= \sqrt{4000^2 + 2500^2} = 4717 \text{ N}$$

Its inclination with the horizontal, $\theta = \tan^{-1}[(R_Y)_C/(R_X)_C]$

$$= \tan^{-1}(2500/4000) = 32^0$$

Example 9

A roller of radius 0.3m and weight 2000N is to be pulled over a kerb of height 0.15m by a horizontal force P applied to the end of a string wound around the circumference of the roller, as shown. Find the magnitude of minimum force required.

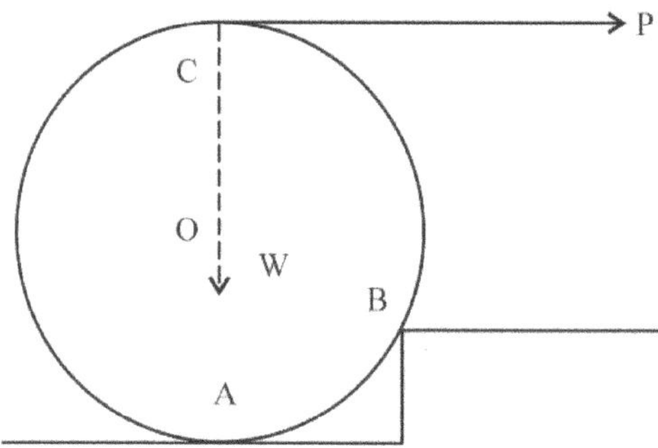

Solution:

When the cylinder is ***on the verge of being lifted up, contact between the roller and the floor ceases and normal reaction at A will not exist***. The cylinder is in equilibrium, therefore, under the influence of three forces - weight 'W', applied force 'P' and the reaction 'R' at B. Three forces in equilibrium must be either concurrent or parallel. W and P are perpendicular, meeting at C. Hence, 'R' must also pass through the point of concurrence C. To find its inclination with the vertical, draw horizontal line through B to intersect the vertical through C at D.

Then, $\angle OCB = \theta = \tan^{-1}(BD/CD)$

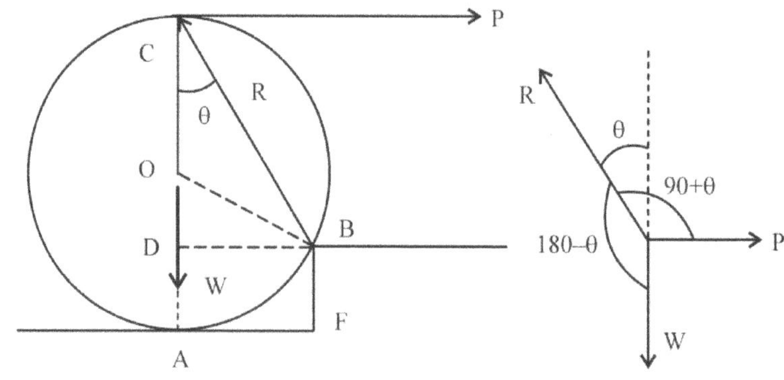

$$CD = CA - DA = 2 \times OC - BF = 2 \times 0.3 - 0.15 = 0.45 \text{ m}$$

From $\Delta ODB,\ DB = \sqrt{(OB^2 - OD^2)} = \sqrt{[OB^2 - (OA - DA)^2]}$

$$= \sqrt{[0.3^2 - (0.3 - 0.15)^2]} = 0.2598 \text{ m}$$

Therefore, $\theta = \tan^{-1}(0.2598 / 0.45) = \tan^{-1}(0.5773) = 30^0$

Using Lame's theorem, $P / \sin(180 - \theta) = W / \sin(90 + \theta)$

$\Rightarrow$ $P = W \sin(180 - \theta) / \sin(90 + \theta) = W \sin\theta/\cos\theta$

$= W \tan\theta = 2000 \times 0.5773 = 1154.7$ N

Example 10

A 675N man stands on the middle rung of a 225N ladder, as shown. Assuming a smooth wall at B and a stop at A to prevent slipping, find the reactions at A and B. AD = 1.8m and BD = 3.6m

Solution:

Weight of the ladder and weight of the man are acting at C, the mid-point of the ladder. Therefore,

$$W_C = 675 + 225 = 900 \text{ N}$$

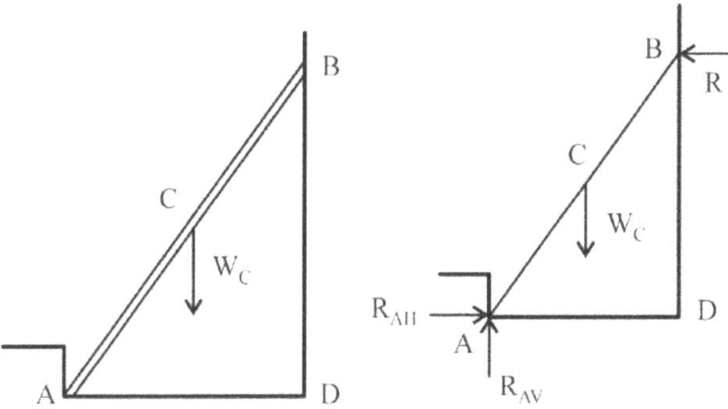

This load is balanced by the reactions at the two ends of the ladder. Normal reaction at the wall, R_B, has to be horizontal while the reaction at the floor can have horizontal and vertical components R_{AH} and R_{AV}.

Then, taking moments about A,

$$\Sigma M_A = R_B \times BD - W_C \times (AD/2) = 0$$

$\Rightarrow$ $R_B = W_C \times (AD/2) / BD = 900 \times 0.9 / 3.6 = 225$ N

$$\Sigma F_X = R_{AH} - R_B = 0 \Rightarrow R_{AH} = 225 \text{ N}$$

$$\Sigma F_Y = R_{AV} - W_C = 0 \Rightarrow R_{AV} = 900 \text{ N}$$

$$R_A = \sqrt{R_{AH}^2 + R_{AV}^2} = 927.7 \text{ N}$$

Example 11

A man weighing 750 N stands at the middle of a uniform ladder weighing 250 N and connected by a string to the corner of wall and floor. Assuming there is no friction between the ladder and wall or floor, calculate the tension in the string and reactions at the wall and floor. Assume string DO is perpendicular to AB.

Solution:

Weight of ladder and weight of man act from the mid-point C of ladder. Therefore, $W_C = 750 + 250 = 1000$ N

$$\Sigma F_X = T\cos 30 - H = 0 \quad \Rightarrow \sqrt{3}\ T = 2\ H \qquad .....(1)$$

$$\Sigma F_Y = V - 1000 - T\sin 30 = 0 \quad \Rightarrow 2\ V - T = 2000 \qquad .....(2)$$

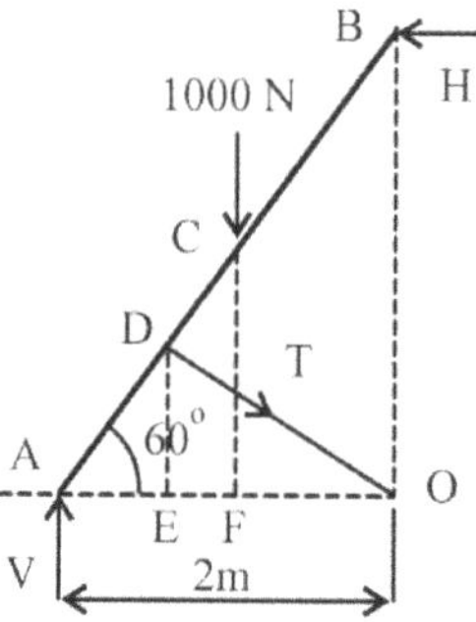

Taking moments about O,

$$\Sigma M_O = H \times OB + 1000 \times OF - V \times OA = 0$$

$$H \times OA\tan 60 + 1000 \times 1 - V \times 2 = 0$$

$$V - \sqrt{3}\ H = 500 \qquad .....(3)$$

Eliminating V from (2) and (3),

$$1000 + 2\sqrt{3}\ H - T = 2000 \qquad .....(4)$$

Substituting for 2H from (1) in (4), $1000 + 3\ T - T = 2000 \Rightarrow T = 500$ N

From (2), $V = (T + 2000) / 2 = 1250$ N

From (1), $H = \sqrt{3}T / 2 = 250\sqrt{3} = 433$ N

Example 12

Two ropes connected to A and B carry a bar CD with a load of W at D and 25kN at C, as shown. Find the value of W so that the bar remains horizontal.

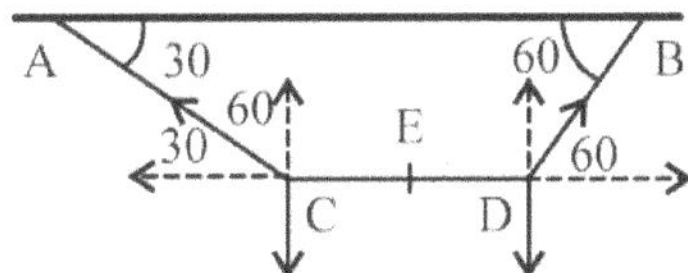

Solution:

This problem was solved earlier, treating this non-concurrent force system as systems of concurrent forces at C and D. The problem is solved here as a system of coplanar non-concurrent forces. Considering equilibrium of point C, since bar CD is horizontal, applied force of 25kN is to be balanced by the vertical component of force in member CA. Similarly, for equilibrium of point D, since bar CD is horizontal, load W should balance vertical component of force in member DB.

The three equations of equilibrium for this non-concurrent force system on the bar CD are

$$\Sigma F_X = -F_{AC} \times \cos 30 + F_{BD} \times \cos 60 = 0$$

or $F_{BD} = F_{AC} \times \cos 30 / \cos 60 = \sqrt{3}\ F_{AC}$ .....(1)

$$\Sigma F_Y = F_{AC} \times \cos 60 + F_{BD} \times \cos 30 - W - 25 = 0$$

Substituting eq (1), $F_{AC} \times (1/2) + \sqrt{3}\ F_{AC} \times (\sqrt{3}/2) = W + 25$

or $F_{AC} = (W + 25) / 2$ .....(2)

and $\Sigma M_E = 25 \times CE - (F_{AC} \times \cos 60) \times CE - W \times DE + (F_{BD} \cos 30) \times DE = 0$

or $25 - F_{AC} \times \cos 60 - W + F_{BD} \times \cos 30 = 0$ since $CE = DE$

Substituting eq (1), $-F_{AC} \times (1/2) + \sqrt{3}\ F_{AC} \times (\sqrt{3}/2) = W - 25$

or $F_{AC} = W - 25$ .....(3)

From eq (2) and eq. (3), $(W + 25) / 2 = W - 25$

or $W + 25 = 2W - 50$ or $2W - W = 25 + 50$

$\Rightarrow$ $W = 75$ kN

Example 13

Two cylinders A and B rest in a channel as shown. The cylinder A is of 100mm dia and weighs 200N while cylinder B is of 180mm dia and weighs 500N. If the bottom width of the box is 180mm with one side vertical and the other inclined at 60^0, determine reactions at the four points of contact C, D, E and F

Solution:

$\angle CGD = 120^0$

ΔGBC and ΔGBD are identical.

Since BG is a common side, CB = BD and $\angle BCG \Rightarrow \angle BDG = 90^o$

$\Rightarrow \quad \angle CGB = \angle BGD = 60^0$

Therefore, $GD = BD / \tan 60 = 90 / \sqrt{3} = 51.96$ mm

$DI = GH - GD - IH = 180 - 51.96 - 50 = 78.04$ mm

$AB = AE + EB = 50 + 90 = 140$ mm

$\text{Cos}\ \theta = DI / AB = 78.04 / 140 = 0.5574$

$\Rightarrow \quad \theta = 56.12^0$ and $\sin\theta = 0.8302$

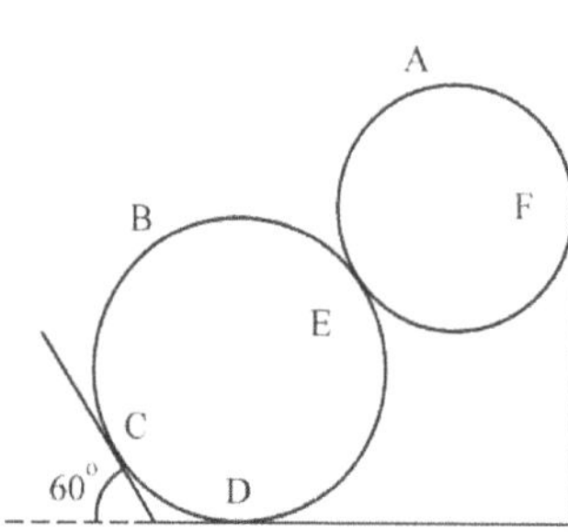

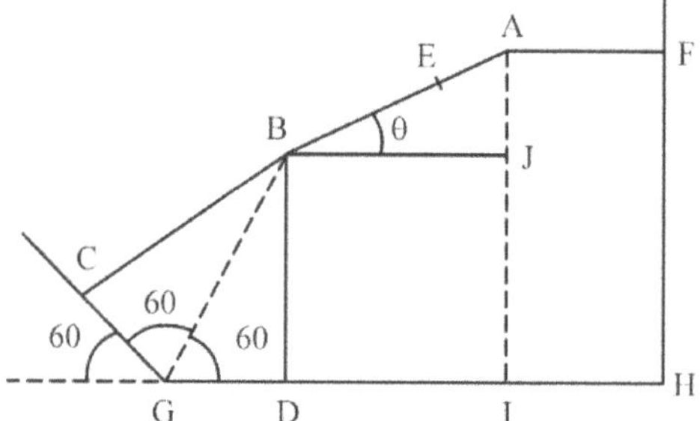

Cylinder A is in equilibrium under the influence of W_A, R_F and R_E.

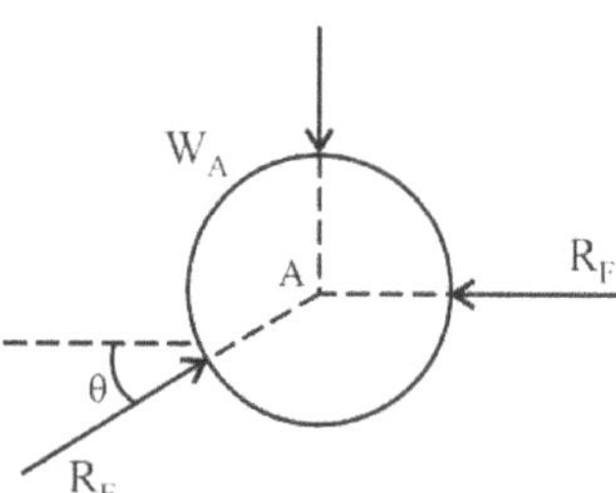

Applying Lami's theorem,

$R_F / \sin(90 + \theta) = R_E / \sin 90$

$= W_A / \sin(180 - \theta)$

$\Rightarrow \quad R_F = W_A (\cos\theta / \sin\theta)$

$= 200 \times (0.5574/0.8302) = 134.27$ N

$R_E = W_A (1 / \sin\theta) = 200 / 0.8302$

$= 240.91$ N

For equilibrium of cylinder B,

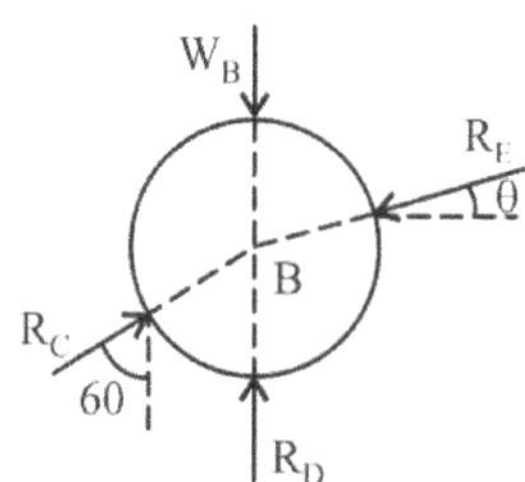

$\sum F_X = R_C \sin 60 - R_E \cos\theta = 0$

$R_C = R_E (\cos\theta / \sin 60) = 155.06$ N

$\sum F_Y = R_C \cos 60 + R_D - W_B - R_E \sin\theta = 0$

$R_D = W_B + R_E \sin\theta - R_C \cos 60 = 622.47$ N

3.2 RESULTANT OF NON-CONCURRENT FORCES

In the case of a set of concurrent forces, direction and magnitude will be sufficient to define the resultant force, since it passes through the point of concurrency of other forces. However, in the case of non-concurrent forces, point of application of resultant force also needs to be calculated using the law of moments, to ensure same resultant moment.

3.2.1 SYSTEM OF PARALLEL FORCES

Resultant of any coplanar force system is the single force acting through any arbitrary point such that $R_x = \Sigma(F_x)_i$ and $R_y = \Sigma(F_y)_i$

Location of the resultant is obtained from $M_R = \Sigma(M_F)_i$ by taking all moments about any point. When all the forces are parallel, the resultant also will have the same direction and need not be calculated.

3.2.2 ANALYTICAL METHOD

Resolve all forces along horizontal and vertical directions and obtain their algebraic sums, $\sum F_X$ and $\sum F_Y$. Resultant force 'R' is obtained by the addition of these two vectors, $R_X = \sum F_X$ and $R_Y = \sum F_Y$. Point of application of the resultant force is obtained by equating moment of the resultant about any point with the algebraic sum of the moments of the system of forces about the same point. Its orientation with X-axis in the chosen X-Y coordinate system is obtained from the relation $\alpha = \tan^{-1}(R_y/R_x)$.

The procedure is illustrated through the following examples.

Example 14

A dam shown in figure, is subjected to three forces pressure load of 50kN on the upstream vertical face AB through E, pressure load of, 30kN on the downstream inclined face CD through F and its own weight of 120kN through G. Determine the resultant force and its point of intersection with base AC.

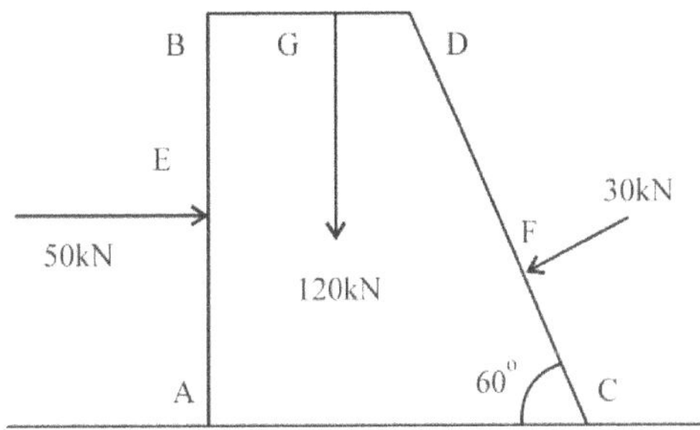

AE = 2m; BG = 2m ; FC = 1.5m; AC = 5m ; $\angle C = 60^0$

Solution:

Since $\angle C = 60^{\circ}$, force on the face CD at F is inclined with horizontal at 30°.

Let R be the resultant, acting at some angle. Its horizontal and vertical components are obtained from

$$R_X = \sum F_X = 50 - 30 \cos 30 = 24 \text{ kN}$$

and $$R_Y = \sum F_Y = -120 - 30 \sin 30 = -135 \text{ kN}$$

Resultant, $$R = \sqrt{(R_X^2 + R_Y^2)} = 137 \text{ kN}$$

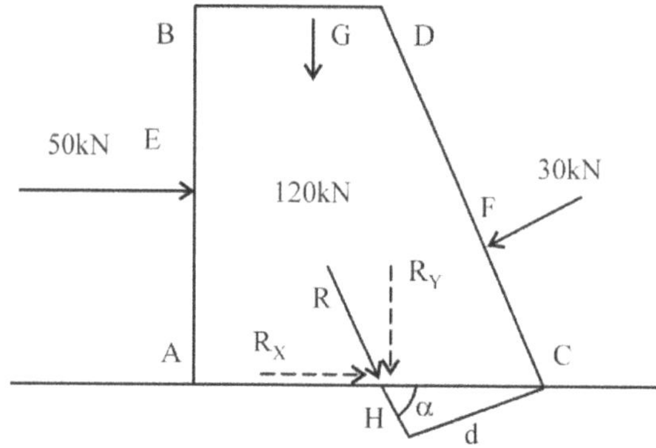

It is oriented at an angle α with the horizontal, given by

$$\alpha = \tan^{-1}(R_Y/R_X) = \tan^{-1}(-135/24) = -73.5^0$$

Since R_X is +ve and R_Y is –ve, direction of R lies in 4th quadrant.

Its location H on the base AC is obtained from the moment equation. If 'd' is the normal distance of the resultant 'R' from C, taking moments of forces and resultant about C,

$$\sum M_F = 120 \times (AC - BG) + 30 \times FC - 50 \times AE$$
$$= 120 \times (5 - 2) + 30 \times 1.5 - 50 \times 2$$
$$= 360 + 45 - 100 = 305$$

$$M_R = R \times d = \Sigma M_F$$

$$\Rightarrow \quad d = 305 / R = 305 / 137 = 2.2263\text{m}$$

Now, $$HC = d / \sin \alpha = d / \sin 73.5 = 2.26 \text{ m}$$

Alternative method: HC can also be found directly, by taking moments about C,

$$\sum M_F = M_R$$

$\sum M_F = 120 \times (AC - BG) + 30 \times FC - 50 \times AE$

$= 120 \times (5 - 2) + 30 \times 1.5 - 50 \times 2 = 305$

$M_R = R_Y \times HC + R_X \times 0$

Therefore, $HC = 305 / R_Y = 305 / 135 = 2.26$ m

Example 15

Four forces of magnitudes 5N, 10N, 15N and 20N are acting respectively along the four sides of a square ABCD of side 'a' as shown. Determine the magnitude, direction and position of the resultant force.

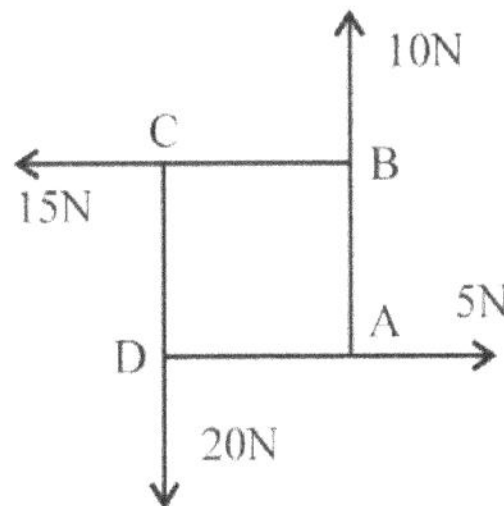

Solution:

The horizontal and vertical components of resultant are obtained from

$R_X = \sum F_X = 5 - 15 = -10$ N

and $R_Y = \sum F_Y = 10 - 20 = -10$ N

Then,

$\sqrt{[R_X^2 + R_Y^2]} = \sqrt{[(-10)^2 + (-10)^2]} = 10\sqrt{2}$N

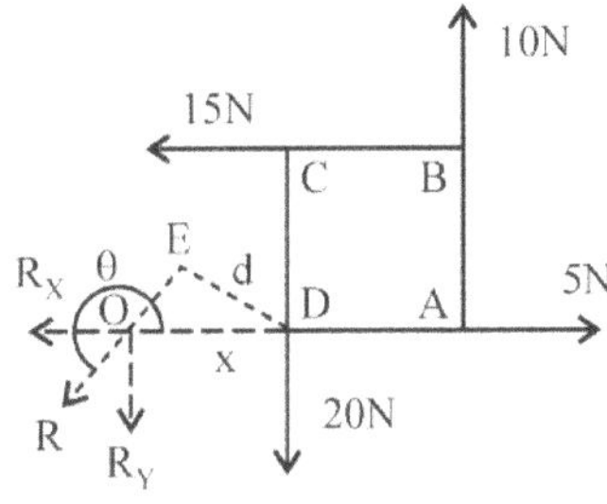

Since R_X is –ve and R_Y is –ve, orientation of R must be in the 3rd quadrant. Its inclination 'θ' with the horizontal is given by

$\theta = \tan^{-1}(R_Y/R_X) = \tan^{-1}[(-10)/(-10)] = 225^0$ from +X axis

Its location can be identified from the moment equation. Let 'x' be the distance of the resultant along x-axis from D and 'd' the normal distance to the resultant from D. Taking moments about D,

$\sum M_D = R \times d = 10\,a + 15\,a = 25\,a$

$\Rightarrow$ $d = (25\,a) / R = (25\,a) / (10\sqrt{2}) = 5a / (2\sqrt{2})$

In Δ EDO, $\angle EOD = \theta - 180 = 45^0$ and so $\angle EDO = 45^0$

Then, $x = d / \cos \angle EDO = d / \cos 45^0 = [5a/(2\sqrt{2})] / (1/\sqrt{2})$

$= 2.5a$ (to the left of D, since $\angle EDO$ is in the 2nd quadrant)

Alternative method: It can also be calculated directly, by taking moments of components of R, as $M_R = \sum M_F$

or $\qquad R_X \times 0 + R_Y \times x = 10\ a + 15\ a = 25\ a$

$\Rightarrow \qquad x = 25\ a / R_Y = 25\ a/(-10) = -2.5a$ along – X axis

Example 16

A particle is acted upon by three forces of magnitudes 50N, 100N and 150N respectively along the sides of an equilateral triangle of side 'a' as shown. Determine the magnitude, direction and position of the resultant force.

Solution:

The geometry can be initialised in many ways, by considering different orientations of the triangle as well as forces in clockwise or counter clockwise direction. The problem is solved here for the particular geometry shown in figure.

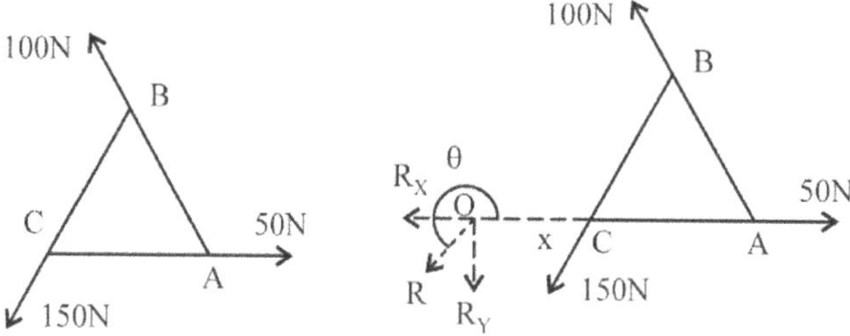

In the equilateral Δ ABC, $\angle A = \angle B = \angle C = 60^0$.

Therefore, the three forces taken along CA, AB and BC are oriented along 0^0, 120^0 and 240^0 w.r.t. +X axis.

The horizontal and vertical components of resultant are obtained from

$$R_X = 50 \cos 0 + 100 \cos 120 + 150 \cos 240$$
$$= 50 \times 1 + 100 \times (-1/2) + 150 \times (-1/2) = -75N$$

and
$$R_Y = 50 \sin 0 + 100 \sin 120 - 150 \sin 240$$
$$= 50 \times 0 + 100 \times (\sqrt{3}/2) + 150 \times (-\sqrt{3}/2) = -43.3N$$

Then, $\sqrt{R_X^2 + R_Y^2} = \sqrt{(-75)^2 + (-43.3)^2} = 86.6N$

Its inclination 'θ' with the horizontal is given by

$\theta = \tan^{-1}(R_Y/R_X) = \tan^{-1}[(-75)/(-43.3)] = 210^0$ from + X axis, in the 3rd quadrant, since R_X and R_Y are –ve

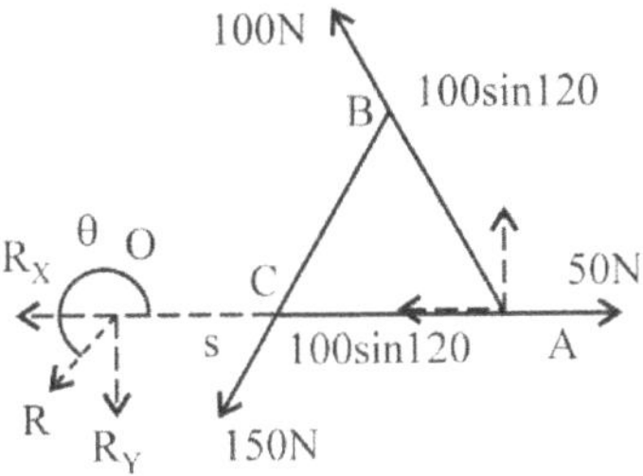

Its location can be determined from the moment equation. Let 'S' be the distance of the resultant along X-axis from C. Taking moments about C,

$$M_R = \sum M_F$$

or $$R_X \times 0 + R_Y \times s = 50 \times 0 + 100 \times CD + 150 \times 0$$

From Δ ACD, CD = AC sin 60 = a ($\sqrt{3}/2$)

Therefore, $s = 100 \times CD / R_Y$

$$= 100 \times (a\ \sqrt{3}/2) / 43.3 = 2\ a$$

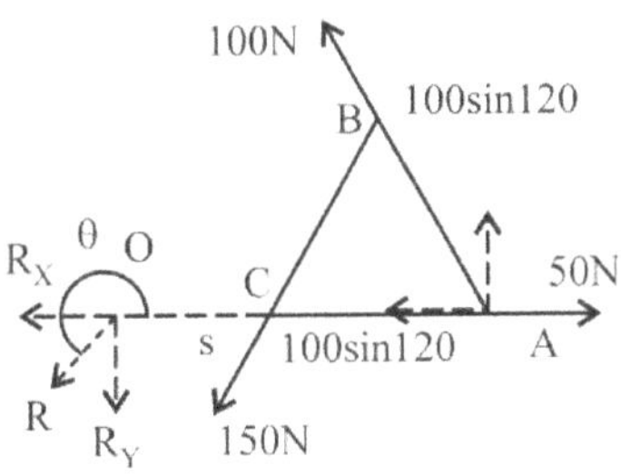

Alternative Method: It can also be calculated by taking moment of components of 100N,

$$M_R = \sum M_F$$

$$R_X \times 0 + R_Y \times s$$

$$= 50 \times 0 + (100 \sin 120 \times AC + 100 \cos 120 \times 0) + 150 \times 0$$

$$R_Y \times s = 100\ (\sqrt{3}/2) \times a$$

$$s = 100\ (\sqrt{3}/2)\ a / R_Y = 50\sqrt{3}\ a / 43.3 = 2\ a$$

SUMMARY

1. **Moment 'M_O' of a force** 'P' about any point 'O' is a vector whose magnitude is the product of force and its normal distance 'd' from the point (or moment center)

   $M_O = P \times d$ In vector approach, $\boldsymbol{M_O} = \boldsymbol{P} \times \boldsymbol{d}$, *a cross product*

It tends to rotate the body about O in the clockwise or counter-clockwise direction.

2. Moment 'M' of a couple (two equal, parallel and non-collinear forces in opposite directions), separated by a distance 'a' is $M = P \times a$ about any point O.
3. Three forces acting on a body <u>can</u> keep it in equilibrium <u>only</u> if they are parallel or concurrent
4. Moment of a force 'P' passing through a point 'A', about a point 'O' at a distance 'd' from 'A' can be obtained either by determining normal distance 'a' or by resolving the force into two components 'P_X' and 'P_Y' along OA and perpendicular to OA and adding the moments of the force components.

   $$M_O = P_X \times y + P_Y \times x = P \times a$$

   where, 'a' is the normal distance of the force from O

   P_X, P_Y are the X and Y components of force P

   and x and y are the X and Y components of distance 'd'
5. **Varignon's theorem (or Law of moments)** – Moment of resultant of a number of coplanar forces about any point is equal to the algebraic sum of moments of individual forces about the same point. $M_R = \sum M_i$
6. For **any** system of coplanar non-concurrent forces, in ***equilibrium***,

   The sum of moments of all the forces acting on the component, about any point, is equal to zero. $M = \sum M_i = 0$. This is the ***3rd equation of equilibrium*** to be satisfied simultaneously along with two equations of equilibrium for force components

PROBLEMS FOR PRACTICE

1. A system of levers is shown supporting a load of 80N at D. Determine the reactions at A and B on the lever. All dimensions are in meters.

   (***Ans:*** $R_C = 53.33$N, $R_B = 124.44$N, $R_A = 71.11$N)

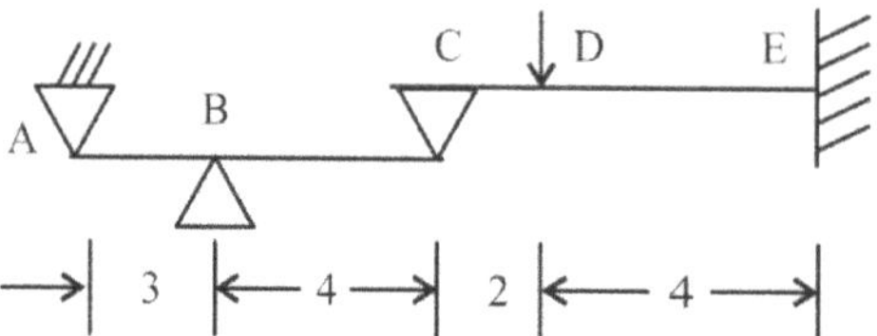

2. A bar weighing 12 N/cm is 300 cm long. Its left end is inserted loosely into a wall of 20 cm thick. A 20 cm pulley of weight 40 N is fixed to the bar on its right end. A rope passes over the pulley. Calculate the reactions at the two ends of the wall in contact with the bar, if the tension in the rope on either side of the pulley is 80N

(*Ans:* R_C = 26200N, R_B = 22400N)

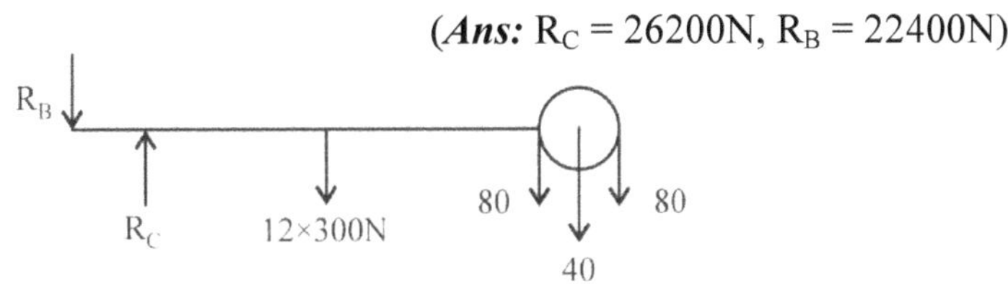

3. Determine the force P required to keep the bell crank lever in equilibrium.

(*Ans:* 52.2N)

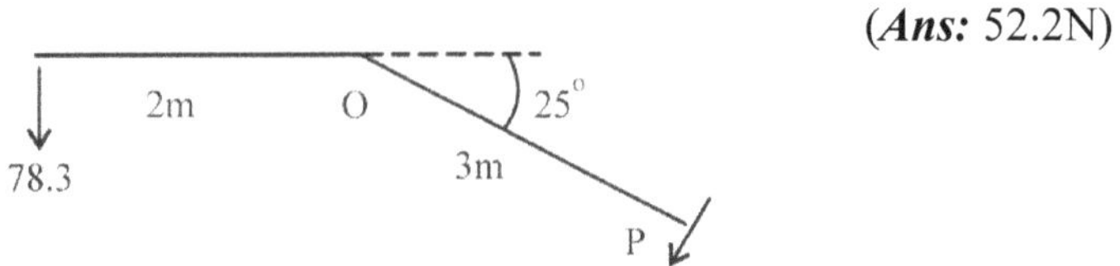

5. Determine resultant of the four forces and one couple that act on the plate as shown. Take AB = 1m, BC = 4m, CD = 2m, DE = 5m.

(*Ans:* R = 146.35N, $\theta = 63.21^0$, x = 4.703m)

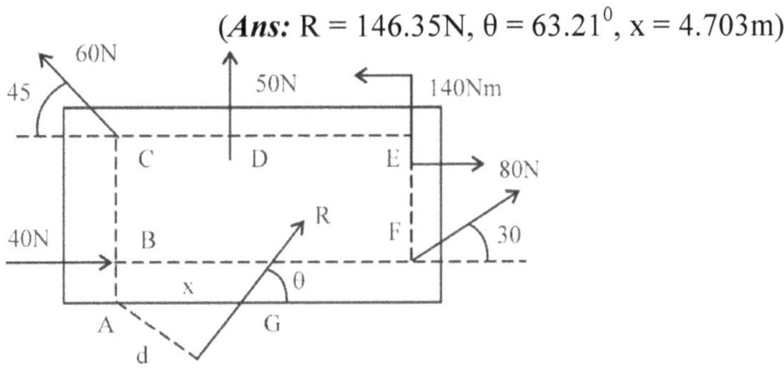

6. Determine the reactions at the supports of the beam, loaded as shown

(*Ans:* R_B = 1.9kN; R_{DH} = – 0.6kN; R_{DV} = –0.1kN)

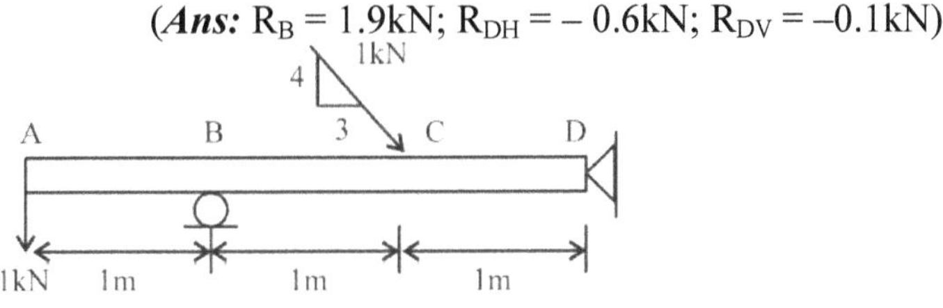

7. Two wheel loads of a tractor crossing a 6m span are shown in figure. Find the distance 'x' of the load 9kN from the support A, if the reaction at A is twice that of B

(*Ans:* 1.5m)

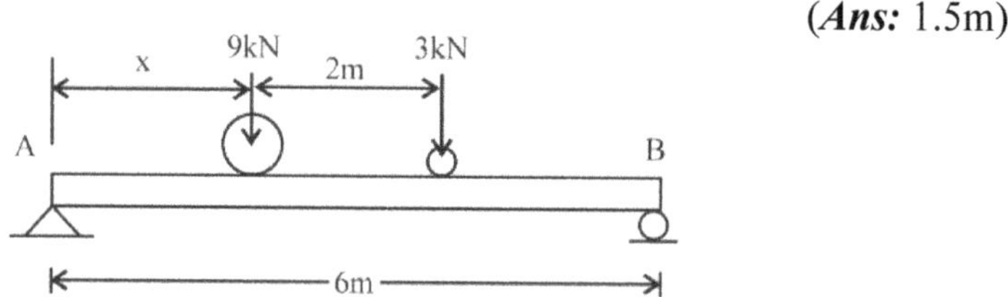

8. A bar 2m long and of negligible weights rests in horizontal position on two smooth inclined planes. Determine the distance 'x' at which the load Q=100N should be placed from end B to keep the bar horizontal.

(*Ans:* 0.804m)

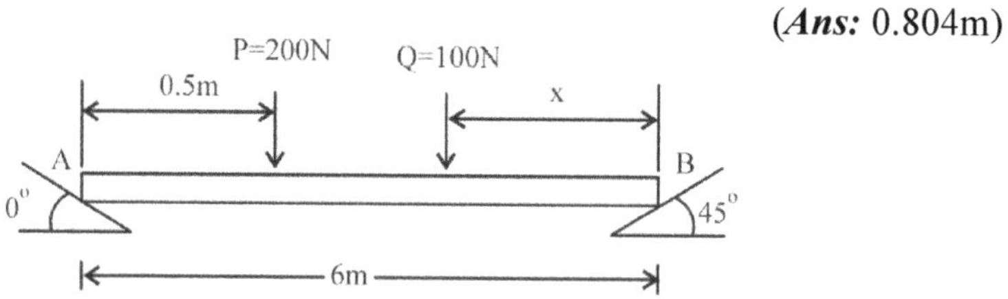

9. Find the reactions R_A and R_B induced at the supports A and B of the right angled bar ACB (AC=1.2m, BC=0.9m) supported as shown and subjected to a vertical load P at the mid-point of AC

(*Ans:* $R_{AX} = R_B = 0.67P$, $R_{AY} = P$, $R_A = 1.2P$, $\theta = 33.8^0$)

B

P

C

10. A cylinder of 1m dia and 1000N weight is lodged between two symmetric cross pieces ACE and BCD, which make an angle of 60^0 and are tied together with a horizontal rope, as shown. AC=BC=1m; CD=CE=2m. Determine the tension in the rope DE assuming a smooth floor.

(*Ans:* 644.34N)

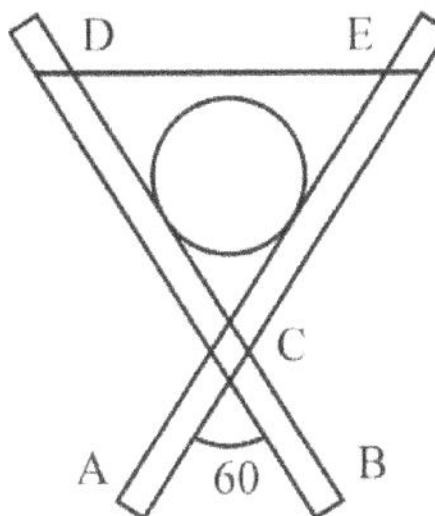

11. A pulley 'A' is supported by two bars AB and AC, which are hinged at B and C to a vertical mast EF. Over the pulley hangs a flexible cable DG which is fastened to the mast at D and carries at the other end G a load of 20kN. Neglecting friction in the pulley, determine the forces in the bars.

(***Ans:*** $F_{AB} = 0$; $F_{AC} = 34.64$kN)

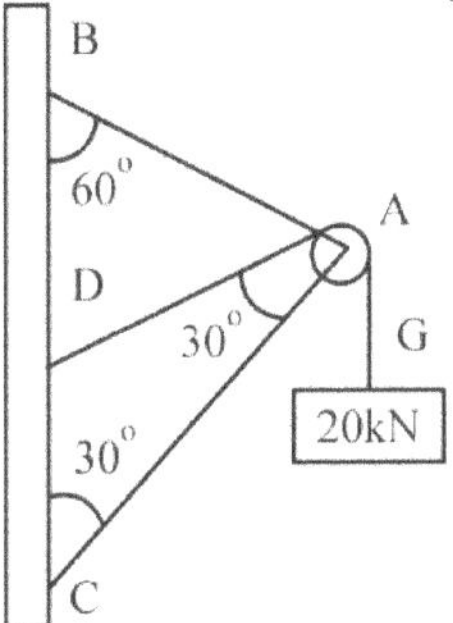

CHAPTER **4**

SPATIAL CONCURRENT FORCES AND VECTOR METHOD

Parallelogram law and **Triangle law** can be used to obtain the resultant of any two concurrent forces, oriented at random in 3-dimensional space. Their ***resultant*** lies in the same plane as that formed by the two forces.

Resolution of a force into two components along two mutually perpendicular directions X and Y of Cartesian coordinate system was explained earlier as.

$$F_X / \cos\theta = F_Y / \sin\theta = F$$

$$F^2 = F_X^{\ 2} + F_Y^{\ 2}; \ \tan\theta = F_Y / F_X$$

where, θ is the angle made by the force F with the X-axis.

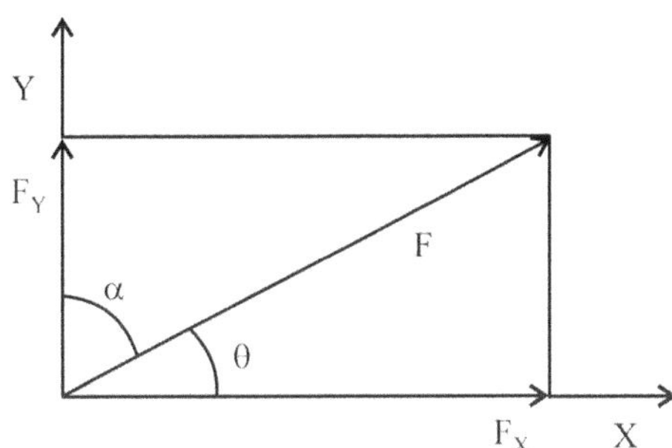

If α is the angle made by the force with the Y-axis,

then, $\theta + \alpha = 90^0$ and $\sin\theta = \cos\alpha$

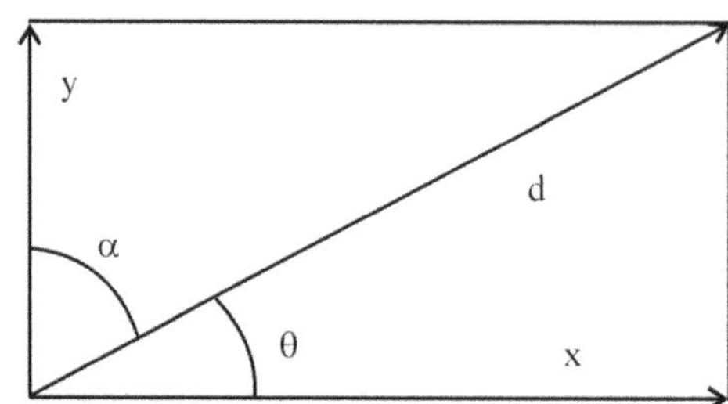

If the force is represented by a line of length 'd' on a paper, then the lengths of force components along X and Y axes are related by

$$F_X / x = F_Y / y = F / d \quad \text{where, } d^2 = x^2 + y^2$$

4.1 RESOLUTION OF A FORCE IN 3-D SPACE

Resolution of forces oriented in 3-dimensional space is an extension of the 2-D forces

- Resolution of a force into three mutually perpendicular components

$$F_x / \cos \theta_x = F_y / \cos \theta_y = F_z / \cos \theta_z = F$$

$$\cos^2\theta_x + \cos^2\theta_y + \cos^2\theta_z = 1$$

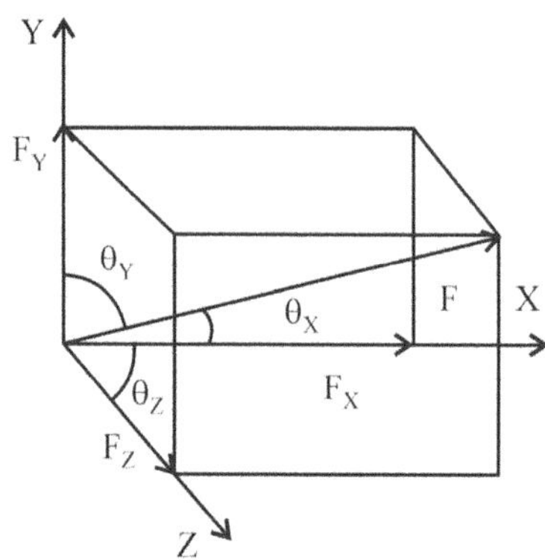

- If the force is represented by a line of length 'd' on a paper, then the lengths of force components along X, Y and Z axes are related by

$$F_x / x = F_y / y = F_z / z = F / d$$

$$\Rightarrow \quad F^2 = F_x^2 + F_y^2 + F_z^2 \; ; \; d^2 = x^2 + y^2 + z^2$$

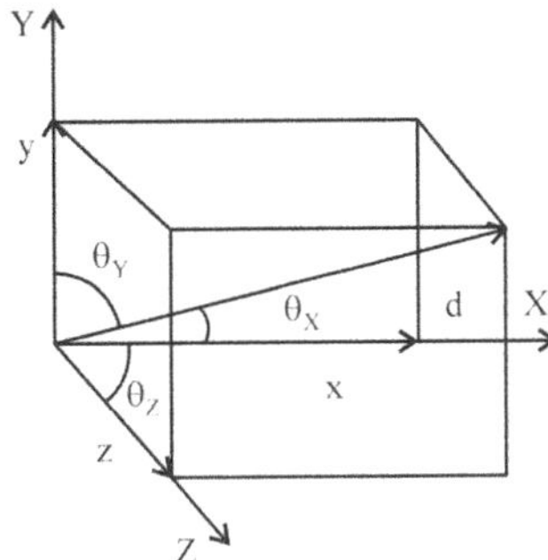

Resultant of three or more concurrent forces in space

$$R_x = \sum F_x ; \quad R_y = \sum F_y ; \quad R_z = \sum F_z$$

$$R^2 = R_x^2 + R_y^2 + R_z^2 = (\sum F_x)^2 + (\sum F_y)^2 + (\sum F_z)^2$$

$$R_x / \cos \theta_x = R_y / \cos \theta_y = R_z / \cos \theta_z = R$$

Resultant of any force system is the single force acting through any arbitrary point and a couple which is the moment sum of the original force system about that point.

O P Q a b ≡ R P R Q R x ≡ R M

$$R = (P \times a + Q \times b) / x \qquad M = R \times x$$

4.2 VECTOR MECHANICS

Every force is represented by its magnitudes along three mutually perpendicular directions, identified by unit vectors. Vectors are represented by bold letters

$$\mathbf{F} = \mathbf{F_x} + \mathbf{F_y} + \mathbf{F_z} = F_x \boldsymbol{i} + F_y \boldsymbol{j} + F_z \boldsymbol{k} ; \quad |F|^2 = F_x^2 + F_y^2 + F_z^2$$

4.2.1 VECTOR ALGEBRA

Two types of vector multiplication are possible

(a) Dot (scalar) product : $\mathbf{a} \cdot \mathbf{b} = a\, b \cos \theta$

- If **b** is a unit vector, $\mathbf{a} \cdot \mathbf{b} = a \cos \theta$ i.e., The magnitude of component of a vector in any direction is the dot product of the vector with a unit vector in the desired direction. This component in vector form is

  $(a \cos \theta)\, \boldsymbol{e}$, where $\boldsymbol{e} = \boldsymbol{b} / |\boldsymbol{b}|$ is the unit vector along $\boldsymbol{b}$.

Thus, $\mathbf{F} \cdot \boldsymbol{i} = F_X$; $\mathbf{F} \cdot \boldsymbol{j} = F_Y$ and $\mathbf{F} \cdot \boldsymbol{k} = F_Z$

where, F_X, F_Y and F_Z are components of force vector $\boldsymbol{F}$ along X, Y and Z directions, represented by unit vectors $\boldsymbol{i}, \boldsymbol{j}$ and $\boldsymbol{k}$

- In orthogonal coordinate system, represented by unit vectors $\boldsymbol{i}, \boldsymbol{j}$ and $\boldsymbol{k}$ in the three directions, dot product of a unit vector with itself is unity (since, $\theta = 0$ and $\cos\theta = 1$). Other dot products will be zero (since, $\theta = 90^0$ and $\cos\theta = 0$)

  $\boldsymbol{i} \cdot \boldsymbol{i} = \boldsymbol{j} \cdot \boldsymbol{j} = \boldsymbol{k} \cdot \boldsymbol{k} = 1$; $\boldsymbol{i} \cdot \boldsymbol{j} = \boldsymbol{j} \cdot \boldsymbol{k} = \boldsymbol{k} \cdot \boldsymbol{i} = 0$

  $\mathbf{a} \cdot \mathbf{b} = (a_x \boldsymbol{i} + a_y \boldsymbol{j} + a_z \boldsymbol{k}) \cdot (b_x \boldsymbol{i} + b_y \boldsymbol{j} + b_z \boldsymbol{k}) = a_x b_x + a_y b_y + a_z b_z$

  and $\mathbf{a} \cdot \mathbf{a} = a_x^2 + a_y^2 + a_z^2 = a^2$

- If $\mathbf{a} \cdot \mathbf{b} = 0$, and $a \neq 0$, $b \neq 0$ then they must be perpendicular to each other.

Properties :	Commutative	$\boldsymbol{a} \cdot \boldsymbol{b} = \boldsymbol{b} \cdot \boldsymbol{a}$
	Associative	$(m\boldsymbol{a}) \cdot \boldsymbol{b} = \boldsymbol{a} \cdot (m\boldsymbol{b}) = m\,(\boldsymbol{a} \cdot \boldsymbol{b})$
	Distributive	$\boldsymbol{c} \cdot (\boldsymbol{a} + \boldsymbol{b}) = \boldsymbol{c} \cdot \boldsymbol{a} + \boldsymbol{c} \cdot \boldsymbol{b}$

(b) Cross (vector) product $\boldsymbol{a} \times \boldsymbol{b} = (a\,b\,\sin\theta)\,\boldsymbol{n}$

where, $\boldsymbol{n}$ is a unit vector perpendicular to both $\boldsymbol{a}$ and $\boldsymbol{b}$, positive direction following right hand rule

$|\boldsymbol{a} \times \boldsymbol{b}| =$ Area of parallelogram with sides 'a' and 'b'

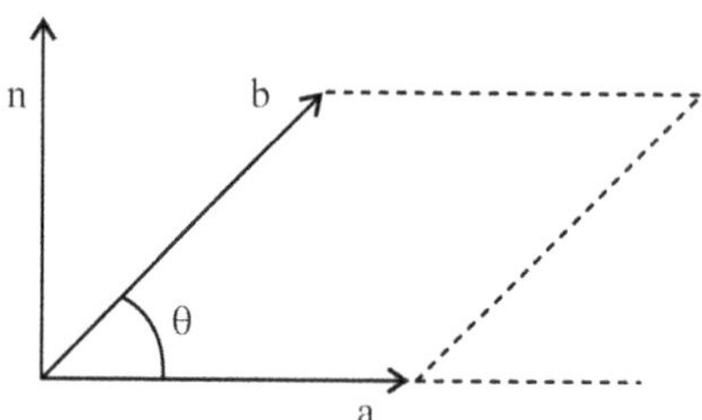

Properties :

<u>***Not***</u> commutative (due to change of direction of $\boldsymbol{n}$) $\boldsymbol{a} \times \boldsymbol{b} = -\,\boldsymbol{b} \times \boldsymbol{a}$

Associative with scalar multiplication $m\,(\boldsymbol{a} \times \boldsymbol{b}) = (m\boldsymbol{a}) \times \boldsymbol{b} = \boldsymbol{a} \times (m\boldsymbol{b})$

<u>***Not***</u> Associative with vector multiplication $\boldsymbol{a} \times (\boldsymbol{b} \times \boldsymbol{c}) \neq (\boldsymbol{a} \times \boldsymbol{b}) \times \boldsymbol{c}$

Distributive $\boldsymbol{c} \times (\mathbf{a} + \mathbf{b}) = \boldsymbol{c} \times \boldsymbol{a} + \boldsymbol{c} \times \boldsymbol{b}$

Cross product of a unit vector with itself is zero. $\boldsymbol{i} \times \boldsymbol{i} = \boldsymbol{j} \times \boldsymbol{j} = \boldsymbol{k} \times \boldsymbol{k} = \mathbf{0}$

Cross product of any two perpendicular unit vectors is the unit vector perpendicular to both these unit vectors

$\boldsymbol{i} \times \boldsymbol{j} = \boldsymbol{k}$; $\boldsymbol{j} \times \boldsymbol{k} = \boldsymbol{i}$; $\boldsymbol{k} \times \boldsymbol{i} = \boldsymbol{j}$

$\boldsymbol{j} \times \boldsymbol{i} = -\boldsymbol{k}$; $\boldsymbol{k} \times \boldsymbol{j} = -\boldsymbol{i}$; $\boldsymbol{i} \times \boldsymbol{k} = -\boldsymbol{j}$

$$\boldsymbol{a} \times \boldsymbol{b} = \boldsymbol{i}\,(a_y b_z - a_z b_y) + \boldsymbol{j}\,(a_z b_x - a_x b_z) + \boldsymbol{k}\,(a_x b_y - a_y b_x) = \begin{vmatrix} \boldsymbol{i} & \boldsymbol{j} & \boldsymbol{k} \\ a_x & a_y & a_z \\ b_x & b_y & b_z \end{vmatrix}$$

A force of magnitude P inclined at an angle θ, is expressed in matrix form as

$$\boldsymbol{P} = P \cos\theta\, \boldsymbol{i} + P \sin\theta\, \boldsymbol{j} = P\,(\cos\theta\, \boldsymbol{i} + \sin\theta\, \boldsymbol{j}) = P\, \boldsymbol{e}$$

where, $\boldsymbol{e} = \cos\theta\, \boldsymbol{i} + \sin\theta\, \boldsymbol{j}$ is called the ***unit vector*** along the direction of force

and $|e| = \sqrt{[(\cos\theta)^2 + (\sin\theta)^2]} = 1$

4.2.2 MOMENT OF A FORCE

Moment of a force about a moment center is the product of the position vector from the moment center to any point on the line of action of the force crossed with the force vector. It is oriented towards the positive normal to the plane of position vector and force vector, following right hand rule.

$$\mathbf{M} = \mathbf{r} \times \mathbf{F} \quad ; \quad |\mathbf{M}| = \mathbf{F} \times \mathbf{d}$$

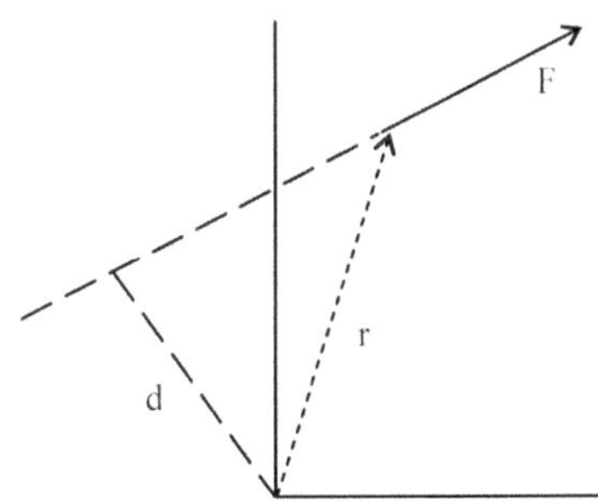

A better understanding of the vector method of calculating moment of a force can be obtained by resolving the force '***F***' and the distance '***r***' in vector form into its components along the coordinate axes and then calculating the moment for these

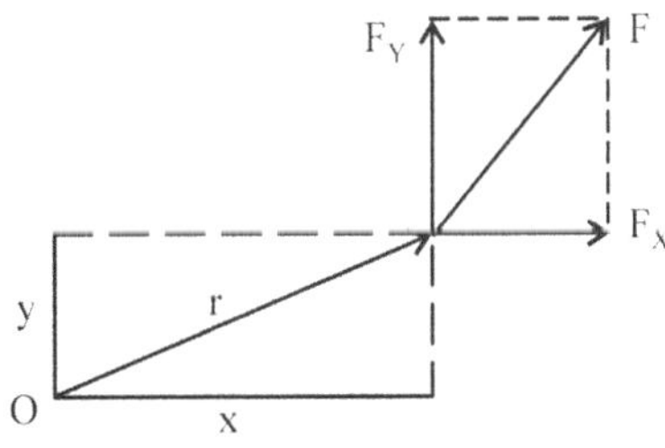

components of force. A two-dimensional representation in graphical form is shown.

(a) Moment about a point O

$$\boldsymbol{M_O} = \boldsymbol{r} \times \boldsymbol{F} = (x\,\boldsymbol{i} + y\,\boldsymbol{j}) \times (F_X\,\boldsymbol{i} + F_y\,\boldsymbol{j})$$

$$= (x \times F_Y - y \times F_X)\,\boldsymbol{k}$$

$$\mathbf{M}_0 = \begin{vmatrix} \mathbf{i} & \mathbf{j} & \mathbf{k} \\ x & y & 0 \\ F_x & F_y & 0 \end{vmatrix}$$

By the non-vector approach, moment about O for the force components,

M_O = Counter clockwise moment due to F_Y

– Clockwise moment due to F_X

$$= x \times F_Y - y \times F_X$$

and is oriented about the axis perpendicular to X-Y plane

Moment of force P about C, $\boldsymbol{M_C^P} = \boldsymbol{r_{CA}} \times \boldsymbol{P}$

(cross product, resulting in a vector))

where, $\boldsymbol{R_{CA}}$ is a position vector from C to A

$$= (X_A - X_C)\,\boldsymbol{i} + (Y_A - Y_C)\,\boldsymbol{j} + (Z_A - Z_C)\,\boldsymbol{k}$$

$$= X_{AC}\,\boldsymbol{i} + Y_{AC}\,\boldsymbol{j} + Z_{AC}\,\boldsymbol{k}$$

and Force vector $\boldsymbol{P} = p\,\mathbf{e_{AB}} = p_x\,\mathbf{i} + p_y\,\mathbf{j} + p_z\,\mathbf{k}$

$$\mathbf{M_C^P} = \begin{vmatrix} \mathbf{i} & \mathbf{j} & \mathbf{k} \\ x_{AC} & y_{AC} & z_{AC} \\ P_x & P_y & P_z \end{vmatrix}$$

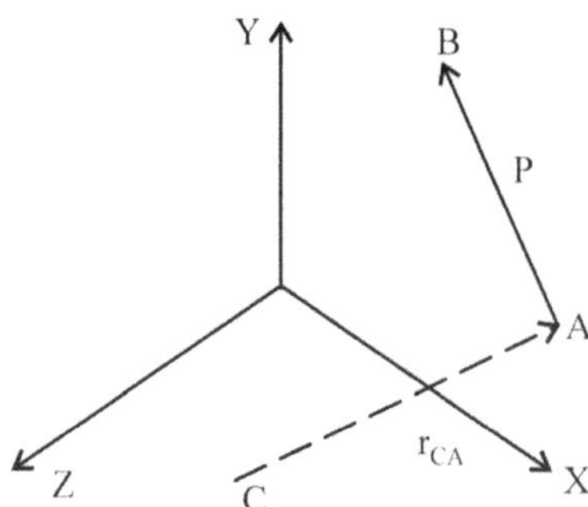

An important advantage with the vector approach is that the distance vector '**r**' can be chosen from 'C' to any point along the line of action of force. This property is very much useful in solving some problems, as shown in the following examples.

(b) Moment about a line OA: It is the component of moment about O along OA

$$|\boldsymbol{M_{OA}}| = \mathbf{e_{OA}} \cdot \mathbf{M_O} = \boldsymbol{e_{OA}} \cdot (\boldsymbol{r} \times \boldsymbol{F})$$

$$= (\text{OA} / |\boldsymbol{OA}|) \cdot (\boldsymbol{r} \times \boldsymbol{F})$$

where $\mathbf{e_{OA}}$ is the unit vector along OA

In vector form, $\boldsymbol{M_{OA}} = |\boldsymbol{M_{OA}}|\ \boldsymbol{e_{OA}}$

Moment of a force 'P' about an axis CD = $\boldsymbol{e}_{CD} \cdot \boldsymbol{M_C}^{P}$

(dot product, resulting in a scalar)

where, $\boldsymbol{e_{CD}}$ is a unit vector from C to D

$$= [\,(X_D - X_C)\,\boldsymbol{i} + (Y_D - Y_C)\,\boldsymbol{j} + (Z_D - Z_C)\,\boldsymbol{k}\,] / |CD|$$

and $|CD| = \sqrt{(X_D - X_C^2) + (Y_D - Y_C^2) + (Z_D + Z_C^2)}$

Couple – A pair of equal, parallel and non-collinear forces in opposite directions

- Moment sum of its forces is constant and independent of any moment center

  $M_O = \boldsymbol{F} \times \boldsymbol{r_2} - \boldsymbol{F} \times \boldsymbol{r_1} = \boldsymbol{F} \times (\boldsymbol{r_2} - \boldsymbol{r_1}) = \boldsymbol{F} \times \boldsymbol{d}$

- Moment of a couple = One of the forces × Perpendicular distance between their lines of action

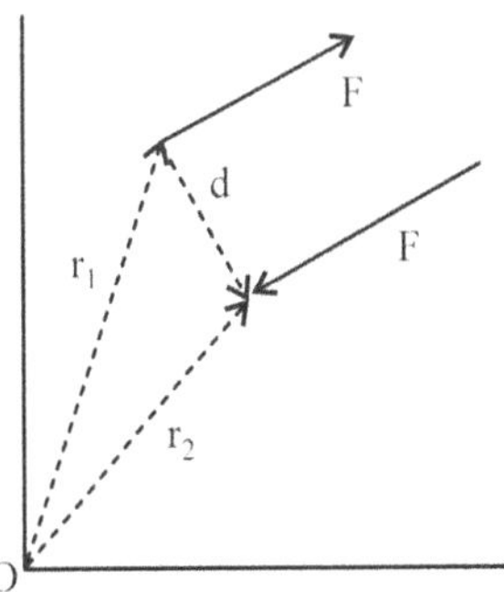

Any force F, not passing through O, produces an effect at O equivalent to a force F through O and a moment of magnitude F × d. This can be better understood by applying two forces parallel to F, opposing each other, at O. The two parallel, non-collinear forces produce a moment at O, in addition to the force at O.

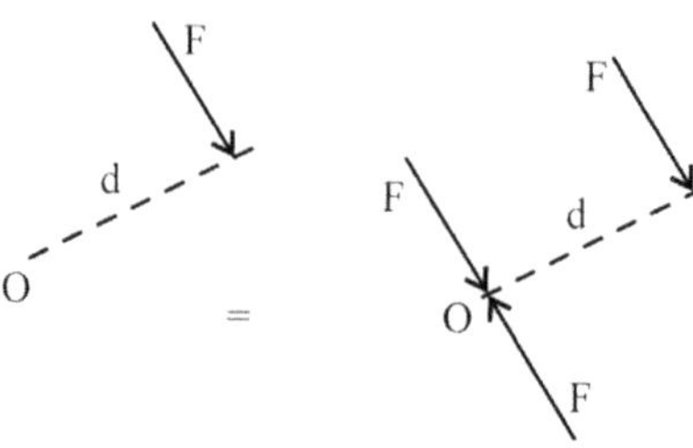

Important Note: Addition of moments of many forces about any single point is done with signs (corresponding to clockwise or counter-clockwise moments) in the conventional method and without explicit use of signs in vector method, since vector includes the direction effect.

System of non-concurrent forces as a force and a couple at a point

If force $\boldsymbol{P}$ is acting along CD and force $\boldsymbol{Q}$ along AB, it is equivalent to a resultant force $\boldsymbol{R}$ and a couple of moment $\boldsymbol{M}$ acting at O

Force vector, $\boldsymbol{P} = p\, \boldsymbol{e}_{CD}$ (scalar multiplication)

Force vector, $\boldsymbol{Q} = q\, \boldsymbol{e}_{AB}$ (scalar multiplication)

Resultant, $\boldsymbol{R} = \boldsymbol{P} + \boldsymbol{Q}$

Moment of couple, $\boldsymbol{M} = \boldsymbol{P} \times \boldsymbol{r}_{OC} + \boldsymbol{Q} \times \boldsymbol{r}_{OA}$

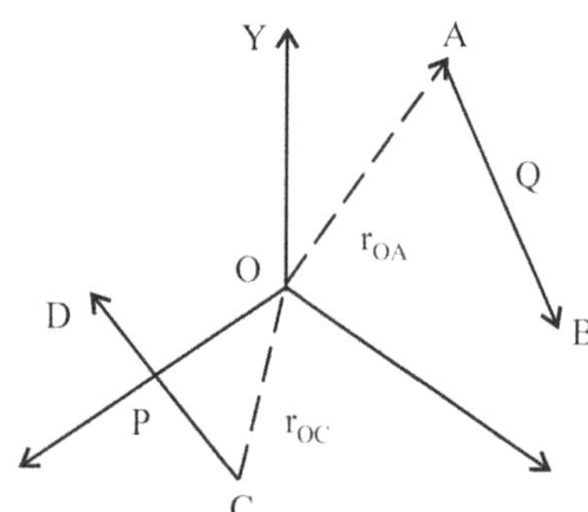

Example 1

A ladle is lifted by means of three sling chains, each 1 m in length. The upper ends of the chains are attached to a ring while the lower ends are attached to three hooks fixed to the ladle forming an equivalent triangle of 1.2 m side. If the weight of the ladle and its contents is 4500 N, find the load taken up by each chain.

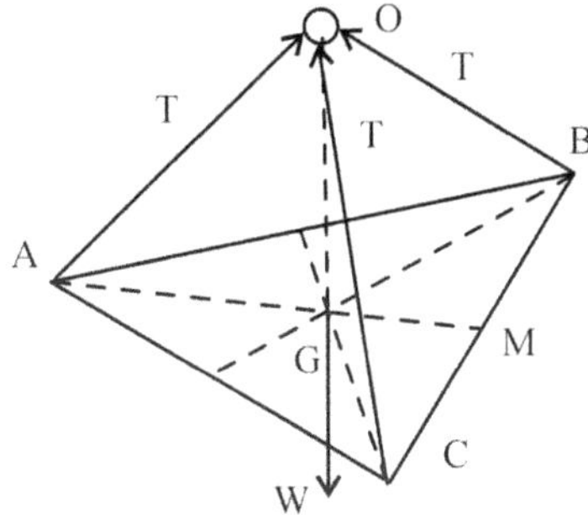

Solution:

If G is the centroid of the equilateral ΔABC of side 1.2 m,

$$AG = (2/3)\ AM = (2/3)\ AB \cos 30^0 = (2/3)\ 0.6\ \sqrt{3}\ \text{m}$$

$$= 0.4\ \sqrt{3}\ \text{m}$$

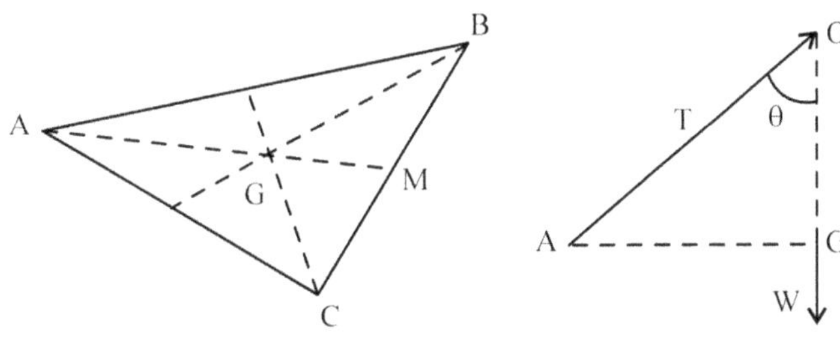

Considering right angled Δ AOG, let θ be the inclination of each chain with the vertical.

Then, $\sin\theta = AG/AO = 0.4\ \sqrt{3}\ /\ 1$

or $\theta = 43^0\ 51'$

Vertical components of forces in the 3 chains should balance the weight of the ladle.

$$3\ T\cos\theta = W = 4500\ \text{N} \qquad \text{or} \qquad T = 2080\ \text{N}$$

Example 2

Three bars, each of length 50 cm, are hinged together at D and to fixed points at A, B and C as shown. Find the forces in the bars for a vertical load P applied at D, if OA = OB = OC = 40 cm.

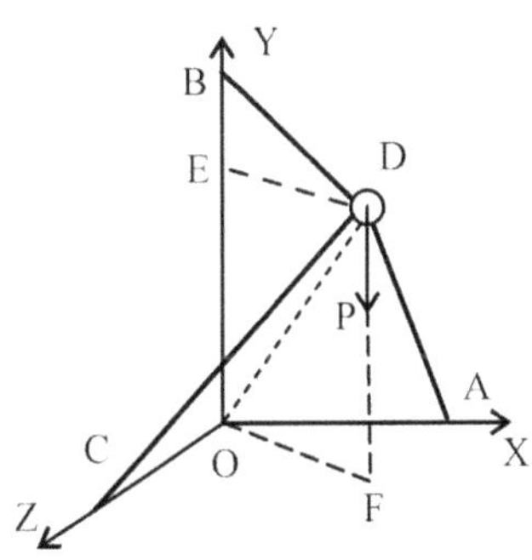

Solution:

Let the line OD be inclined to the three axes at α, β and γ.

These are related by $\cos^2 \alpha + \cos^2 \beta + \cos^2 \gamma = 1$

Due to symmetry of the system (since OA = OB = OC & DA = DB = DC),

$$\alpha = \beta = \gamma = \cos^{-1}(1/\sqrt{3}) = 54.74^0$$

In Δ OBD, as per sine rule of a triangle, $BD/\sin\beta = OD/\sin\theta = OB/\sin\varphi$

Then, $\varphi = \sin^{-1}(OB \times \sin\beta / BD) = 40.77^0$

and $\theta = 180 - \beta - \varphi = 84.49^0$

Method-I : Let the forces in members AD, BD and CD be respectively be F_1, F_2 and F_3 along DA, DB and DC respectively. Then, force F_2 in member BD can be resolved along and perpendicular to DE as $F_2 \sin\theta$ and $F_2 \cos\theta$. Due to the symmetry of the system, projection of the component $F_2 \sin\theta$ on X–Z plane (OF) is equally inclined to X and Z axes. Hence, components of F_2 along X and Z axes are

$$F_{2X} = -(F_2 \sin\theta)\cos 45 \quad \text{and} \quad F_{2Z} = -(F_2 \sin\theta)\sin 45$$

Similarly,

$$F_{1X} = F_1 \cos\theta;\ F_{1Y} = -(F_1 \sin\theta)\cos 45;\ F_{1Z} = -(F_1 \sin\theta)\sin 45$$

$$F_{3X} = -(F_3 \sin\theta)\cos 45\ ;\ F_{3Y} = -(F_3 \sin\theta)\sin 45\ ;\ F_{3Z} = F_3 \cos\theta$$

For equilibrium of point D,

$$\sum F_X = F_1 \cos\theta - (F_2 \sin\theta)\cos 45 - (F_3 \sin\theta)\cos 45 = 0$$

$$\sum F_Y = -(F_1 \sin\theta)\cos 45 + F_2 \cos\theta - (F_3 \sin\theta)\sin 45 - P = 0$$

$$\sum F_Z = -(F_1 \sin\theta)\sin 45 - (F_2 \sin\theta)\sin 45 + F_3 \cos\theta = 0$$

Solving these three simultaneous equations,

$$F_1 = -0.67\,P\ ;\ F_2 = 0.58\,P\ ;\ F_3 = -0.67\,P$$

Method-II (Vector approach)**:** Using the angle 'θ' calculated above, we get coordinates OG, OE and OH of D along X, Y and Z axes as (35.2,35.2,35.2). Let the forces in members AD, BD and CD be F_1, F_2 and F_3 along DA, DB and DC respectively.

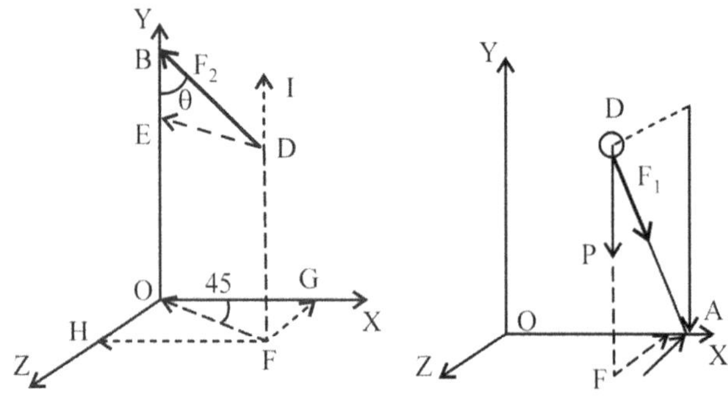

The components of DB in the plane OFDB are DI or EB along Y-axis and DE or FO. The component FO can then be resolved into FH along X-axis and FG along Z-axis.

Thus, $DI = BD \cos\theta = 4.8$; $FO = BD \sin\theta$

Also, FH or $OG = (BD \sin\theta) \cos 45 = -35.2$

& FG or $OH = (BD \sin\theta) \sin 45 = -35.2$

Using the relation, $F_x / x = F_y / y = F_z / z = F / d$, for $F = F_2$,

$$(F_2)_X = F_2 \times x / d = F_2 \times OG / DB = F_2 \times (-35.2 / 50)$$

Similarly, $(F_2)_Y = F_2 \times y / d = F_2 \times DI / DB = F_2 \times (4.8 / 50)$

and $(F_2)_Z = F_2 \times z / d = F_2 \times OH / DB = F_2 \times (-35.2 / 50)$

With these components, force vector can be written as

$$\mathbf{F_2} = (-35.2\,\boldsymbol{i} + 4.8\,\boldsymbol{j} - 35.2\,\boldsymbol{k}) / 50$$

Similarly, $\mathbf{F_1} = (4.8\,\boldsymbol{i} - 35.2\,\boldsymbol{j} - 35.2\,\boldsymbol{k}) / 50$

and $\mathbf{F_3} = (-35.2\,\boldsymbol{i} - 35.2\,\boldsymbol{j} + 4.8\,\boldsymbol{k}) / 50$

For equilibrium of point D, $F_1 + F_2 + F_3 = P$

Coefficients of '*i*' in these force vectors,

$$\sum F_X = (4.8\,F_1 - 35.2\,F_2 - 35.2\,F_3) / 50 = 0$$

Coefficients of '*j*' in these force vectors,

$$\sum F_Y = (-35.2\,F_1 + 4.8\,F_2 - 35.2\,F_3) / 50 = P$$

Coefficients of '*k*' in these force vectors,

$$\sum F_Z = (-35.2\,F_1 - 35.2\,F_2 + 4.8\,F_3) / 50 = 0$$

Solving these three simultaneous equations,

$$F_1 = -0.67\,P ; \quad F_2 = 0.58\,P ; \quad F_3 = -0.67\,P$$

Example 3

A pulley is supported from the face of a vertical wall by two braces AB and AC together with a tie bar AD as shown. A flexible cord EAG, fastened to the wall at E, passes over the pulley and carries a load P at its free end. Find the tensile force 'T_{AD}' produced in the tie bar AD if P = 100 N ; a = 60 cm; b = 40 cm and c = 75 cm.

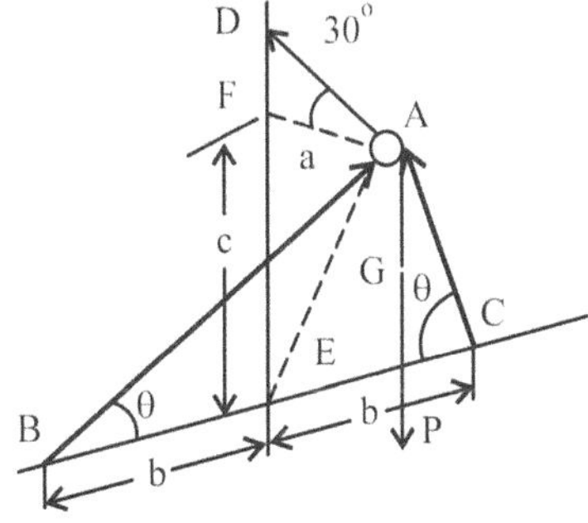

Solution:

In ΔAEF, $AE = \sqrt{AF^2 + EF^2} = \sqrt{(a^2 + c^2)} = 96cm$

In ΔABE, $AB = AC = \sqrt{AE^2 + BE^2} = \sqrt{\left(a^2 + c^2\right) + b^2} = 104cm$

$$\tan\theta = AE/BE \text{ or } AE/CE = 96/40 = 2.4 \Rightarrow \theta = 67.4^0$$

Forces in the two braces AB and AC are symmetric about the plane through AED. Therefore, they can be replaced by their resultant along EA given by,

$$F_{EA} = F_{BA}\sin\theta + F_{CA}\sin\theta = 2\,F_{BA}\sin\theta$$

Considering the vertical plane through A, E and D, pulley A with load P is in equilibrium due to the forces F_{AD}, F_{EA} and tension T in the chord EAG

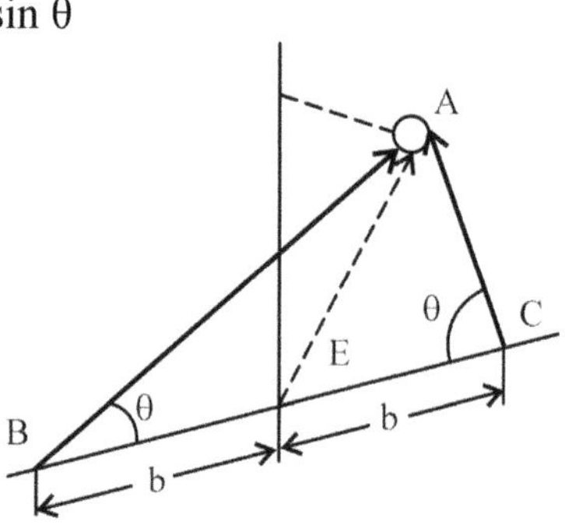

For equilibrium of G, T = P and the same tension acts along AE also

The equations of equilibrium for horizontal and vertical forces at A are

$$F_{EA}\sin\alpha - F_{AD}\cos 30 - T\sin\alpha = 0 \quad .....(1)$$

and $$F_{EA}\cos\alpha + F_{AD}\sin 30 - T\cos\alpha = T \quad .....(2)$$

where, $\sin\alpha = FA/EA = 60/96 = 0.625$

and $\cos\alpha = 0.7806$

=> $$F_{EA} - F_{AD}\cos 30/\sin\alpha = T$$

or $$F_{EA} - 1.38564\,F_{AD} = 100$$

$$F_{EA} + F_{AD}\sin 30/\cos\alpha = T(1 + 1/\cos\alpha)$$

or $$F_{EA} + 0.6405F_{AD} = 228$$

Subtracting, $F_{AD}\,[0.6405 + 1.38564] = 228 - 100$

$$F_{AD} = 128/2.026 = \boldsymbol{63.17N}$$

and $$F_{EA} = 187.53N$$

$$F_{BA} = F_{CA} = F_{EA}/(2\sin\theta) = \boldsymbol{101.56N}$$

Example 4

Three equal spheres of radius 'r' rest on a smooth horizontal surface, touching one another. A fourth sphere of same size is placed on top of these three, to form a pyramid and the three lower spheres are now held together by an encircling string. Calculate tension in the string.

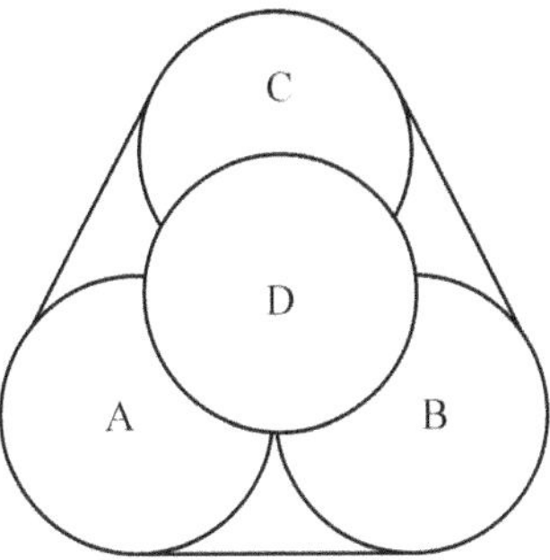

Solution:

Top view of the centers of the three lower spheres forms an equilateral triangle ABC. Projection of the center of the top sphere 'D', being equidistant to the vertices A, B and C, lies at the centroid 'O' of ΔABC. AB = BC = CA = AD = BD = CD = 2r

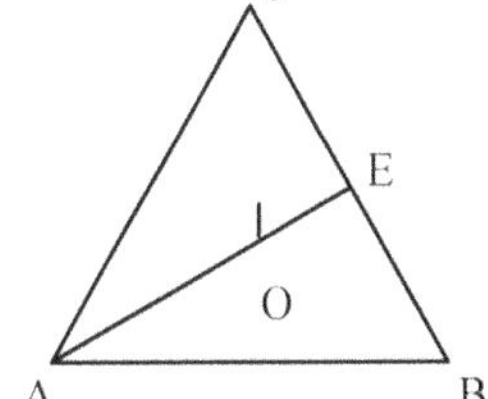

$$\angle CAB = 60^0 \quad \text{and} \ \angle CAE = \angle BAE = 30^0$$

The perpendicular on BC,

$$AE = AB \cos 30 = 2r\,(\sqrt{3}/2) = r\sqrt{3}$$

and $$AO = (2/3)\,AE = (2/3)\,r\sqrt{3} = 2\,r/(\sqrt{3})$$

Now, considering triangle AOD,

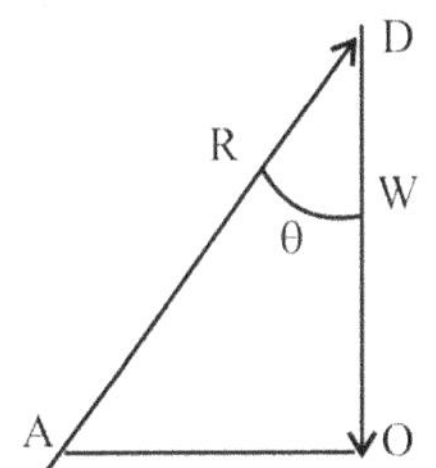

$$\angle DAO = \cos^{-1}(AO/AD) = \cos^{-1}(1/\sqrt{3})$$

$$= 54^0 44'$$

$$\theta = \angle ADO = 90 - 54^0 44' = 35^0 16'$$

Due to symmetry, weight 'W' of the top cylinder acting along DO is supported equally by reactions 'R' acting along AD, BD and CD. Thus, resolving forces 'R' along OD,

$$3\,R\cos\theta = W \quad \text{or} \quad R = 0.3657\,W$$

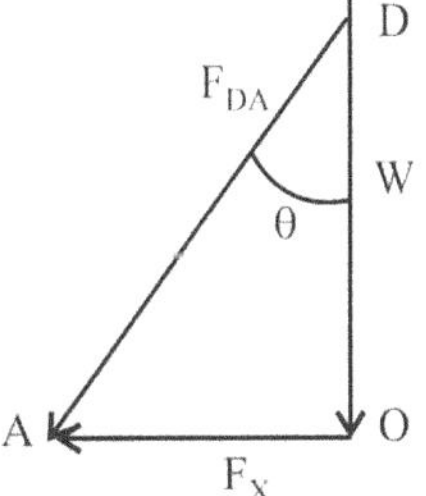

Force F_{DA} exerted by the weight of D on A is equal and opposite to the reaction R. Horizontal component F_X of F_{DA} is balanced by the tension in the string, on each side of the sphere.

Thus, $2\ T \cos 30 = F_X = F_{DA} \sin\theta$

or $T = F_{DA} \sin\theta / (2 \cos 30) = 0.1225\ W$

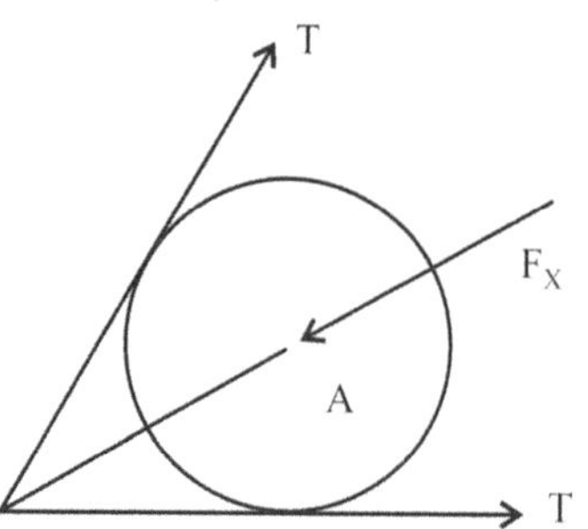

Example 5

Calculate tension in the guy wire CD and the compression in the struts AC and BC of the shear leg derrick, as shown, for the load P acting at C. Take OE = 30cm, EC = 120cm, AO = OB = 40cm and $\angle CDO = 30^0$

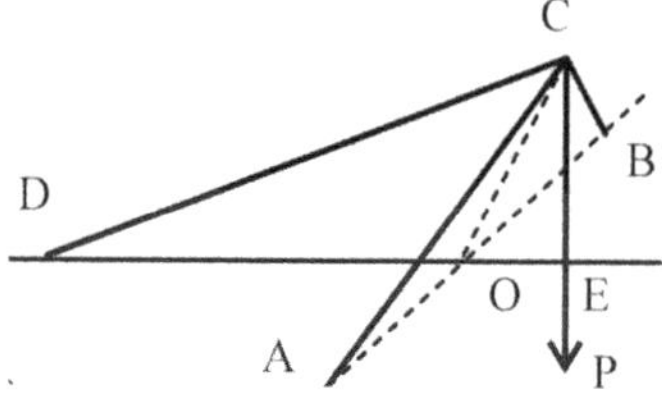

Solution:

Method-1: Components of forces approach

The problem is analysed in two stages

- First by imagining a member OC (to represent resultant of forces in members AC and BC) and calculating force in this member for the equilibrium of point C
- In the second stage, forces in members AC and BC can be calculated such that the resultant of these two forces is equal in magnitude to the force in the imaginary member OC and has same direction

Stage-I : Let $\angle COE = \theta$; $\tan\theta = 120/30 \Rightarrow \theta = 76^0$

Applying Lami's theorem at point C for the forces, F_{CD}, F_{OC} and P,

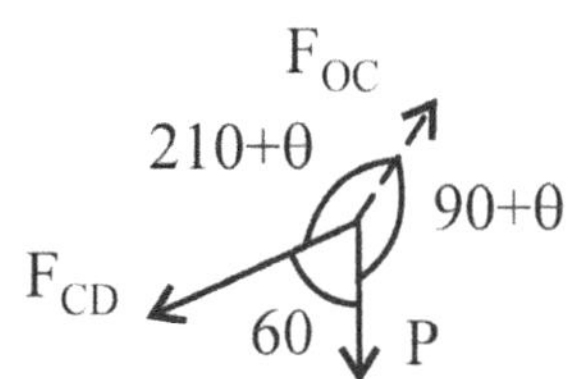

$$F_{CD}/\sin(90+\theta) = P/\sin(210+\theta)$$

$\Rightarrow$ $F_{CD} = 0.337\ P$

and $F_{OC}/\sin 60 = P/\sin(210+\theta)$

$\Rightarrow$ $F_{OC} = 1.203\ P$

Stage-II : $OC^2 = OE^2 + EC^2 = 30^2 + 120^2$

$\Rightarrow$ $OC = 123.7\text{cm}$

Let $\angle ACO = \angle BCO = \alpha$; $\tan\alpha = AO/OC = 40/123.7$

$\Rightarrow$ $\alpha = 17.9^0$

Taking X-axis along OB and Y-axis along OC,

$$\sum F_X = F_{AC}\sin\alpha - F_{BC}\sin\alpha = 0 \quad \Rightarrow \quad F_{AC} = F_{BC}$$

$$\sum F_Y = F_{AC}\cos\alpha + F_{BC}\cos\alpha = F_{OC}$$

or $F_{AC} = F_{BC} = F_{OC} / (2\cos\alpha) = 1.203\,P / (2\cos 17.9) = 0.632\,P$

Method-2: Vector approach

From triangle CDE, $DC = EC / \sin 30 = 240$ cm

Considering O as the origin, X-axis along OE, Y-axis along OB and Z-axis parallel to EC through O, coordinates of the end points of the members are

A(0,–40,0), B(0,40,0), C(30,0,120) and D(–210,0,0)

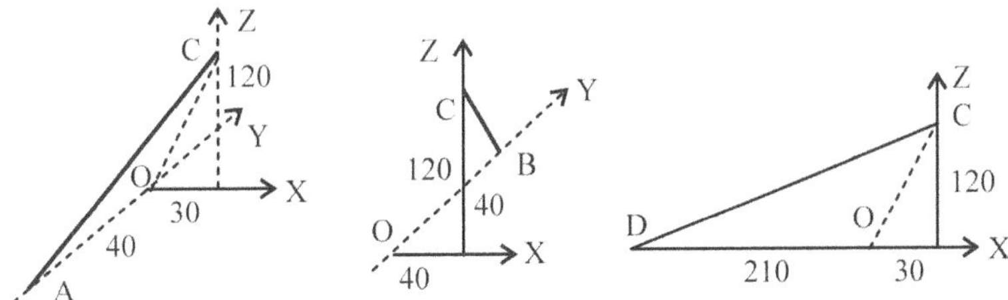

Position vectors of members are, therefore

$AC = (X_C - X_A)\,\boldsymbol{i} + (Y_C - Y_A)\,\boldsymbol{j} + (Z_C - Z_A)\,\boldsymbol{k} = 30\,\boldsymbol{i} + 40\,\boldsymbol{j} + 120\,\boldsymbol{k}$

$BC = (X_C - X_B)\,\boldsymbol{i} + (Y_C - Y_B)\,\boldsymbol{j} + (Z_C - Z_B)\,\boldsymbol{k} = 30\,\boldsymbol{i} - 40\,\boldsymbol{j} + 120\,\boldsymbol{k}$

$DC = (X_C - X_D)\,\boldsymbol{i} + (Y_C - Y_D)\,\boldsymbol{j} + (Z_C - Z_D)\,\boldsymbol{k} = 240\,\boldsymbol{i} + 0\,\boldsymbol{j} + 120\,\boldsymbol{k}$

Unit vectors along these directions are given by,

$$\mathbf{e}_{AC} = (30\boldsymbol{i} + 40\boldsymbol{j} + 120\boldsymbol{k}) / \sqrt{(30^2 + 40^2 + 120^2)}$$

$$= (3/13)\,\boldsymbol{i} + (4/13)\,\boldsymbol{j} + (12/13)\,\boldsymbol{k}$$

$$\mathbf{e}_{BC} = (30\boldsymbol{i} - 40\boldsymbol{j} + 120\boldsymbol{k}) / \sqrt{(30^2 + 40^2 + 120^2)}$$

$$= (3/13)\,\boldsymbol{i} - (4/13)\,\boldsymbol{j} + (12/13)\,\boldsymbol{k}$$

$$\mathbf{e}_{DC} = (240\boldsymbol{i} + 0\boldsymbol{j} + 120\boldsymbol{k}) / \sqrt{(240^2 + 0^2 + 120^2)}$$

$$= (2/\sqrt{5})\,\boldsymbol{i} + 0\,\boldsymbol{j} + (1/\sqrt{5})\,\boldsymbol{k}$$

Let the magnitudes of forces along AC, BC and DC be F_1, F_2 and F_3 respectively

Applying condition of equilibrium at C, $F_1\,e_{AC} + F_2\,e_{BC} + F_1\,e_{DC} - P\,\boldsymbol{k} = 0$

$$F_1\,[\,(3/13)\,\boldsymbol{i} + (4/13)\,\boldsymbol{j} + (12/13)\,\boldsymbol{k}\,] + F_2\,[\,(3/13)\,\boldsymbol{i} - (4/13)\,\boldsymbol{j} + (12/13)\,\boldsymbol{k}\,]$$

$$+ F_3\,[\,(2/\sqrt{5})\,\boldsymbol{i} + 0\,\boldsymbol{j} + (1/\sqrt{5})\,\boldsymbol{k}\,] - P\,\boldsymbol{k} = 0$$

This equation is satisfied only when coefficients of unit vectors $\boldsymbol{i}$, j and k are zero

i.e., $(3/13)\,F_1 + (3/13)\,F_2 + (2/\sqrt{5})\,F_3 = 0$

$$(4/13)\,F_1 - (4/13)\,F_2 + 0\,F_3 = 0$$

$$(12/13)\,F_1 + (12/13)\,F_2 + (1/\sqrt{5})\,F_3 - P = 0$$

$\Rightarrow$ $F_1 = F_2 = 0.632\,P$ and $F_3 = -\,0.337\,P$

Example 6

A concentrated load of 5 kN is applied downwards at the common point A of a 3–D truss, consisting of members A–B, A–C and A–D. All the ends are hinged and the truss is fixed at B, C and D. If the coordinates (in cm) of the end points are A(40,0,0), B(0,0,30), C(0,0,–30) and D(0,–30,0). Calculate forces in the three members.

Solution:

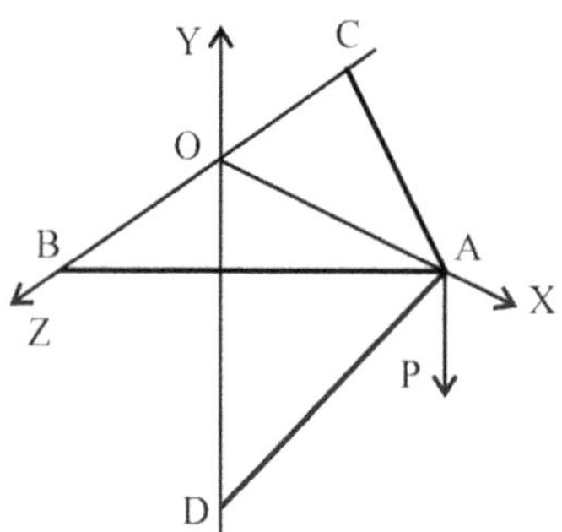

Method-1: Components of forces approach - The problem is analysed in two stages

- First by imagining a member OA (to represent resultant of forces in members AB and AC) and calculating forces in OA and AD for equilibrium of point A
- In the second stage, forces in members AB and AC can be calculated such that the resultant of these two forces is equal in magnitude to the force in the imaginary member OA and has same direction

Stage-I: Let $\angle ADO = \theta$; $\tan\theta = AO/OD = 40/30 = 4/3$

Therefore, $\sin\theta = 4/5$ and $\cos\theta = 3/5$

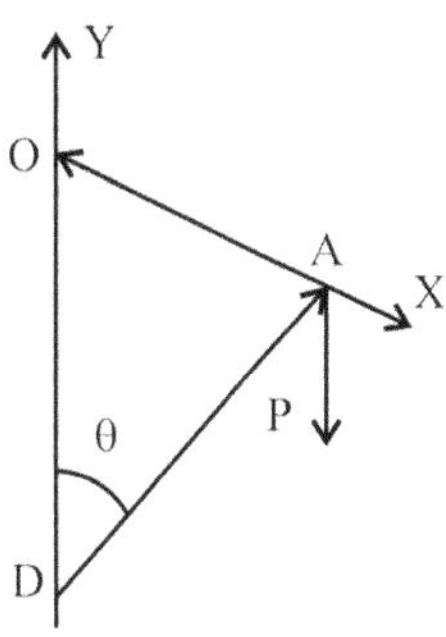

For equilibrium of point A,

$\sum F_Y = F_{AD} \cos\theta - P = 0$

$\Rightarrow$ $\mathbf{F_{AD}} = P/\cos\theta = 5/(3/5) = 25/3 = \mathbf{8.33\ kN}$

$\sum F_X = F_{AD} \sin\theta - F_{AO} = 0$

$\Rightarrow$ $\mathbf{F_{AO}} = F_{AD} \sin\theta = (25/3) \times (4/5) = 20/3 = 6.67$ kN

Stage-II : Let /ABO = /ACO = α ;

tan α = AO/BO = 40/30

Therefore, sin α = 4/5 and cos α = 3/5

If F_{AO} is the resultant of forces F_{AB} and F_{AC},

$\sum F_X = F_{AC} \sin\alpha + F_{AB} \sin\alpha = F_{AO}$

$\Rightarrow$ $\mathbf{F_{AC}} = \mathbf{F_{AB}} = F_{AO} / 2 \sin\alpha = 6.67 / [2 \times (4/5)] = \mathbf{4.167\ kN}$

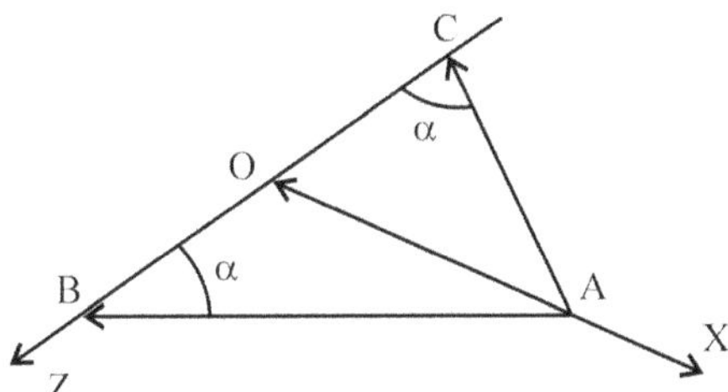

Equilibrium equations at point A are written with the assumed directions for member forces –

Forces in AB and AC are tensile and force in AD is compressive

Method-2: Vector approach

Coordinates of the four points of the truss are A(40, 0, 0), B(0, 0, 30), C(0, 0, –30) and D(0,–30,0).

Position vectors of members are, therefore

$$AB = (X_B - X_A)\,\boldsymbol{i} + (Y_B - Y_A)\,\boldsymbol{j} + (Z_B - Z_A)\,\boldsymbol{k} = -40\,\boldsymbol{i} + 0\,\boldsymbol{j} + 30\,\boldsymbol{k}$$
$$AC = (X_C - X_A)\,\boldsymbol{i} + (Y_C - Y_A)\,\boldsymbol{j} + (Z_C - Z_A)\,\boldsymbol{k} = -40\,\boldsymbol{i} + 0\,\boldsymbol{j} - 30\,\boldsymbol{k}$$
$$AD = (X_D - X_A)\,\boldsymbol{i} + (Y_D - Y_A)\,\boldsymbol{j} + (Z_D - Z_A)\,\boldsymbol{k} = -40\,\boldsymbol{i} - 30\,\boldsymbol{j} + 0\,\boldsymbol{k}$$

Length of each member is given by $\sqrt{(40^2 + 30^2)} = 50$ cm

Unit vectors along these directions are given by,

$$\mathbf{e_{AB}} = (-40\boldsymbol{i} + 0\boldsymbol{j} + 30\boldsymbol{k}) / 50 = -(4/5)\,\boldsymbol{i} + 0\,\boldsymbol{j} + (3/5)\,\boldsymbol{k}$$
$$\mathbf{e_{AC}} = (-40\boldsymbol{i} + 0\boldsymbol{j} - 30\boldsymbol{k}) / 50 = -(4/5)\,\boldsymbol{i} + 0\,\boldsymbol{j} - (3/5)\,\boldsymbol{k}$$
$$\mathbf{e_{AD}} = (-40\boldsymbol{i} - 30\boldsymbol{j} + 0\boldsymbol{k}) / 50 = -(4/5)\,\boldsymbol{i} - (3/5)\,\boldsymbol{j} + 0\,\boldsymbol{k}$$

Let the magnitudes of forces along AC, BC and DC be F_1, F_2 and F_3 respectively

Applying condition of equilibrium at A,

$$F_{AB}\,\boldsymbol{e_{AB}} + F_{AC}\,\boldsymbol{e_{AC}} + F_{AD}\,\boldsymbol{e_{AD}} - \boldsymbol{P} = 0$$

$$F_{AB}\,[-(4/5)\,\boldsymbol{i} + (0)\,\boldsymbol{j} - (3/5)\,\boldsymbol{k}\,] + F_{AC}\,[-(4/5)\,\boldsymbol{i} + (0)\,\boldsymbol{j} - (3/5)\,\boldsymbol{k}\,]$$

$$+ F_{AD}\,[-(4/5)\,\boldsymbol{i} - \mathbf{(3/5)}\,\boldsymbol{j} + (0)\,\boldsymbol{k}\,] - 5\,\mathbf{j} = 0$$

Since load **P** acts along –ve Y-direction, $\boldsymbol{P} = -P\,\boldsymbol{j} = -5\,\boldsymbol{j}$

This equation is satisfied only when coefficients of unit vectors ***i***, ***j*** and ***k*** are all zero

i.e., $[-(4/5)\,F_{AB} - (4/5)\,F_{AC} - (4/5)\,F_{AD}\,]\,\boldsymbol{i} = 0$

$[\,(0)\,F_{AB} + (0)\,F_{AC} - (3/5)\,F_{AD} - 5]\,\boldsymbol{j} = 0$

$[\,(3/5)\,F_{AB} - (3/5)\,F_{AC} + (0)\,F_{AD}\,]\,k = 0$

$\Rightarrow$ $F_{AD} = -25/3 = -8.33$ kN (Compressive)

and $F_{AB} = F_{AC} = -F_{AD}/2 = +\,4.167$ kN (tensile)

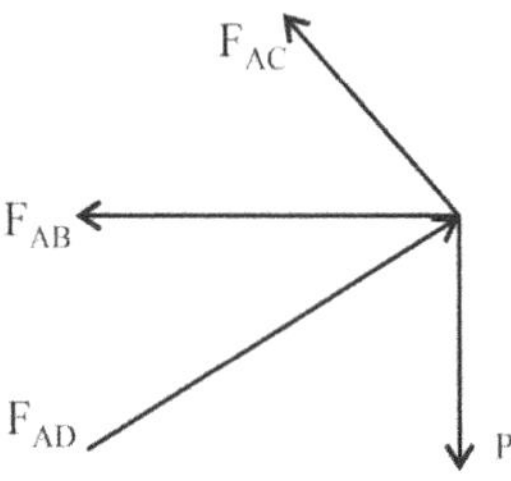

Example 7

Lines of action of three concurrent forces, meeting at point O(0,0,0), pass through A(–1,2,4), B(3,0,–3) and C(2,–2,4) and have magnitudes 40N, 10N and 30N respectively. Calculate magnitude and direction of resultant.

Solution:

Position vectors along the lines of action of the three forces are

$$AO = (X_O - X_A)\,\boldsymbol{i} + (Y_O - Y_A)\,\boldsymbol{j} + (Z_O - Z_A)\,\boldsymbol{k} = \boldsymbol{i} - 2\,\boldsymbol{j} - 4\,\boldsymbol{k}$$

$$BO = (X_O - X_B)\,\boldsymbol{i} + (Y_O - Y_B)\,\boldsymbol{j} + (Z_O - Z_B)\,\boldsymbol{k} = -3\,\boldsymbol{i} + 0\,\boldsymbol{j} + 3\,\boldsymbol{k}$$

and $$CO = (X_O - X_C)\,\boldsymbol{i} + (Y_O - Y_C)\,\boldsymbol{j} + (Z_O - Z_C)\,\boldsymbol{k} = -2\,\boldsymbol{i} + 2\,\boldsymbol{j} - 4\,\boldsymbol{k}$$

$$|AO| = \sqrt{[1^2]+(-2)^2+(-4)^2} = \sqrt{21}$$

$$|BO| = \sqrt{[(-3)^2]+0^2+3^2} = \sqrt{18}$$

and $$|CO| = \sqrt{[(-2)^2]+2^2+(-4)^2]} = \sqrt{24}$$

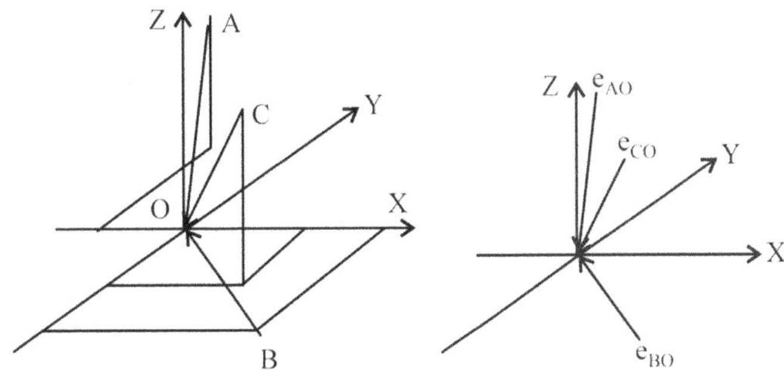

Unit vectors along the three forces are

$$\boldsymbol{e_{AO}} = AO \,/\, |AO| = (+\boldsymbol{i} - 2\boldsymbol{j} - 4\,\boldsymbol{k}) \,/\, \sqrt{21}$$

$$\boldsymbol{e_{BO}} = BO \,/\, |BO| = (-3\,\boldsymbol{i} + 0\,\boldsymbol{j} + 3\,\boldsymbol{k}) \,/\, \sqrt{18}$$

and $$\boldsymbol{e_{CO}} = CO \,/\, |CO| = (-2\,\boldsymbol{i} + 2\,\boldsymbol{j} - 4\,\boldsymbol{k}) \,/\, \sqrt{24}$$

Then, $$\boldsymbol{F_{AO}} = 40\,\boldsymbol{e_{AO}} = 40(\,\boldsymbol{i} - 2\,\boldsymbol{j} - 4\,\boldsymbol{k}) \,/\, \sqrt{21}$$

$$\boldsymbol{F_{BO}} = 10\,\boldsymbol{e_{BO}} = 10(-3\,\boldsymbol{i} + 0\,\boldsymbol{j} + 3\,\boldsymbol{k}) \,/\, \sqrt{18}$$

and $$\boldsymbol{F_{CO}} = 30\,\boldsymbol{e_{CO}} = 30(-2\,\boldsymbol{i} + 2\,\boldsymbol{j} - 4\,\boldsymbol{k}) \,/\, \sqrt{24}$$

Resultant, $$\boldsymbol{R} = \boldsymbol{F_{OA}} + \boldsymbol{F_{OB}} + \boldsymbol{F_{OC}}$$

$$= [\,(1/\sqrt{21}) - (3/\sqrt{18}\,) - (2/\sqrt{24}\,)\,]\,\boldsymbol{i} + [-(2\sqrt{21}\,) + 0 + (2/\sqrt{24}\,)\,]\,\boldsymbol{j} + [-(4/\sqrt{21}\,) + (3/\sqrt{18}\,) - (4/\sqrt{24}\,)\,]\,\boldsymbol{k}$$

$$= -\,0.8972\,i\; - 0.0282\,j\; - 0.982\,k$$

Magnitude of resultant, $|R| = \sqrt{[(-0.8972)^2 + (-0.0282)^2 + (-0.988)^2]} = 1.77$

Direction of the resultant is specified by

its angle with X-axis, $\alpha = \cos^{-1}(-0.8972\,/\,|\mathbf{R}|\,) = \cos^{-1}(-0.5069) = 59.54^0$

its angle with Y-axis, $\beta = \cos^{-1}(-0.0282\,/\,|\mathbf{R}|\,) = \cos^{-1}(-0.0159) = 89.1^0$

its angle with Z-axis, $\gamma = \cos^{-1}(-0.982\,/\,|\mathbf{R}|\,) = \cos^{-1}(-0.5548) = 56.3$

Example 8

Coordinates of initial and terminal points of a vector are A(3,1,–2) and B(4,–7,10) respectively. Determine magnitude of vector and its inclination with the axes.

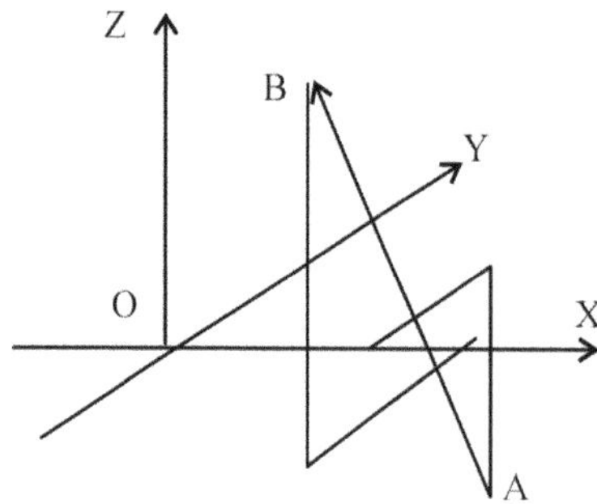

Solution:

Vector, $\boldsymbol{R_{AB}} = (X_B - X_A)\,\boldsymbol{i} + (Y_B - Y_A)\,\boldsymbol{j} + (Z_B - Z_A)\,\boldsymbol{k}$

$= (4 - 3)\,\boldsymbol{i} + (-7 - 1)\,\boldsymbol{j} + [10 - (-2)]\,\boldsymbol{k}$

$= \boldsymbol{i} - 8\,\boldsymbol{j} + 12\,\boldsymbol{k}$

Magnitude of the vector,

$$|\mathbf{R_{AB}}| = \sqrt{[(1)^2 + (-8)^2 + (12)^2]} = 14.46$$

Direction of the resultant is specified by

its angle with X-axis, $\alpha = \cos^{-1}(1/|\mathbf{R}|) = 86^0$

its angle with Y-axis, $\beta = \cos^{-1}(-8/|\mathbf{R}|) = 123.6^0$

its angle with Z-axis, $\gamma = \cos^{-1}(12/|\mathbf{R}|) = 33.91^0$

Example 9

A particle is acted upon simultaneously by the following forces - 150N inclined at 60^0 to the North of East ; 200N towards the North ; 250N towards North-West ; 300N inclined at 50^0 to the South of West and 350N towards South-East. Determine the magnitude and direction of the resultant.

Solution:

These forces are represented in vector form respectively by

$$(150 \cos 60)\,\boldsymbol{i} + (150 \cos 30)\,\boldsymbol{j}$$

$0\,\boldsymbol{i} + 200\,\boldsymbol{j}$

$-(250\cos 45)\,\boldsymbol{i} + (250\cos 45)\,\boldsymbol{j}$

$-(300\cos 50)\,\boldsymbol{i} - (300\cos 40)\,\boldsymbol{j}$

and $(350\cos 45)\,\boldsymbol{i} - (350\cos 45)\,\boldsymbol{j}$

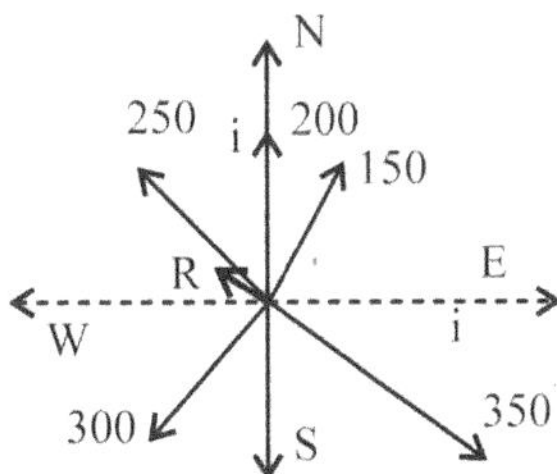

Resultant, $\boldsymbol{R} = \sum F_X\ \boldsymbol{i} + \sum F_Y\,\boldsymbol{j}$

$= (150\cos 60 + 0 - 250\cos 45 - 300\cos 50 + 350\cos 45)\,\boldsymbol{i}$

$+ (150\cos 30 + 200 + 250\cos 45 - 300\cos 40 - 350\cos 45)\,\boldsymbol{j}$

$= -47.13\,\boldsymbol{i} + 29.38\,\boldsymbol{j}$

Magnitude of the resultant, $|R| = \sqrt{[(-47.13)^2 + (29.38)^2} = 55.54$

Inclination of the resultant with respect to East direction is given by

$\theta = \cos^{-1}(\sum F_X / |R|\ ) = \cos^{-1}(-47.13 / 55.54) = 148.06^0$

Example 10

A force of 100 N passes through A(2,0,4) and B(3,3,5). Find (a) moment about origin and (b) moment about C(–2,3,4)

Solution:

(a) Vector AB $= (3-2)\,\boldsymbol{i} + (3-0)\,\boldsymbol{j} + (5-4)\,\boldsymbol{k} = \boldsymbol{i} + 3\boldsymbol{j} + \boldsymbol{k}$

Unit vector along AB, $\boldsymbol{e} = (\boldsymbol{i} + 3\boldsymbol{j} + \boldsymbol{k}) / \sqrt{(1^2 + 3^2 + 1^2)}$

$= (\boldsymbol{i} + 3\boldsymbol{j} + \boldsymbol{k}) / \sqrt{11}$

Force vector along AB, $\boldsymbol{F_{AB}} = 100\,\boldsymbol{e} = (100 / \sqrt{11}\ )(\boldsymbol{i} + 3\boldsymbol{j} + \boldsymbol{k})$

Position vector, $\boldsymbol{r_{OA}} = (2-0)\,\boldsymbol{i} + (0-0)\,\boldsymbol{j} + (4-0)\,\boldsymbol{k} = 2\,\boldsymbol{i} + 4\,\boldsymbol{k}$

Moment about O, $\boldsymbol{M_O} = \boldsymbol{r_{OA}} \times \boldsymbol{F_{AB}} = (100/\sqrt{11}) \begin{vmatrix} \boldsymbol{i} & \boldsymbol{j} & \boldsymbol{k} \\ 2 & 0 & 4 \\ 1 & 3 & 1 \end{vmatrix}$

$$= (100/\sqrt{11}\ )\,(-12i + 2j + 6k)$$

or

Position vector, $\mathbf{r_{OB}} = (3-0)\,\boldsymbol{i} + (3-0)\,\boldsymbol{j} + (5-0)\,\boldsymbol{k} = 3\,\boldsymbol{i} + 3\,\boldsymbol{j} + 5\,\boldsymbol{k}$

Moment about O, $\boldsymbol{M_O} = \boldsymbol{r_{OB}} \times \boldsymbol{F_{AB}} = (100/\sqrt{11}) \begin{vmatrix} \boldsymbol{i} & \boldsymbol{j} & \boldsymbol{k} \\ 3 & 3 & 5 \\ 1 & 3 & 1 \end{vmatrix}$

$$= (100/\sqrt{11})\,[(3-15)\boldsymbol{i} + (5\text{-}3)\boldsymbol{j} + (9\text{-}3)\boldsymbol{k}]$$

$$= (100\,/\,\sqrt{11})\,(-12\boldsymbol{i} + 2\boldsymbol{j} + 6\boldsymbol{k})$$

Magnitude of moment, $|M_O| = (100\,/\,\sqrt{11}\ )\ \sqrt{(12^2 + 2^2 + 6^2)}$

$$= 408.93 \text{ Nm}$$

(b) Position vector, $\boldsymbol{r_{CA}} = (2+2)\,\boldsymbol{i} + (0-3)\,\boldsymbol{j} + (4-4)\,\boldsymbol{k} = 4\,\boldsymbol{i} - 3\,\boldsymbol{j}$

Moment about C, $\mathbf{M_C} = \boldsymbol{r_{CA}} \times \boldsymbol{F_{AB}} = (100/\sqrt{11}) \begin{vmatrix} \boldsymbol{i} & \boldsymbol{j} & \boldsymbol{k} \\ 4 & -3 & 0 \\ 1 & 3 & 1 \end{vmatrix}$

$$= (100\,/\sqrt{11}\ )\,(-\,3\boldsymbol{i} - 4\boldsymbol{j} + 15\boldsymbol{k})$$

Magnitude of moment, $|\,M_C\,| = (100\,/\sqrt{11}\ )\ \sqrt{(3^2 + 4^2 + 15^2)}$

$$= 476.25 \text{ Nm}$$

Example 11

Two cables are used, as shown, to support a rod on which a sign board is hung. If the tensions P and F in the cables are 450N and 950N and the sign board weighs 200N, what is the resultant moment about O. Take OC = CD = 4m ; AG = 1m ; GB = 3.2m ; OG = 8m

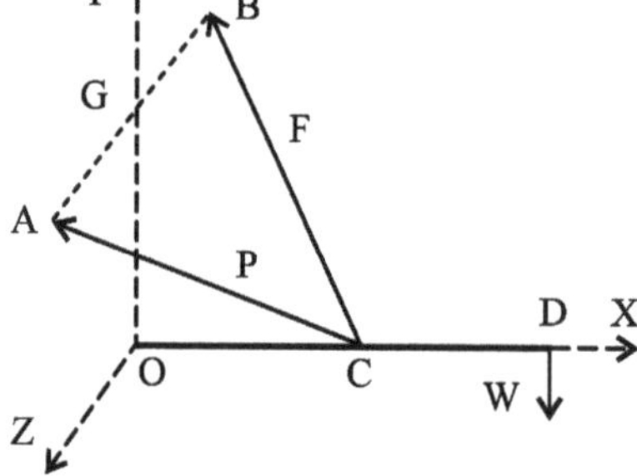

Solution:

Coordinates of the points are :

A(0,8,1), B(0,8,–3.2), C(4,0,0) and D(8,0,0))

Position vector, $\boldsymbol{r_{CA}} = (0-4)\boldsymbol{i} + (8-0)\boldsymbol{j} + (1-0)\boldsymbol{k}$

$$= -\,4\boldsymbol{i} + 8\boldsymbol{j} + \boldsymbol{k}$$

Unit vector along CA,

$$e_{CA} = (-4\boldsymbol{i} + 8\boldsymbol{j} + \boldsymbol{k}) / \sqrt{(4^2 + 8^2 + 1^2)} = (-4\boldsymbol{i} + 8\boldsymbol{j} + \boldsymbol{k}) / 9$$

Then, Force vector, $\boldsymbol{P} = 450\, \boldsymbol{e}_{CA} = 450\,(-4\boldsymbol{i} + 8\boldsymbol{j} + \boldsymbol{k}) / 9$

$$= -200\,\boldsymbol{i} + 400\,\boldsymbol{j} + 50\,\boldsymbol{k}$$

Similarly, other position vectors are $\boldsymbol{r}_{OC} = 4\,\boldsymbol{i}$ & $\boldsymbol{r}_{OD} = 8\,\boldsymbol{i}$

and other force vectors are

$$F = -400\,\boldsymbol{i} + 800\,\boldsymbol{j} - 320\,\boldsymbol{k} \quad \& \quad W = -200\,\boldsymbol{j}$$

Resultant Moment about O,

$$\mathbf{M_O} = \mathbf{r_{OC}} \times \mathbf{P} + \mathbf{r_{OC}} \times \mathbf{F} + \mathbf{r_{OD}} \times \mathbf{W}$$

$$= \begin{vmatrix} i & j & k \\ 4 & 0 & 0 \\ -200 & 400 & 50 \end{vmatrix} + \begin{vmatrix} i & j & k \\ 4 & 0 & 0 \\ -400 & 800 & -320 \end{vmatrix} + \begin{vmatrix} i & j & k \\ 8 & 0 & 0 \\ 0 & -200 & 0 \end{vmatrix}$$

$$= [-4 \times 50\,\boldsymbol{j} + 4 \times 400\,\boldsymbol{k}] + [-4 \times (-320)\,\boldsymbol{j} + 4 \times 800\,\boldsymbol{k}]$$

$$+ [8 \times (-200)\,\boldsymbol{k}] = 1080\,\boldsymbol{j} + 3200\,\boldsymbol{k}$$

$$M_O = \sqrt{(1080)^2 + (3200)^2} = 3377.3\text{Nm}$$

Example 12

A force F intersects X-axis at C(2,0). If the moment of this force about A(0,4) is 200 Nm counter clockwise and about B(7,0) is 100 Nm clockwise, determine its Y intercept

Solution :

Since all points are identified by two coordinates only, let us consider them in X–Y plane.

Position vector, $\boldsymbol{r}_{AC} = (X_C - X_A)\,\boldsymbol{i} + (Y_C - Y_A)\,\boldsymbol{j} = (2 - 0)\,\boldsymbol{i} + (0 - 4)\,\boldsymbol{j} = 2\,\boldsymbol{i} - 4\,\boldsymbol{j}$

and $\boldsymbol{r}_{BC} = (X_C - X_B)\,\boldsymbol{i} + (Y_C - Y_B)\,\boldsymbol{j} = (2 - 7)\,\boldsymbol{i} + (0 - 0)\,\boldsymbol{j} = -5\,\boldsymbol{i}$

Let Force vector F be represented by $\boldsymbol{F} = F_X\,\boldsymbol{i} + F_Y\,\boldsymbol{j}$

Then,

$$\boldsymbol{M}_B = \boldsymbol{r}_{BC} \times \boldsymbol{F} = \begin{vmatrix} i & j & k \\ -5 & 0 & 0 \\ F_X & F_Y & 0 \end{vmatrix} = (-5\,F_Y)\,\boldsymbol{k}$$

$$|M_B| = -5\,F_Y = -100\text{Nm} \quad \text{or} \quad F_Y = +20\text{N}$$

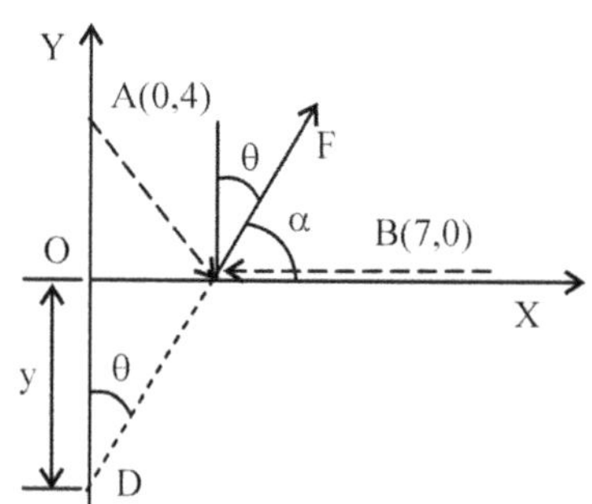

$$M_A = r_{AC} \times F = \begin{vmatrix} i & j & k \\ 2 & -4 & 0 \\ F_X & F_Y & 0 \end{vmatrix}$$

$$= [2\,F_Y - (-4)\,F_X]\,k$$

$M_A \,| = 2\,F_Y + 4\,F_X = 40 + 4\,F_X = 200\text{Nm}$

or $\quad F_X = +\,40\text{N}$

Y-intercept of the force can be calculated by different methods

Method-1: Since moment of a force about a point is same with position vector meeting any point on the line of action of the force, choosing D as the point of intersection of the force with the Y-axis at y from the origin O, position vector AD is given by

$$r_{AD} = (-\,y - 4)\,j = -\,(y + 4)\,j$$

$$M_A = r_{AD} \times F = \begin{vmatrix} i & j & k \\ 0 & -(y+4) & 0 \\ F_X & F_Y & 0 \end{vmatrix} = F_X\,(y + 4)\,k = 40\,(y + 4)\,k$$

$$|M_A| = 40\,(y + 4) = 200 \quad \text{or} \quad y = 1\text{ m}$$

Method-2: Considering the angle θ made by the force vector with the vertical, $\tan\theta = F_X / F_Y$. Also, from Δ ACD, $\tan\theta = OC / OD = 2 / y$

Therefore, $y = OD = 2 / \tan\theta = 2 \times (F_Y / F_X) = 2 \times (20/40) = 1\text{ m}$

Method-3: Considering the angle α made by the force vector with the horizontal,

Equation of the straight line, representing force vector, is given by

$$y = m\,x + C'$$

where, $\quad m = \tan\alpha = F_Y / F_X = 20/40 = 0.5$ is the slope of the line

and $\quad C'$ = intercept of the line on y-axis

At point C, substituting its coordinates in the equation of the straight line,

$0 = 0.5 \times 2 + C' \Rightarrow y$ or $C' = -\,1$ m is the intercept along –ve Y-axis at D

Example 13

A boom is supported by a ball and socket joint at C and by the cables BE and AD as shown. Forces 'F' and 'P' act respectively from B to E and A to D. Taking $F_m = 10\text{N/m}$ and $P_m = 20\text{ N/m}$, determine

(a) Component of F perpendicular to plane DAC
(b) Component of P perpendicular to plane EAC and
(c) Length of common perpendicular between BE and AD

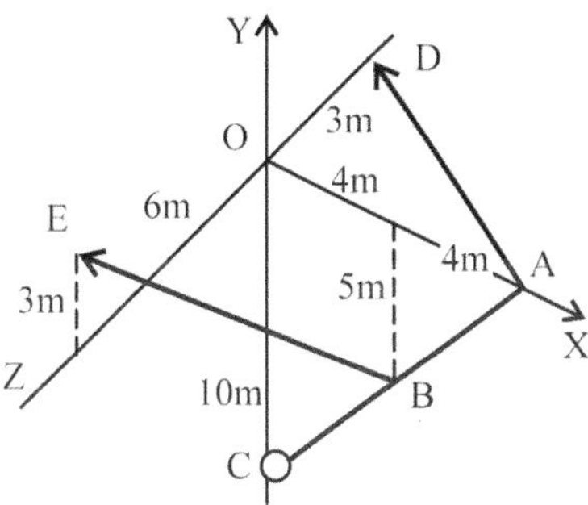

Solution:

From the coordinates A(8,0,0), B(4,–5,0), C(0,–10,0), D(0,0,–3) and E(0,3,6),

$$\mathbf{AC} = (X_C\,\boldsymbol{i} + Y_C\,\boldsymbol{j} + Z_C\,\boldsymbol{k}) - (X_A\,\boldsymbol{i} + Y_A\,\boldsymbol{j} + Z_A\,\boldsymbol{k})$$

$$= (0\,\boldsymbol{i} - 10\,\boldsymbol{j} + 0\,\boldsymbol{k}) - (8\,\boldsymbol{i} + 0\,\boldsymbol{j} + 0\,\boldsymbol{k}) = -8\,\boldsymbol{i} - 10\,\boldsymbol{j}$$

$$e_{AC} = \mathbf{AC} / |\mathbf{AC}| = (-8\,\boldsymbol{i} - 10\,\boldsymbol{j}) / \sqrt{(8^2 + 10^2)} = (-8i - 10\boldsymbol{j}) / 12.806$$

$$e_{CA} = (8\,\boldsymbol{i} + 10\,\boldsymbol{j}) / 12.806$$ full square root in the above line

$$\mathbf{BE} = (0\,\boldsymbol{i} + 3\,\boldsymbol{j} + 6\,\boldsymbol{k}) - (4\,\boldsymbol{i} - 5\,\boldsymbol{j} + 0\,\boldsymbol{k}) = -4\,\boldsymbol{i} + 8\,\boldsymbol{j} + 6\,\boldsymbol{k}$$

$$e_{BE} = \mathbf{BE} / |\mathbf{BE}| = (-4\,\boldsymbol{i} + 8\,\boldsymbol{j} + 6\,\boldsymbol{k}) / \sqrt{(4^2 + 8^2 + 6^2)}$$

$$= (-4\,\boldsymbol{i} + 8\,\boldsymbol{j} + 6\,\boldsymbol{k}) / 10.77$$

$$\mathbf{CE} = (0\,\boldsymbol{i} + 3\,\boldsymbol{j} + 6\,\boldsymbol{k}) - (0\,\boldsymbol{i} - 10\,\boldsymbol{j} + 0\,\boldsymbol{k}) = +13\,\boldsymbol{j} + 6\,\boldsymbol{k}$$

$$e_{CE} = \mathbf{CE} / |\mathbf{CE}| = (13\,\boldsymbol{j} + 6\,\boldsymbol{k}) / \sqrt{(13^2 + 6^2) = (13\boldsymbol{j} + 6\boldsymbol{k})} / 14.32$$

$$\mathbf{AD} = (0\,\boldsymbol{i} + 0\,\boldsymbol{j} - 3\,\boldsymbol{k}) - (8\,\boldsymbol{i} + 0\,\boldsymbol{j} + 0\,\boldsymbol{k}) = -8\,\boldsymbol{i} - 3\,\boldsymbol{k}$$

$$e_{AD} = \mathbf{AD} / |\mathbf{AD}| = (-8\,\boldsymbol{i} - 3\,\boldsymbol{k}) / \sqrt{(8^2 + 3^2)} = (-8\,\boldsymbol{i} - 3\,\boldsymbol{k}) / 8.544$$

In the absence of any applied load, forces in the two members BE and AD are specified in terms of the member stiffness (F_m and P_m) and changes in lengths of members (f and p) along the two members respectively. Thus,

$$\boldsymbol{F} = |\mathbf{F}|\, e_{BE} = 10\text{ f } (-4\,\boldsymbol{i} + 8\,\boldsymbol{j} + 6\,\boldsymbol{k}) / 10.77$$

and $$\boldsymbol{P} = |\mathbf{P}|\, e_{AD} = 20\text{ p } (-8\,\boldsymbol{i} - 3\,\boldsymbol{k}) / 8.544$$

(a) Perpendicular to plane DAC is indicated by the unit vector given by $e_{AD} \times e_{AC}$

Component of F perpendicular to plane DAC is obtained from

$= \boldsymbol{F}.\ (\boldsymbol{e}_{AD} \times \boldsymbol{e}_{AC})$

$= (10f/10.77)\,(-4\,\boldsymbol{i} + 8\,\boldsymbol{j} + 6\,\boldsymbol{k})\,.\,[\ \{(-8\,\boldsymbol{i} - 3\,\boldsymbol{k}) / 8.544\} \times \{(-8\,\boldsymbol{i} - 10\,\boldsymbol{j}) / 12.806\}\ ]$

$$= \frac{10f}{10.77 \times 8.544 \times 12.806} \begin{vmatrix} -4 & 8 & 6 \\ -8 & 0 & -3 \\ -8 & -10 & 0 \end{vmatrix} = 6.721\ f$$

(b) Perpendicular to plane EAC is indicated by the unit vector given by $e_{CE} \times e_{CA}$

Component of P perpendicular to plane EAC is obtained from

$= \boldsymbol{F}.\ (\boldsymbol{e}_{CE} \times \boldsymbol{e}_{CA})$

$= (20p/8.544)\,(-8\,\boldsymbol{i} - 3\,\boldsymbol{k})\,.\,[\ \{(13\,\boldsymbol{j} + 6\,\boldsymbol{k}) / 14.32\} \times \{(8\,\boldsymbol{i} + 10\,\boldsymbol{j}) / 12.806\}\ ]$

$$= \frac{20\ p}{8.544 \times 14.32 \times 12.806} \begin{vmatrix} -8 & 0 & -3 \\ 0 & 13 & 6 \\ 8 & 10 & 0 \end{vmatrix} = 10.11\ p$$

(c) Length of common perpendicular between BE and AD is given by the projection (or dot product) of BA or ED on the unit vector along the common normal to these two lines.

Line $\mathbf{BE} = (0 - 4)\,\boldsymbol{i} + [3 - (-5)]\,\boldsymbol{j} + (6 - 0)\,\boldsymbol{k} = -4\,\boldsymbol{i} + 8\,\boldsymbol{j} + 6\,\boldsymbol{k}$

Let $\mathbf{R} = a\,\boldsymbol{i} + b\,\boldsymbol{j} + c\,\boldsymbol{k}$ be the line perpendicular to **BE**

Then $\mathbf{R}\,.\,\mathbf{BE} = 0 \Rightarrow -4a + 8b + 6c = 0$ .....(1)

Line $\mathbf{AD} = (0 - 8)\,\boldsymbol{i} + (0 - 0)\,\boldsymbol{j} + (-3 - 0)\,\boldsymbol{k} = -8\,\boldsymbol{i} - 3\,\boldsymbol{k}$

If $\mathbf{R} = a\,\boldsymbol{i} + b\,\boldsymbol{j} + c\,\boldsymbol{k}$ is also perpendicular to **AD**

Then $\mathbf{R}\,.\,\mathbf{AD} = 0 \Rightarrow -8a - 3c = 0$ .....(2)

Equations (1) and (2) are simultaneously satisfied only when,

$$a / 3 = b / 7.5 = c / (-8)$$

If $L = \sqrt{(3^2 + 7.5^2 + 8^2)} = 11.3688$, then unit vector along the common normal

$$\mathbf{e_R} = (3/L)\,\boldsymbol{i} + (7.5/L)\,\boldsymbol{j} + (-8/L)\,\boldsymbol{k}$$

Length of common perpendicular between BE and AD

= Projection of **BA** on $\mathbf{e_R}$ or Projection of **ED** on $\mathbf{e_R}$

= $\mathbf{BA} \cdot \mathbf{e_R}$ or $\mathbf{ED} \cdot \mathbf{e_R}$

$= [(8-4)\,\boldsymbol{i} + \{0-(-5)\}\,\boldsymbol{j} + (0-0)\,\boldsymbol{k}] \cdot [(3/L)\,\boldsymbol{i} + (7.5/L)\,\boldsymbol{j} + (-8/L)\,\boldsymbol{k}]$

or $[(0-0)\,\boldsymbol{i} + (0-3)\,\boldsymbol{j} + (-3-6)\,\boldsymbol{k}] \cdot [(3/L)\,\boldsymbol{i} + (7.5/L)\,\boldsymbol{j} + (-8/L)\,\boldsymbol{k}]$

$= (4\,\boldsymbol{i} + 5\,\boldsymbol{j}) \cdot (3\,\boldsymbol{i} + 7.5\,\boldsymbol{j} - 8\,\boldsymbol{k})/L$

or $(-3\,\boldsymbol{j} - 9\,\boldsymbol{k}) \cdot (3\,\boldsymbol{i} + 7.5\,\boldsymbol{j} - 8\,\boldsymbol{k})/L$

$= (12 + 37.5)/L$ or $(-22.5 + 72)/L$

$= 49.5 / 11.3688 = 4.354$ m

Example 14

Determine component of vector A $(3\,\boldsymbol{i} + 2\,\boldsymbol{j} - 5\boldsymbol{k})$ along vector B $(4\,\boldsymbol{i} - 3\,\boldsymbol{j})$

Solution:

Length of vector B, $|B| = \sqrt{(4)^2 + (-3)^2} = 5$

Unit vector along B, $\boldsymbol{e} = \boldsymbol{B}/|B| = (4/5)\,\boldsymbol{i} + (-3/5)\,\boldsymbol{j} = 0.8\,\boldsymbol{i} - 0.6\,\boldsymbol{j}$

Magnitude of component of vector A along vector B

$= \boldsymbol{A} \cdot \boldsymbol{e} = (3\,\boldsymbol{i} + 2\,\boldsymbol{j} - 5\boldsymbol{k}) \cdot (0.8\,\boldsymbol{i} - 0.6\,\boldsymbol{j}) = 3 \times 0.8 - 2 \times 0.6 = 1.2$

Component of vector A along B,

$= (\mathbf{A.e})\,\mathbf{e} = 1.2\,\mathbf{e} = 1.2\,(0.8\,\boldsymbol{i} - 0.6\,\boldsymbol{j}) = 0.96\,\boldsymbol{i} - 0.72\,\boldsymbol{j}$

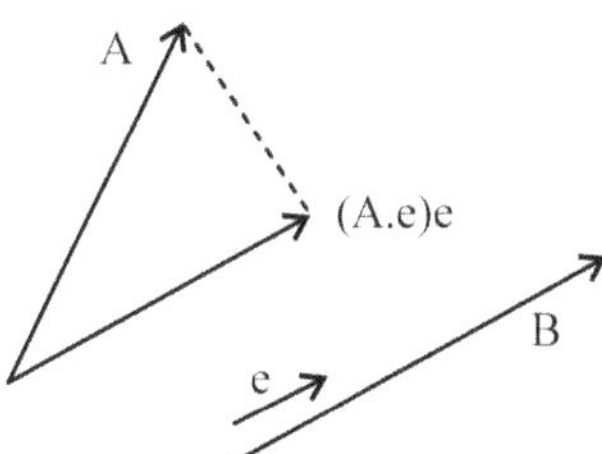

SUMMARY

1. Any force vector can be represented as $\mathbf{F} = F_x\,\boldsymbol{i} + F_y\,\boldsymbol{j} + F_z\,\boldsymbol{k}$ where, F_X, F_Y and F_Z are the components of the force along the three coordinate axes and $\boldsymbol{i}$, $\boldsymbol{j}$ and $\boldsymbol{k}$ are the unit vectors along these coordinate directions

Its magnitude is given by $|F| = \sqrt{[F_x^2 + F_y^2 + F_z^2]}$

2. The magnitude of component of a vector in any direction is the **dot product** of the vector with a unit vector in the desired direction.

   $|F_e| = \mathbf{F} \cdot \mathbf{e}$ or $F_X = \mathbf{F} \cdot \mathbf{i}$, ...

3. **Moment of a force** about a moment center is the **cross product** of the position vector from the moment center to any point on the line of action of the force and the force vector. It is oriented towards the positive normal to the plane of position vector and force vector, following right hand rule.

$$\boldsymbol{M} = \boldsymbol{r} \times \boldsymbol{F} = \begin{vmatrix} \boldsymbol{i} & \boldsymbol{j} & \boldsymbol{k} \\ x & y & z \\ F_X & F_Y & F_Z \end{vmatrix}$$ where, x, y and z are

components of vector **r** along the three coordinate directions

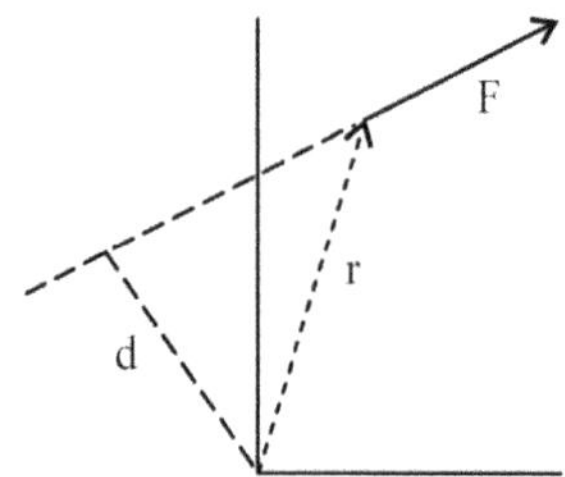

4. Space forces are generalised by treating them as a combination coplanar forces in two or more appropriate planes, by the components of forces approach.

5. In the vector approach, each of the concurrent forces is written as a vector in 3-D space. At a point in equilibrium, vector sum of the concurrent forces is equal to zero, which is satisfied only when the components along each of the three coordinate directions is zero.

PROBLEMS FOR PRACTICE

1. Determine the resultant of a system of concurrent forces having the following magnitudes and passing through origin and the other indicated points. P = 3000N through (12, 6, – 4); Q = 5000N through (– 3, – 4, 12) and F = 2800N through (6, – 3, – 6)

   (***Ans :*** $R = 3284\,\boldsymbol{i} - 1186\,\boldsymbol{j} + 1892\,\boldsymbol{k}$; $|\mathbf{R}| = 3971.3$N)

2. The x, y, z components of a force F are 36kN, 24kN and 24kN respectively. Find the component of the force along the line joining A(1,2, –3) and B(–1, –2,2)

(*Ans :* $(32\,\boldsymbol{i} + 64\,\boldsymbol{j} - 80\boldsymbol{k}) / 15$ and its magnitude is $-16/\sqrt{5}$)

3. Find the shortest distance from the origin to the line passing through the points A(–5,3,7) and B(9,11,0)

4. A force $\boldsymbol{F} = 4\,\boldsymbol{i} + 3\,\boldsymbol{j} + 2\boldsymbol{k}$ is applied at a point whose position vector from 'O' is given by $\boldsymbol{r} = \boldsymbol{i} + 2\,\boldsymbol{j} + 3\boldsymbol{k}$. What is the resulting moment about 'O'.

(*Ans :* $\mathbf{M_O} = -5\,\boldsymbol{i} + 10\,\boldsymbol{j} - 5\boldsymbol{k}$; $|\mathbf{M_O}| = 12.25$)

5. Find the resultant of concurrent force system consisting of F=541N along AB, P = 671N along AC and T = 735N along AD. Coordinates of the points are A(0,6,0), B(–3,1.5,0), C(0,0,3) and D(3,0, –3)

(*Ans :* $0\,\boldsymbol{i} - 1650\,\boldsymbol{j} + 0\,\boldsymbol{k}$)

6. In the tripod shown, force F acts along AB, P along AC and T along AD. Coordinates of points are A(0,4,0), B(–1,0,2), C(–1,0, –1) and D(2,0,1)

   (a) Write each force in terms of force multiplier F_m, P_m and T_m

   (b) What is the component of F along the direction of P

   (c) What is the angle between F and P

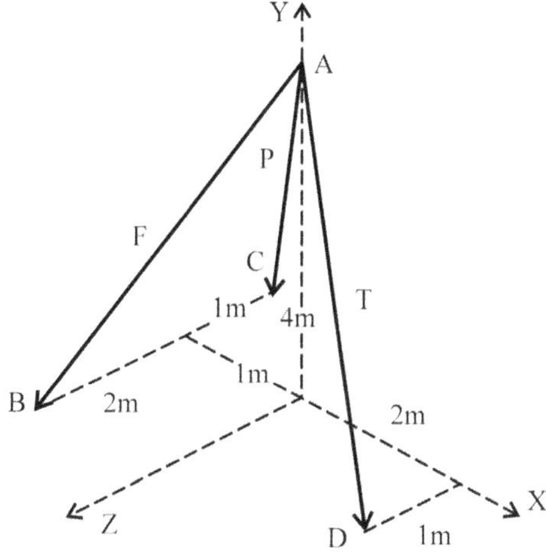

7. A boom AC is supported by ball and socket joint at C and cables BD and AE. The force multiplier of P acting from A to E is 200N/m and that of F acting from B to D is 300N/m. Determine
   (a) Position vector from C to A and
   (b) components of forces P and F along AC

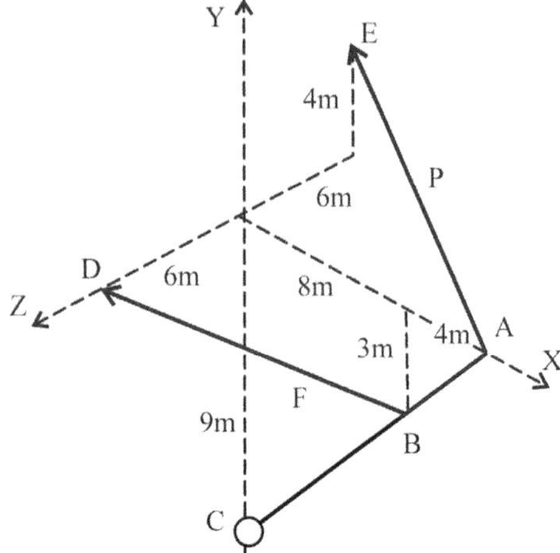

(***Ans:*** CA = 12***i*** + 9***j***; F on AC = 1380N; P on AC = 1440N)

CHAPTER 5

FRICTION

Surface of every object will have certain roughness due to the peaks and troughs (Fig. 5.1). The roughness depends on the type of material (wood, brick, glass,..) and manufacturing process by which a component is generated (casting, forging, machining,..). For example, a saw-cut wooden block may have surface as shown in the figure. After a planing operation, it becomes smooth. After rubbing with an emery paper, it becomes smoother. It still has uneven surface. A polishing operation fills the voids and makes it smoothest. Similarly, a casting operation produces rough surface, which can be smoothened by turning or planing, polishing with an emery paper and finally lapping (used for contact surfaces of valves to ensure leak tightness) in that order.

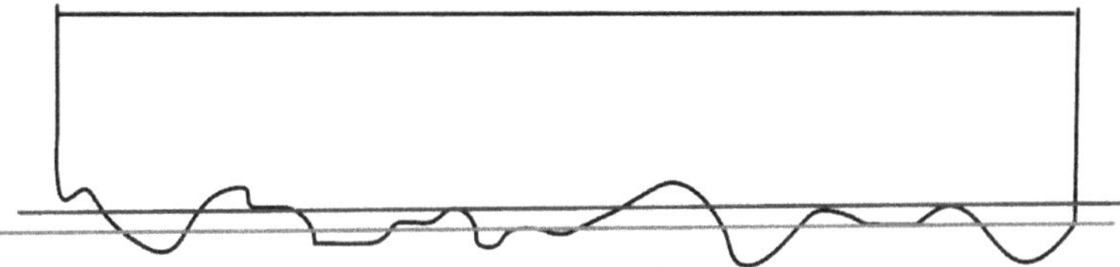

FIGURE 5.1 Surface of every object

5.1 FRICTIONAL RESISTANCE

When one body tries to move over the other, friction between the two surfaces is generated due to inter-meshing of the peaks and troughs of the two contact surfaces. Hence, the ***frictional resistance is a function of surface roughness and comes into play whenever one body tends to move over another body***. It can be minimized by using a solid or liquid lubricant between the two bodies in contact and/or by smoothening of the surfaces.

This can further be subdivided into ***sliding friction*** (same surfaces in continuous contact) and ***rolling friction*** (point or line contact between the rolling body and the supporting surface, with the contact point/line of the rolling body varying with time). In this chapter, we limit our discussion to sliding friction between non-lubricated surfaces (also called, **dry friction**).

5.2 LAWS OF FRICTION

Relative motion between two bodies is governed by the ***four laws of friction***.

- Frictional force 'F' is the resistance to movement between two bodies in contact ; Comes into play only when a body is made to move over another ; Acts parallel to the surface of contact and opposite to the direction of motion ----- ***First law of friction***
- Frictional force is just sufficient to prevent the body from moving. i.e., as the load increases, frictional force increases, upto a limit ----- ***Second law of friction***
- Limiting frictional resistance F_{max} bears a constant ratio with the normal reaction, 'N' (***Coulomb's law of dry friction***) and is independent of area of contact.

  $F_{max} \; \alpha \; N$ or $F_{max} = \mu_S N$ ----- ***Third law of friction***

  where $\mu = \tan \alpha_{max}$ is called the ***coefficient of friction***

  and α_{max} is called the ***angle of friction***

- When motion takes place, kinetic frictional resistance 'F_k' will be less than the limiting value. $F_k < F_{max}$ ---- ***Fourth law of friction***

  Usually, $\mu_k \approx 0.75 \; \mu_S$

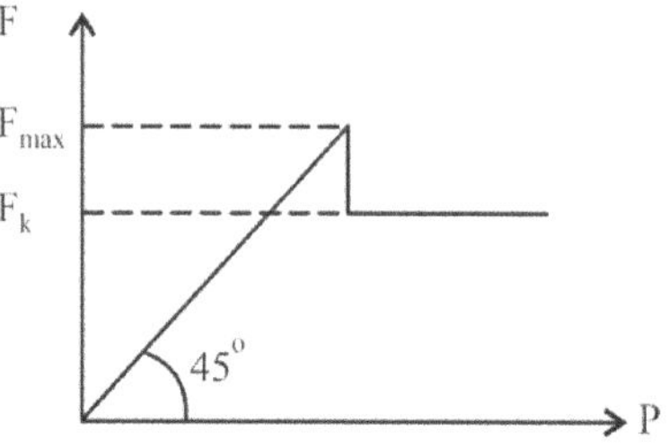

FIGURE 5.2 Laws of friction

These four laws can be summarized by the graph between frictional force 'F' and applied load 'P' (Fig. 5.2).

For static equilibrium,

$W = N$ and $P = F$; $\tan \alpha = F / N$

and $\mu_S = \tan \alpha_{max} = F_{max} / N$

or $F_{max} = \mu_S N$ and $F_k = \mu_k N$

α is the angle between the normal reaction N and the resultant reaction R (Fig. 5.3).

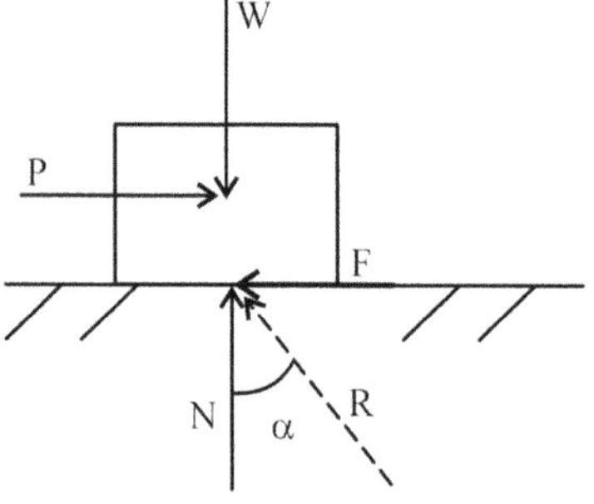

FIGURE 5.3 Angle of friction

5.3 CONE OF FRICTION

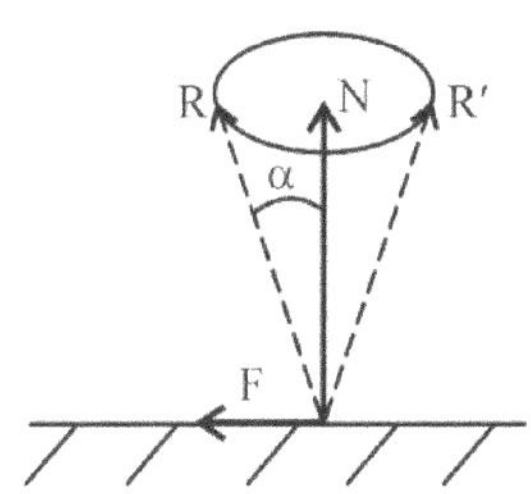

Line of action of resultant force R for a body moving on a flat surface lies on the surface of an inverted cone, for various directions of motion, with a semi-vertex angle α. For example, if a body is moving over a flat surface to the right, friction force F acts to the left and their resultant is along R while for motion in the opposite direction, the resultant is along R′.

5.4 ANGLE OF REPOSE 'θ'

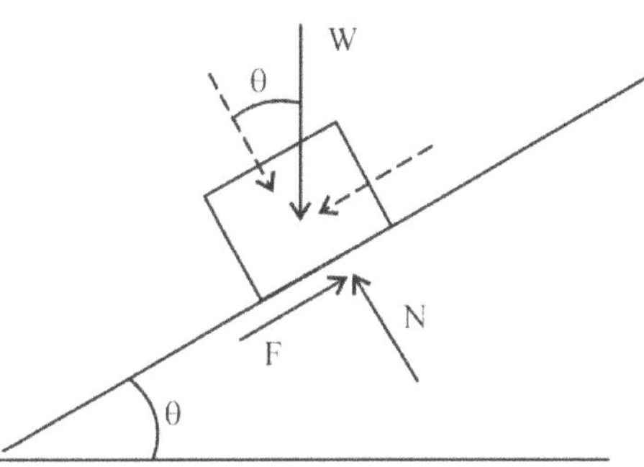

It is the maximum angle of inclined plane beyond which a body placed on the inclined plane slides down due to its own weight. If the vertical weight is resolved along and perpendicular to the inclined plane, the normal component will induce normal reaction 'N' at the the contact surface. The component of W along the inclined plane will tend to move the body down the plane, which induces frictional resistance 'F'. Weight 'W' is also inclined at the same angle 'θ' with the normal to the inclined plane as the inclined plane with the horizontal. If the body is in equilibrium,

$$N = W \cos\theta \ ; \ F_{max} = W \sin\theta$$

Therefore, $F_{max} / N = \tan\theta = \tan\alpha_{max} = \mu$ or $\theta = \alpha_{max}$

5.5 INCLINED PLANE

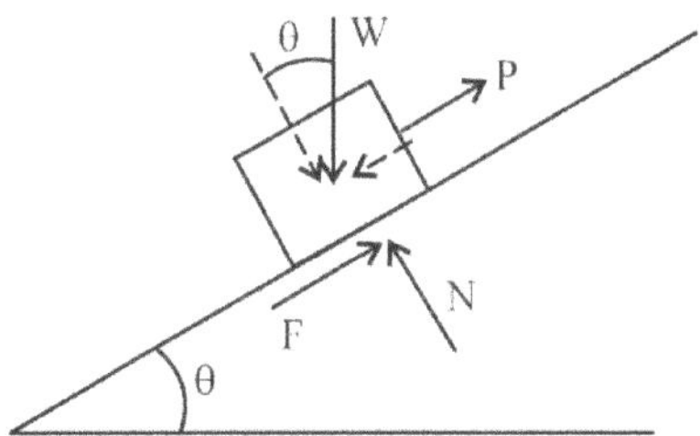

If a force P, acting parallel to the inclined plane, is just sufficient to keep a body of weight W from <u>*sliding down*</u>, then the frictional force will be in the direction of force P. Resolving all forces along and perpendicular to the inclined plane, equations of equilibrium are

$$N = W \cos \theta$$

and $\mu N + P = W \sin \theta$ with $F_{max} = \mu N$

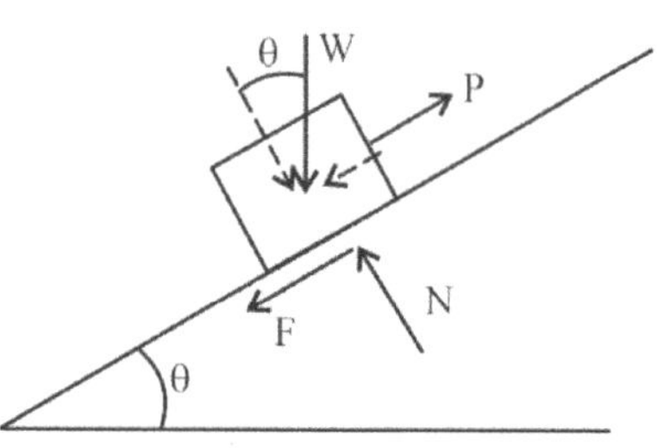

Therefore,

$$P = W \sin \theta - \mu N = W \times (\sin \theta - \mu \cos \theta)$$

Similarly, if the force P is just tending to *move the body upwards*, then the frictional force will be in the opposite direction to P. Resolving all forces along and perpendicular to the inclined plane, equations of equilibrium are

$$N = W \cos \theta \quad \text{and} \quad \mu N + W \sin \theta = P$$

Therefore, $P = W \times (\sin \theta + \mu \cos \theta)$

If the force P is just tending to *move the body upwards*, but acting along the horizontal direction, then the frictional force will be acting downwards along the inclined plane. Resolving all forces along and perpendicular to the inclined plane, equations of equilibrium are

$$N = W \cos \theta + P \sin \theta$$

and $F + W \sin \theta - P \cos \theta = 0$ with $F = \mu N$

Therefore, $P = W \times (\sin \theta + \mu \cos \theta) / (\cos \theta - \mu \sin \theta)$

by substituting $\mu = \tan \alpha$ and simplifying, $P = W \tan (\theta + \alpha)$

Example 1

A ladder AB of length 13 m and weighing 250 N is placed against a vertical wall with its lower end resting on the floor at 5m away from the wall. The coefficient of friction between the ladder and the floor is 0.3 Calculate actual friction force and limiting friction force at the point of contact between the ladder and the floor.

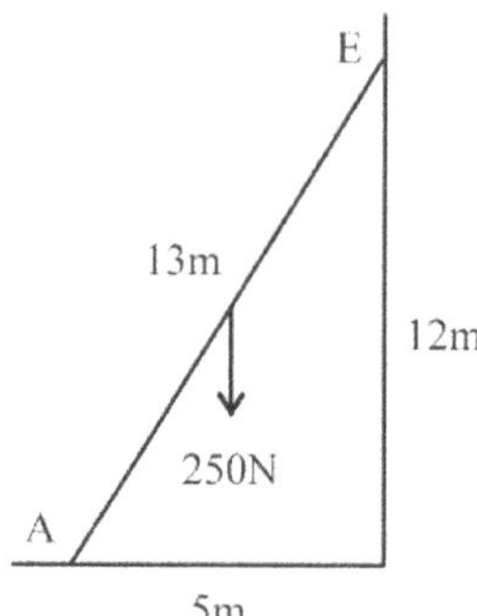

Solution:

Since coefficient of friction between the wall and the ladder is not specified, smooth wall is assumed. Hence, only normal reaction 'H' is considered at B. Normal reaction 'N' and frictional resistance 'F' are considered at A. When the ladder slides, end A of the ladder moves to the left and, hence, frictional resistance at A is taken to the right. Taking moments about A,

$\sum M_A = H \times BC - 250 \times AE = H \times 12 - 250 \times (5/2) = 0$

From the similar triangles ADE and ABC,

$AE/AC = AD/AB = ½$

or $AE = (1/2)\ AC = 5/2$

$BC = \sqrt{(AB^2 - AC^2)} = \sqrt{(13^2 - 5^2)} = 12m$

Then, $H = 250 \times AE / BC$

$= 250 \times (5/2) / 12 = 52.08\ N$

Equations of equilibrium are

$\sum F_X = F - H = 0 \quad \Rightarrow F = H = 52.08N$

and $\sum F_Y = N - 250 = 0 \Rightarrow N = 250N$

Limiting friction, $F_{max} = \mu\ N = 0.3 \times 250 = 75N$

Since $F < F_{max}$, the ladder does not slide.

Example 2

A ladder AB of length 5 m and weighing 250 N is placed against a vertical wall at an angle of 60^0 with the floor. The coefficient of friction between the ladder and the floor as well as between ladder and the wall is 0.3 How far can a person weighing 500 N go up the ladder before the ladder starts sliding down.

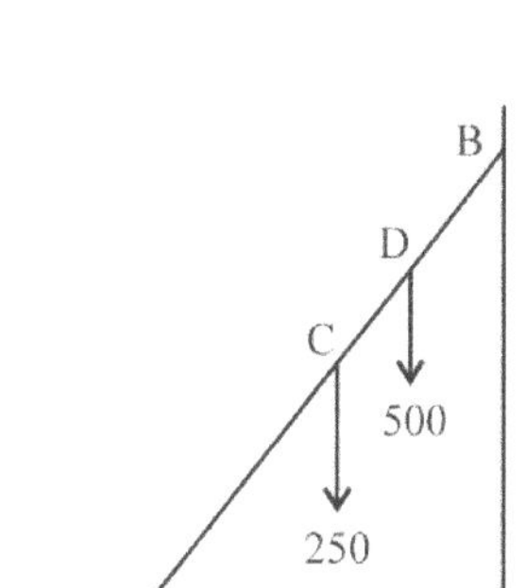

Solution:

Since coefficient of friction is specified at both the ends of the ladder, frictional resistance is considered along with normal reactions at both the ends A and B of the ladder. When the ladder slides, B will come down and A will move to the left. So, the frictional resistance is considered upwards at B and to the right at A. The equations of equilibrium are

$\sum F_X = \mu N - H = 0\ ;\ \sum F_Y = N + \mu H - (250 + 500) = 0$

or $H/\mu + \mu H = H(1+ \mu^2) / \mu = 750$

$\Rightarrow$ $H = 750\ \mu / (1 + \mu^2) = 750 \times 0.3 / (1 + 0.3^2) = 206.42N$

Let D be the max position a person go up the ladder.

Taking moments about A,

$\sum M_A = H \times BE + \mu H \times NE - 250 \times NF - 500 \times NG = 0$

$$206.42 \times AB \sin 60 + 0.3 \times 206.42 \times AB \cos 60$$
$$- 250 \times (5/2) \cos 60 - 500 \times AD \cos 60 = 0$$

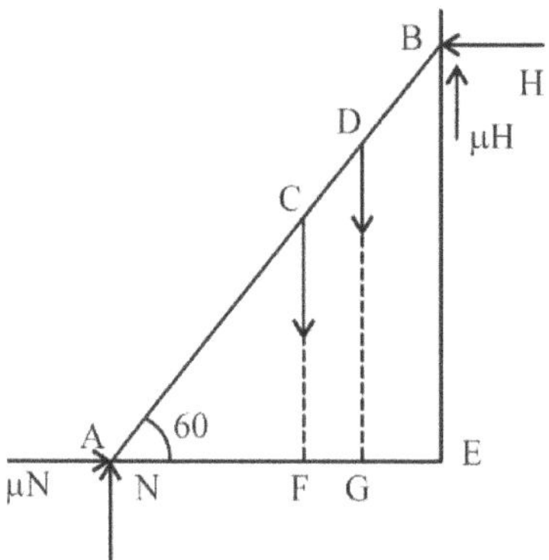

$$\Rightarrow \quad AD = [206.42 \times 5 \tan 60 + 0.3 \times 206.42 \times 5 - 250 \times (5/2)\,] / 500$$
$$= [\,1787.65 + 309.63 - 625\,] / 500 = 1472.28 / 500 = 2.9446\text{m}$$

Example 3

A ladder AB of weight 'W' rests on horizontal floor and leans on vertical wall. Calculate the minimum angle θ which keeps the ladder stable, if the coefficient of friction between the ladder and the floor as well as between ladder and the wall is μ. What is the minimum value of μ required, if $\theta < 45^0$.

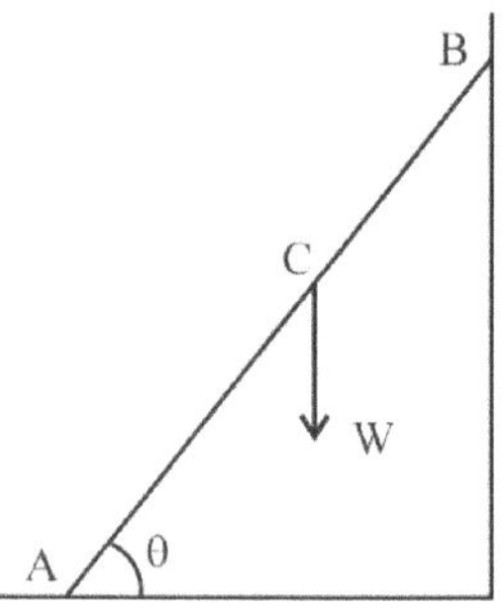

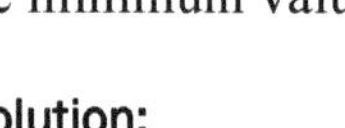
Solution:

Let AB = L. Since coefficient of friction is specified at both the ends of the ladder, frictional resistance is considered along with normal reactions at both the ends A and B of the ladder. When the ladder slides, B will come down and A will move to the left. So, the frictional resistance is considered upwards at B and to the right at A. The equations of equilibrium are

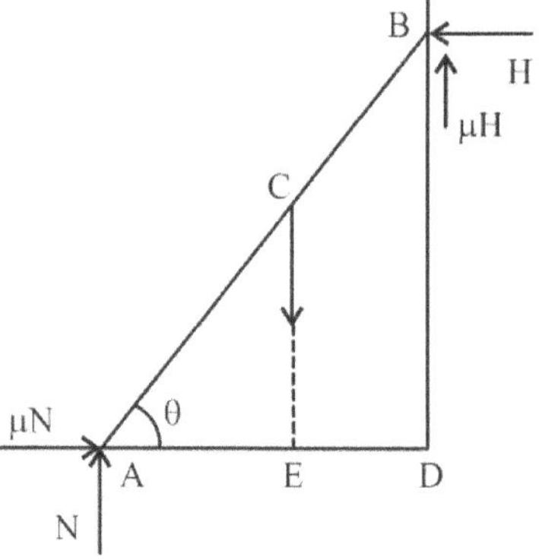

$$\sum F_X = \mu N - H = 0$$

and $$\sum F_Y = N + \mu H - W = 0$$

$$H/\mu + \mu H = H \times (1 + \mu^2) / \mu = W$$

or $$H = \mu W / (1 + \mu^2)$$

Taking moments about A,

$$\sum M_A = H \times BD + \mu H \times AD - W \times AE = 0$$

$$H \times AB \sin\theta + \mu H \times AB \cos\theta - W \times AC \cos\theta = 0$$

and $\quad AC = AB/2$

Dividing by AB,

$$\mu \times W \sin\theta / (1 + \mu^2) + \mu^2 \times W \cos\theta / (1 + \mu^2) - (W/2)\cos\theta = 0$$

Dividing by $W/(1 + \mu^2)$,

$$\mu \sin\theta + \mu^2 \cos\theta - (1 + \mu^2) \times (1/2)\cos\theta = 0$$

$$\Rightarrow \quad \mu \sin\theta - (1 - \mu^2) \times (1/2)\cos\theta = 0$$

$$\Rightarrow \quad \tan\theta = (1 - \mu^2) / 2\mu$$

If $\theta = 45^0$, then $\tan\theta = 1$ or $\mu^2 + 2\mu - 1 = 0 \Rightarrow \mu = 0.414$

$\Rightarrow$ If $\theta < 45^0$, $\mu > 0.414$

Example 4

Two identical blocks A and B, weighing W each, are supported by a rod inclined at 45^0 to the horizontal. If the blocks are in limiting equilibrium, find the coefficient of friction, assuming it to be the same at the floor as well as at the wall.

Solution:

Let the length of the rod be 'L'. The weight of block B will induce a normal reaction 'S' at B. Weight of block A through the rod tends to move the block B to the left, which induces frictional resistance μS to the right at B. The system of non-concurrent forces will satisfy moment equilibrium at B, only if there is normal reaction R at A. Weight of block A tending to move it down induces frictional resistance μR at A. These forces are in equilibrium only if

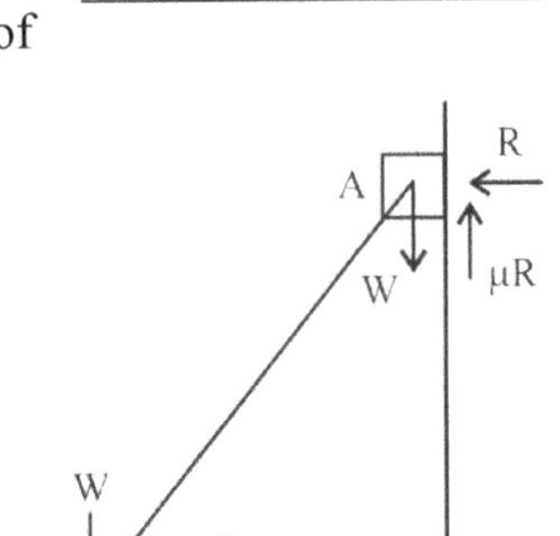

$$\Sigma F_X = \mu S - R = 0$$

and

$$\Sigma F_Y = S + \mu R - 2W = 0$$

Therefore, $S = 2W - \mu R = 2W - \mu \times \mu S$

$$\Rightarrow \quad S \times (\mu^2 + 1) = 2W \quad \text{or} \quad s = 2W / (\mu^2 + 1)$$

and $\quad R = \mu S = 2\mu W / (\mu^2 + 1)$ .....(1)

Taking moments about B, $\sum M_B = R \times AC + \mu R \times BC - W \times BC = 0$

$\Rightarrow \quad R + \mu R - W = 0$

since $AC = L \sin 45^0$, $BC = L \cos 45^0$ and $AC = BC$

$\Rightarrow \quad R \times (\mu + 1) = W \quad$ or $\quad R = W / (\mu + 1)$ .....(2)

Equating R from 1 & 2, $2 \mu W/(\mu^2 + 1) = W/(\mu + 1)$

$\Rightarrow \quad 2\mu \times (\mu + 1) = (\mu^2 + 1) \quad$ or $\quad \mu^2 + 2\mu - 1 = 0$

$\Rightarrow \quad \mu = \left[-2 \pm \sqrt{\{2^2 - 4(-1)\}} / 2\right] = -1 \pm \sqrt{2}$

Since μ can not be –ve, $\mu = -1 + \sqrt{2} = 0.414$

Alternative Method : Replacing R and μR at A by their resultant R′, replacing S and μS at B by their resultant S′, and considering force T in the rod AB, forces at A and B form two systems of concurrent forces and Lami's theorem can be applied to find the unknowns. If $\mu = \tan \phi$, applying Lami's theorem at B,

$$T / \sin(180 - \phi) = W / \sin(135 + \phi)$$

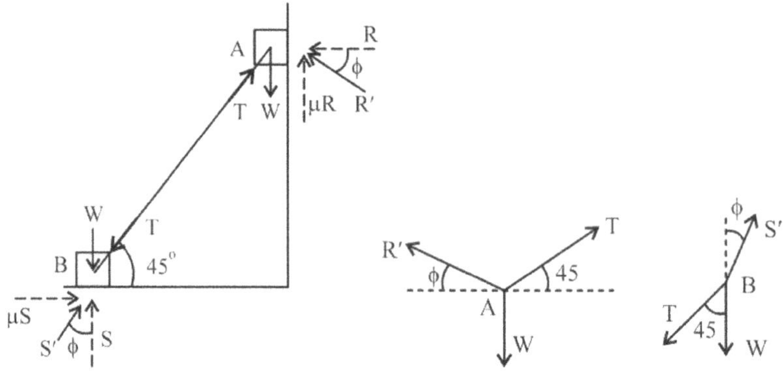

Similarly, applying Lami's theorem at A,

$$T / \sin(90 + \phi) = W / \sin(135 - \phi)$$

Since tension in the rod at A and B is same,

$$T = W \sin(180 - \phi)/\sin(135 + \phi) = W \sin(90 + \phi)/\sin(135 - \phi)$$

$\Rightarrow \quad \sin(180 - \phi) \times \sin(135 - \phi) = \sin(90 + \phi) \times \sin(135 + \phi)$

$\Rightarrow \quad \sin \phi \times \sin [180 - (45 + \phi)] = \cos \phi \times \sin [180 - (45 - \phi)]$

$\Rightarrow \quad \sin \phi \times \sin (45 + \phi) = \cos \phi \times \sin (45 - \phi)$

$\Rightarrow \quad \sin \phi (\sin 45 \cos \phi + \cos 45 \sin \phi)$

$\quad = \cos \phi (\sin 45 \cos \phi - \cos 45 \sin \phi)$

Since sin 45 = cos 45, dividing by $\cos^2\phi$, we get $\tan\phi\,(1 + \tan\phi) = 1 - \tan\phi$

$\Rightarrow \quad \tan^2\phi + 2\tan\phi - 1 = 0 \quad \text{or} \quad \mu^2 + 2\mu - 1 = 0$

$\Rightarrow \quad \mu = [-2 \pm \sqrt{2^2 - 4 \times 1 \times (-1)\}}] / 2 = -1 \pm \sqrt{2}$

Since μ can not be –ve, $\mu = -1 + \sqrt{2} = 0.414$

Example 5

A body of weight 500 N is resting on a plane inclined at 30^0 to the horizontal. Find the maximum and minimum forces to be applied on the body parallel to the inclined plane at limiting equilibrium of the body. Assume μ = 0.2

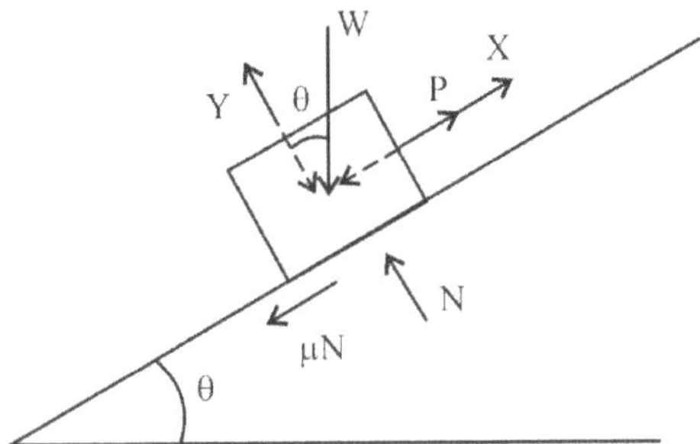

Solution:

Maximum force will tend to move the body upwards whereas minimum force tends to prevent the body from sliding down. Both correspond to limiting equilibrium. Taking X and Y axes along and perpendicular to the inclined plane, weight of the body W can be resolved into W sin θ and W cos θ along X and Y axes respectively.

(i) **Body moving up** : Then, friction force will be acting downwards to oppose upward motion of the body. Then, equations of equilibrium are

$$\Sigma F_X = P_{Max} - \mu N - W\sin\theta = 0 \quad \text{and} \quad \Sigma F_Y = N - W\cos\theta = 0$$

$$\Rightarrow \quad P_{Max} = \mu N + W\sin\theta = \mu\, W\cos\theta + W\sin\theta = 336.6\ \text{N}$$

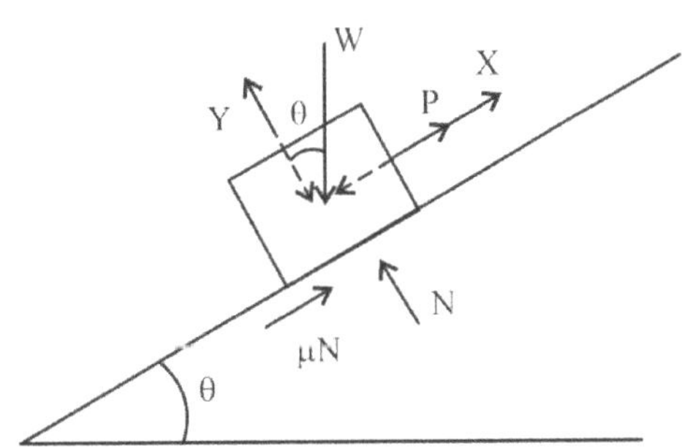

(ii) Body sliding down : Then, friction force will be acting upwards to oppose downward sliding of the body. Then, equations of equilibrium are

$$\Sigma F_X = P_{Min} + \mu N - W \sin\theta = 0 \text{ and } \Sigma F_Y = N - W \cos\theta = 0$$

$$\Rightarrow \quad P_{Min} = -\mu N + W \sin\theta = -\mu W \cos\theta + W \sin\theta = 163.4 \text{ N}$$

Example 6

Find the direction 'α' along which force 'P = W sin(θ + φ)' can move a body of weight 'W' upwards over a plane inclined at 'θ' to the horizontal, where μ = tan φ

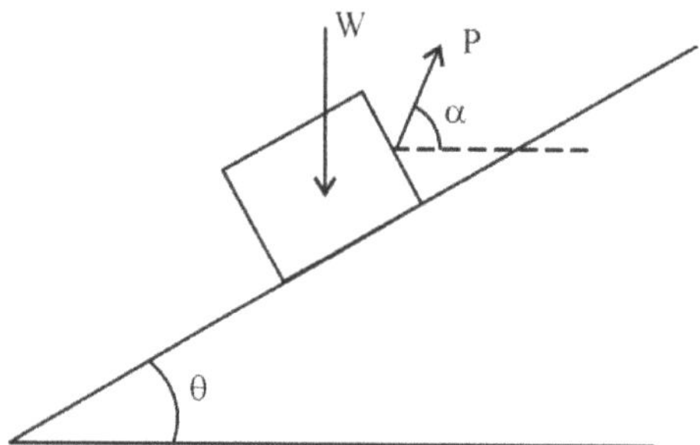

Solution:

Resolving all the forces in horizontal (X) and vertical (Y) directions, equations of equilibrium are

$$\Sigma F_X = P \cos\alpha - N \sin\theta - \mu N \cos\theta = 0 \qquad \text{.....(1)}$$

and

$$\Sigma F_Y = P \sin\alpha + N \cos\theta - \mu N \sin\theta - W = 0 \qquad \text{.....(2)}$$

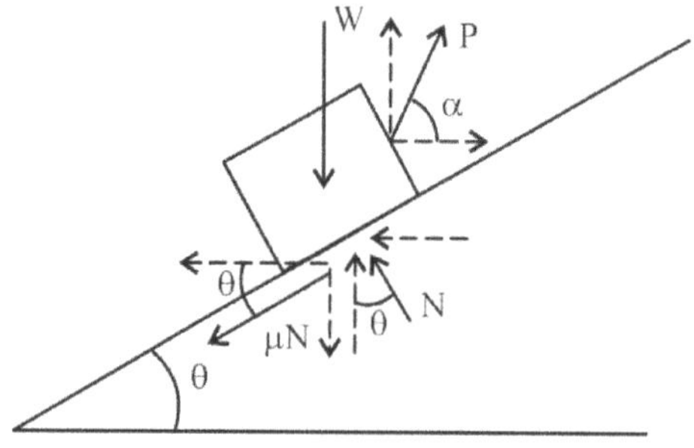

Separating terms with N in both the equations and dividing eq (2) by eq (1)

$$(P \sin\alpha - W) / P \cos\alpha = N(-\cos\theta + \mu \sin\theta) / N(\sin\theta + \mu \cos\theta)$$

Substituting $P = W \sin(\theta + \varphi)$ and $\mu = \tan\varphi$, canceling N on the right side

$$[W \sin(\theta + \varphi) \times \sin\alpha - W] / [W \sin(\theta + \varphi) \times \cos\alpha]$$

$$= [-\cos\theta \times \cos\varphi + \sin\theta \times \sin\varphi] / [\sin\theta \times \cos\varphi + \cos\theta \times \sin\varphi]$$

$$= -\cos(\theta + \varphi) / \sin(\theta + \varphi)$$
$$[\sin(\theta + \varphi) \times \sin\alpha - 1] \times \sin(\theta + \varphi)$$
$$= -\cos(\theta + \varphi) \times [\sin(\theta + \varphi) \times \cos\alpha]$$

Dividing by $\sin(\theta + \varphi)$ on both sides,

$$\sin(\theta + \varphi) \times \sin\alpha + \cos(\theta + \varphi) \times \cos\alpha - 1 = 0$$

$$\cos(\theta + \varphi - \alpha) - 1 = 0 \quad \text{or} \quad \theta + \varphi - \alpha = 0 \quad \Rightarrow \quad \theta + \varphi = \alpha$$

Example 7

A semi-circular cylinder resting on a horizontal plane is pulled by a horizontal force P as shown. Find the angle θ, the diameter makes with the horizontal before it starts sliding.

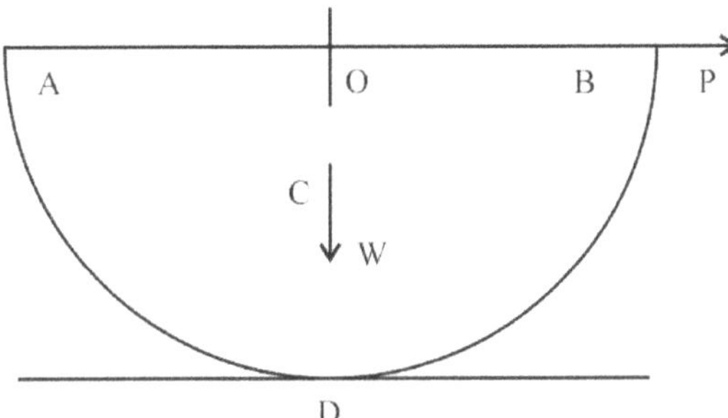

Solution:

C.G. of the semicircular cylinder lies at C, where, $OC = 4r / 3\pi$

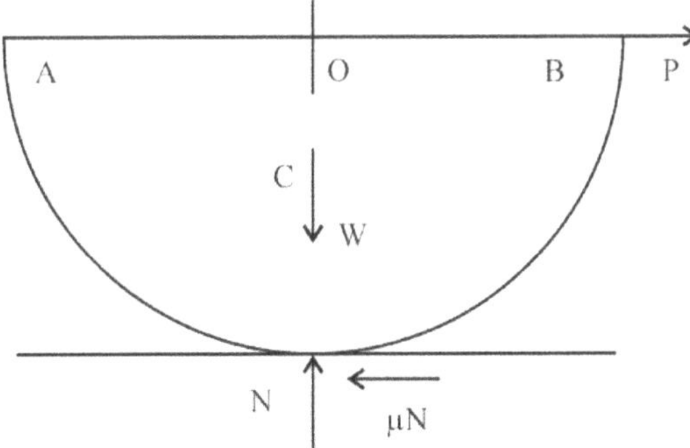

Let us first consider the case when the cylinder is resting on the floor with the diameter parallel to the floor and a horizontal force P applied on it at B. At the contact point D, normal reaction N is developed to balance weight W and frictional force μN is developed to balance the applied load P.

For equilibrium of the body, these forces should satisfy the equations

$$\Sigma F_X = P - \mu N = 0 \quad \text{and} \quad \Sigma F_Y = N - W = 0$$

However, since μN and P are not collinear, a moment is developed which is not balanced.

This moment rotates the body in a clockwise direction till a counter balancing moment is generated by the displacement of centroid away from the line of action of normal reaction N, as shown. Let the inclination of diameter AB with the horizontal be θ.

Force equilibrium conditions remain the same.

Therefore, $N = W$ and $\mu N = P$

Then, taking moments about O,

$$\Sigma M_O = W \times CE + P \times OF - \mu N \times OD = 0$$

$\Rightarrow$ $$W \times OC \sin\theta + P \times OB \sin\theta - \mu N \times OB = 0$$

or $$N \times (4r/3\pi) \sin\theta + \mu N \times (r \sin\theta) = \mu N \times r$$

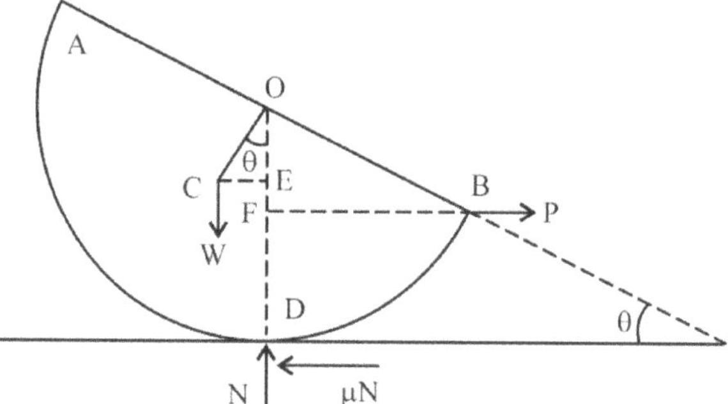

Dividing by $N \times r$ throughout,

$$[\,(4/3\pi) + \mu\,] \times \sin\theta = \mu$$

or $$\sin\theta = 3\,\pi\mu \,/\, (4 + 3\,\pi\,\mu)$$

Example 8

A body is resting on a plane, inclined at 15^0 to the horizontal. A force of 200 N applied on the body, parallel to the plane, tends to move the body upwards. The same body, resting on a plane inclined at 20^0, requires force of 230 N to tend to move the body upwards. Calculate weight of the body and coefficient of friction

Solution:

Resolving the forces along and perpendicular to the plane,

$$\Sigma F_X = P - \mu N - W \sin\theta = 0 \quad \text{and} \quad \Sigma F_Y = N - W \cos\theta = 0$$

or $$P = \mu N + W \sin \theta = W \times (\mu \cos \theta + \sin \theta)$$

Using this relation for the 2 cases,

$$200 = W \times (\mu \cos 15 + \sin 15)$$

and $$230 = W \times (\mu \cos 20 + \sin 20)$$

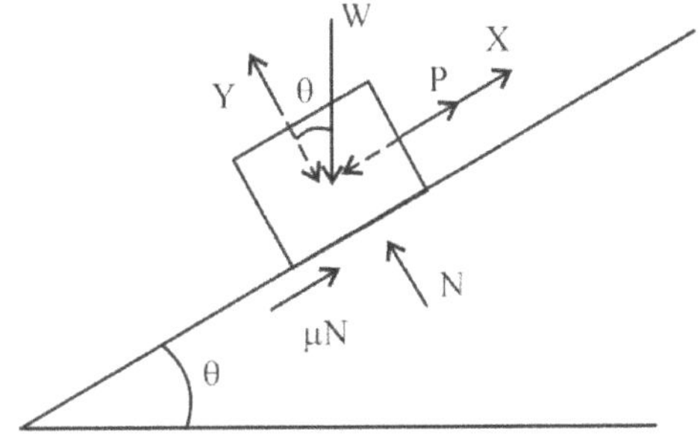

Dividing these two equations, we get,

$$(\mu \cos 15 + \sin 15) / (\mu \cos 20 + \sin 20) = 200/230$$

or $$(\mu \cos 15 + \sin 15) \times 230 = (\mu \cos 20 + \sin 20) \times 200$$

$\Rightarrow$ $$\mu \times (230 \cos 15 - 200 \cos 20) = 200 \sin 20 - 230 \sin 15$$

$\Rightarrow$ $$\mu = 0.26$$

Then, $W = 200 / (\mu \cos 15 + \sin 15)$ or $230 / (\mu \cos 20 + \sin 20) = 392.3$ N

Example 9

Determine force P in limiting equilibrium condition, just before two blocks connected with a rope as shown start moving.

$$W_1 = 300 \text{ N}; \; W_2 = 50 \text{ N} \text{ and } \mu = 0.25$$

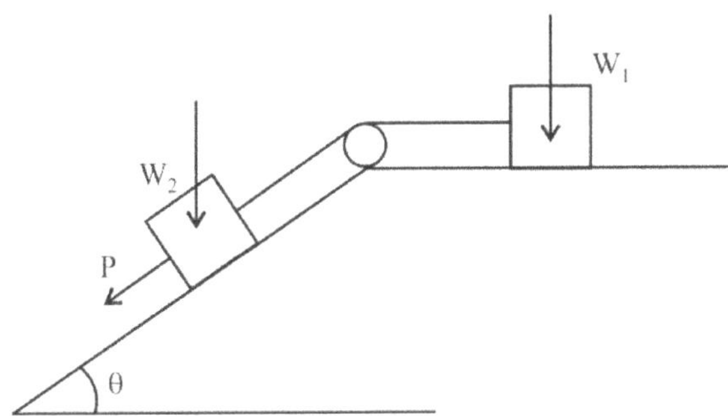

Solution:

Let N_1 and N_2 be the reactions normal to the supporting plane at the two blocks respectively, opposing the components of weight normal to the plane. The force

P tending to move the blocks down the inclined plane induces tension in the rope and frictional resistances of μN_1 and μN_2 between the two blocks and the supporting planes. The problem can be solved for equilibrium of the two blocks separately.

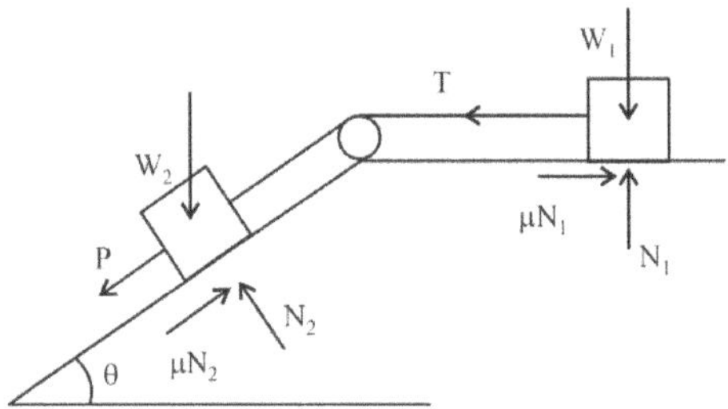

Applying equilibrium conditions for block W_1,

$$\Sigma F_Y = N_1 - W_1 = 0 \quad \text{and} \quad \Sigma F_X = \mu N_1 - T = 0$$

$\Rightarrow$ $$T = \mu N_1 = \mu W_1 = 0.25 \times 300 = 75 \text{ N}$$

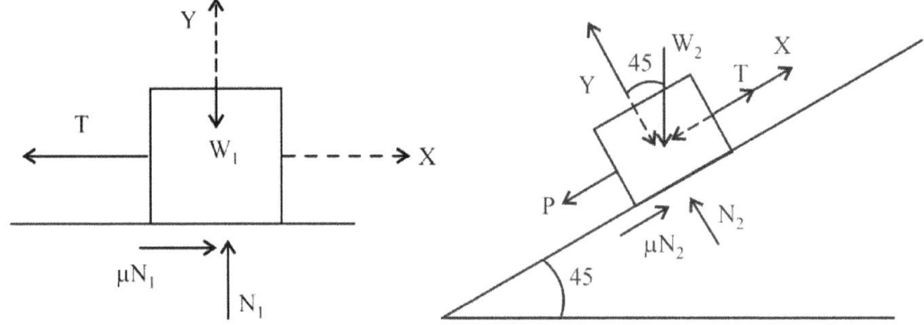

Applying equilibrium conditions for block W_2, with the chosen coordinate system as shown,

$$\Sigma F_Y = N_2 - W_2 \cos 45 = 0 \quad \text{or} \quad N_2 = W_2 \cos 45$$

and $$\Sigma F_X = T + \mu N_2 - P - W_2 \sin 45 = 0$$

or $$P = T + \mu N_2 - W_2 \sin 45$$

$$= T + \mu W_2 \cos 45 - W_2 \sin 45 \text{ (Substituting for } N_2 \text{ from eq. 1)}$$

$$= 75 + 0.25 \times 50 \times 0.707 - 50 \times 0.707 = 48.4875 \text{ N}$$

Example 10

A smooth circular cylinder of radius 'r' and weight 'W' is placed on two semi-circular cylinders of radius 'r' and weight 'W/2', placed 'L' apart between their centers. Find the greatest L at which equilibrium is possible. Neglect friction between circular cylinder and semi-circular cylinders. Assume friction coefficient of 0.5 between semi-circular cylinders and the horizontal plane on which they rest.

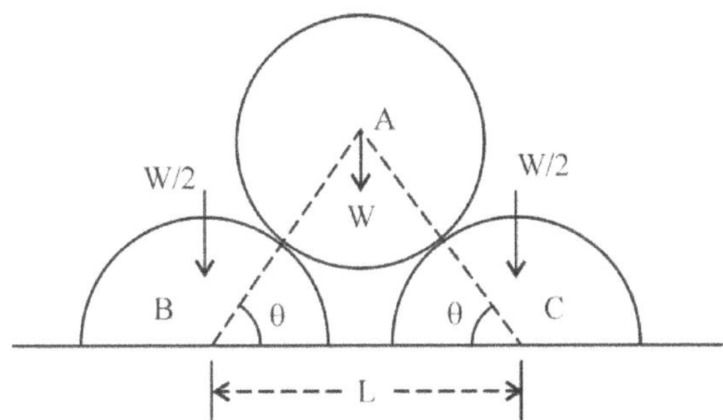

Solution:

Cylinder A is supported by the reactions R from the two semi-circular cylinders at the points of contact D and E along the respective normals BA and CA. Due to symmetry of the geometry, reactions at D and E are equal in magnitude and oriented symmetrically w.r.t. a vertical line through A.

Equations of equilibrium for the cylinder A are

$$\Sigma F_X = R \cos\theta \text{ (at D)} - R \cos\theta \text{ (at E)} = 0$$

and $$\Sigma F_Y = R \sin\theta \text{ (at D)} + R \sin\theta \text{ (at E)} - W = 0$$

$\Rightarrow$ $$R = W / (2 \sin\theta)$$

From the geometry, $\cos\theta = (BC/2) / AB = (L/2) / 2r = L / 4r$

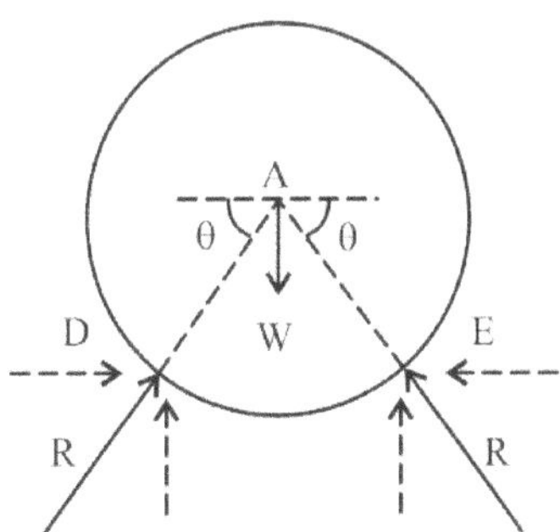

The effect of cylinder A is felt on semi-circular cylinders B and C as forces R acting along DB and EC. Since the geometry is symmetric, only one semi-circular cylinder B is considered in the analysis. Vertical component of force R and weight W/2 are balanced by normal (vertical) reaction S at B. The horizontal component of force R can only be balanced by the frictional resistance μS at B as shown.

Equations of equilibrium are

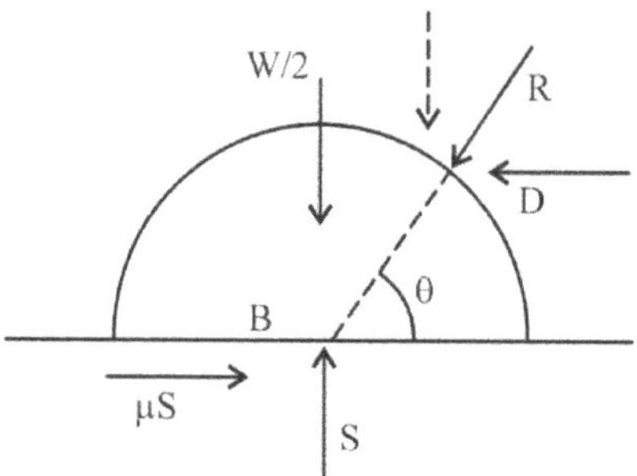

$$\Sigma F_Y = S - W/2 - R \sin\theta = 0$$

and $\Sigma F_X = \mu S - R\cos\theta = 0$

For equilibrium of B, force R exerted by cylinder A through D is equal to the reaction by the semi-circular cylinder at D to support weight of the cylinder A.

From the above equations,

$$W/2 = S - R\sin\theta = (R\cos\theta)/\mu - R\sin\theta$$

$$= [W/(2\sin\theta)] \times [\cos\theta/\mu - \sin\theta]$$

Cancelling W/2 both sides, $\Rightarrow 1 = (\cos\theta - \mu\sin\theta)/\mu\sin\theta$

or $\mu\sin\theta = \cos\theta - \mu\sin\theta \Rightarrow 2\mu\sin\theta = \cos\theta$

$\Rightarrow \sin\theta = \cos\theta$ since $\mu = 1/2$

$\Rightarrow \theta = 45^0$

Then, $L = 2\,(AB\cos 45) = 2\,(2r/\sqrt{2}) = 2.828\,r$

Example 11

Block A of weight 320N, resting on a rough horizontal floor, supports block B of weight 160N. The two blocks are connected to a string passing over round smooth pulley, as shown. Find the horizontal force P to be applied on the block A so as to just move it towards right. Find also the tension in the string, assuming $\mu = 0.25$ at all contact surfaces.

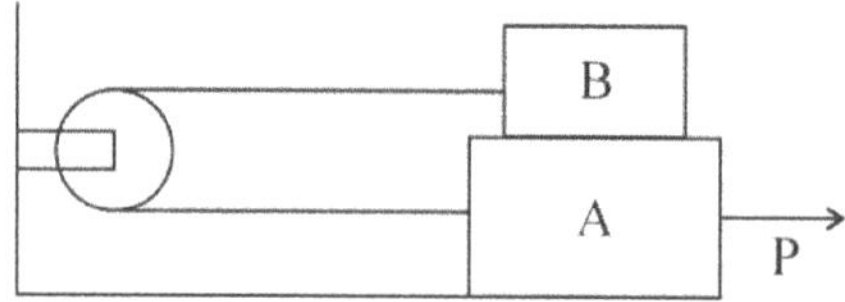

Solution:

When block A is pulled by force P, rope connected to A and B will be in tension. Also, load P has to overcome frictional resistance between block A and the floor as well as between blocks A and B. Tension in the rope has to overcome frictional resistance to sliding of block A to the right relative to block B.

If N is the normal reaction between the two blocks, equations of equilibrium for block B are $\Sigma F_X = \mu N - T = 0$ and $\Sigma F_Y = N - W_B = 0$

$\Rightarrow \quad N = W_B = 160N$ and $T = \mu N = 0.25 \times 160 = 40N$

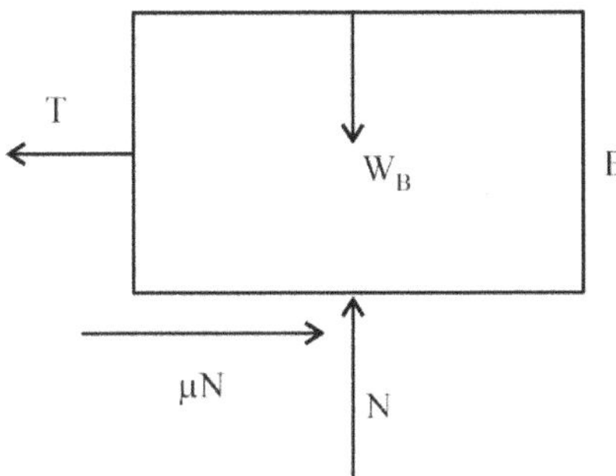

Whenever two blocks tend to move one over the other, normal reaction and frictional resistance from block A to block B will be equal and opposite to those from block B to block A, to ensure force equilibrium at the contact point between the blocks

If N′ is the normal reaction between block A and the floor, equations of equilibrium are

$$\Sigma F_X = P - \mu N - \mu N' - T = 0 \quad \text{and} \quad \Sigma F_Y = N' - N - W_A = 0$$

$$\Rightarrow \quad N' = N + W_A = 160 + 320 = 480N$$

$$\Rightarrow \quad P = \mu N + \mu N' + T = 0.25 \times 160 + 0.25 \times 480 + 40 = 200\text{ N}$$

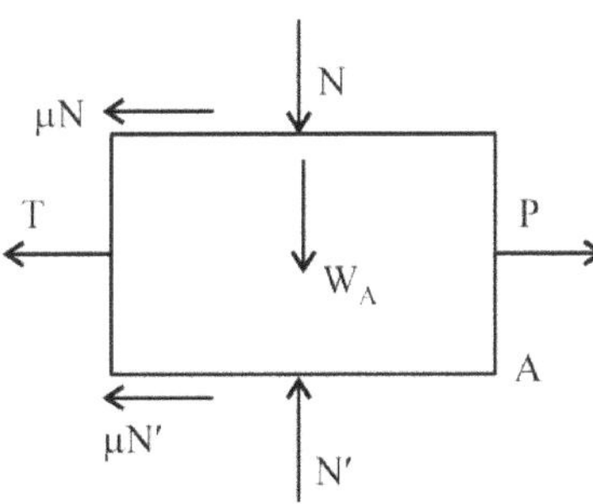

Example 12

A block resting on a 10^0 wedge on horizontal floor and leaning on a vertical wall weighs 1500N. The block is raised by applying a horizontal force to the wedge. Calculate minimum force required, assuming coefficient of friction between all surfaces μ as 0.3. Neglect weight of the wedge.

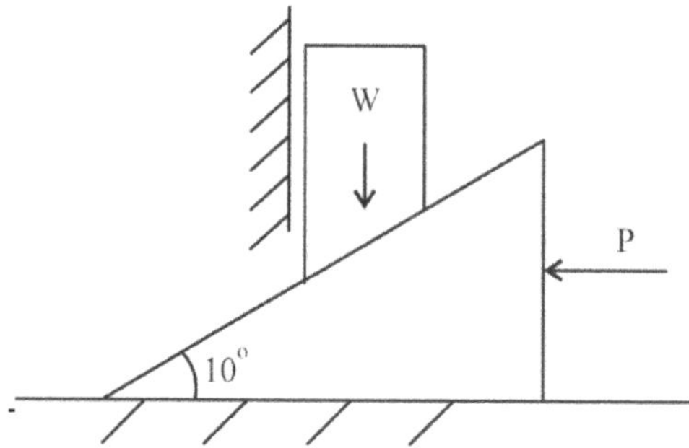

Solution:

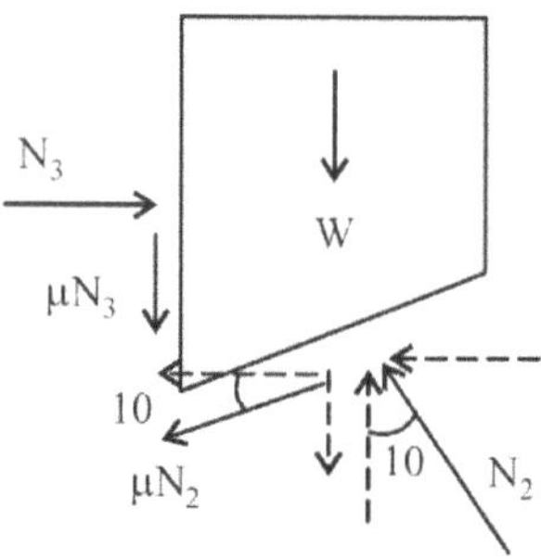

Weight of the block is supported by the vertical component of the normal reaction from the wedge. Its horizontal component is balanced by the normal reaction from the wall. In addition, there are frictional resistances from the wedge and the wall to the upward motion of the block, as shown in the free body diagram of the block. We can consider the block to be so small (in the absence of dimensions) that the ***forces can be assumed to be concurrent***. Then, equations of equilibrium for the block are

$$\Sigma F_X = N_3 - N_2 \sin 10 - \mu N_2 \cos 10 = 0$$

and $$\Sigma F_Y = N_2 \cos 10 - \mu N_2 \sin 10 - \mu N_3 - W = 0$$

$\Rightarrow$ $$0.4691N_2 - N_3 = 0 \quad \text{and} \quad 0.9327N_2 - 0.3N_3 = 1500N$$

or $$3.109N_2 - N_3 = 5000N$$

Eliminating N_3 from these two equations,

$$(3.109 - 0.4691) \times N_2 = 5000N$$

or $N_2 = 1894N$

and $N_3 = 0.4691N_2 = 0.4691 \times 1894$

$= 888.5N$

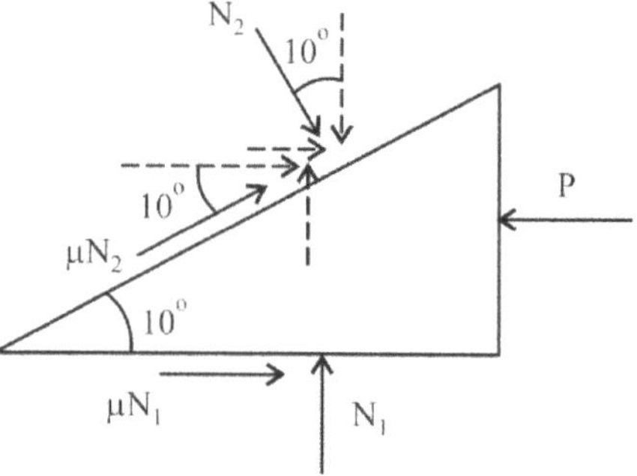

Considering the wedge to be so small that the forces can be assumed concurrent, equations of equilibrium are

$$\Sigma F_X = \mu N_1 + \mu N_2 \cos 10 + N_2 \sin 10 - P = 0$$

and

$\Sigma F_Y = N_1 + \mu N_2 \sin 10 - N_2 \cos 10$

$= 0$

$\Rightarrow N_1 = N_2 \cos 10 - \mu N_2 \sin 10$

$= 1766.6N$

and $P = \mu N_1 + \mu N_2 \cos 10 + N_2 \sin 10 = 1418.4N$

Alternative method: Let $\mu = \tan \varphi$ or $\varphi = 16.7^0$

Then, the pairs of normal reaction and frictional resistance on different surfaces (N_1, μN_1), (N_2, μN_2) and (N_3, μN_3) can be replaced respectively by R_1, R_2 and R_3.

With three forces acting on the block, Lami's theorem can be applied for the equilibrium of the block as

$$W / \sin(100 + 2\varphi) = R_2 / \sin(90 - \varphi) = R_3 / \sin(170 - \varphi)$$

$$R_2 = 1977.45N\ ;\ R_3 = 927.54N$$

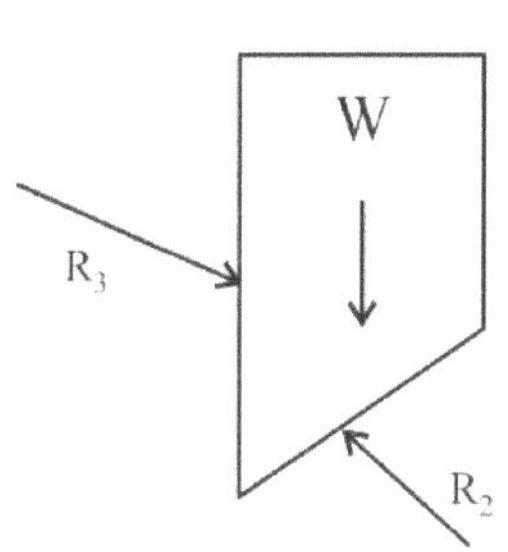

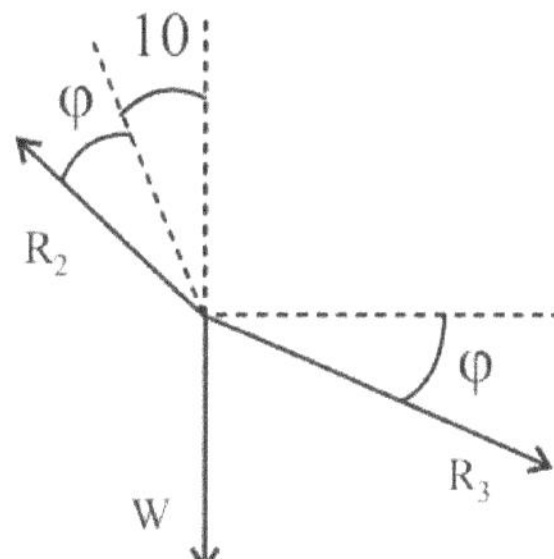

Similarly, with three forces acting on the wedge, Lami's theorem can be applied for the equilibrium of the wedge as

$$P / \sin(170 - 2\varphi) = R_2 / \sin(90 + \varphi) = R_1 / \sin(100 + \varphi)$$

$$\Rightarrow \qquad R_1 = 1844.5N \quad \text{and} \quad P = 1418.4N$$

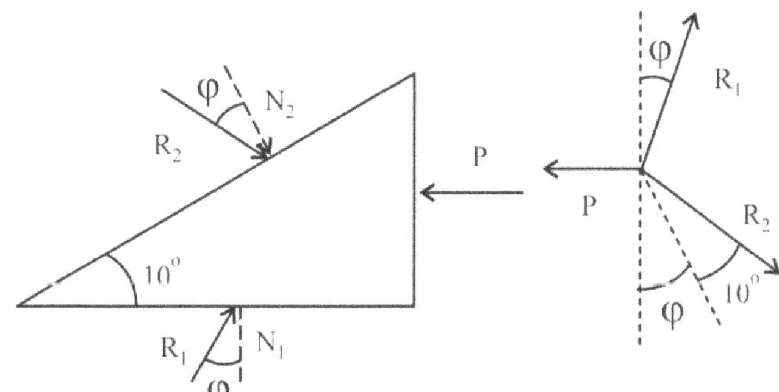

Example 13

A uniform ladder of weight 'W' rests with one end against a rough inclined plane of angle α with the horizontal and the other end against a smooth vertical wall. If the ladder be inclined at θ with the horizontal and is in limiting equilibrium, prove that $2 \tan \theta = \cot (\lambda - \alpha)$, where λ is the angle of limiting friction between the ladder and the inclined plane.

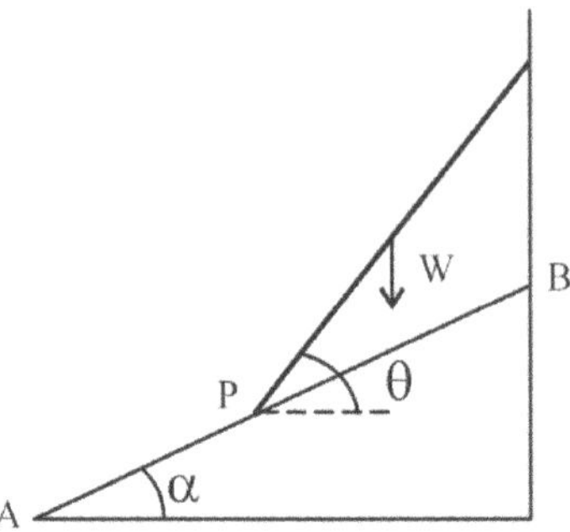

Solution:

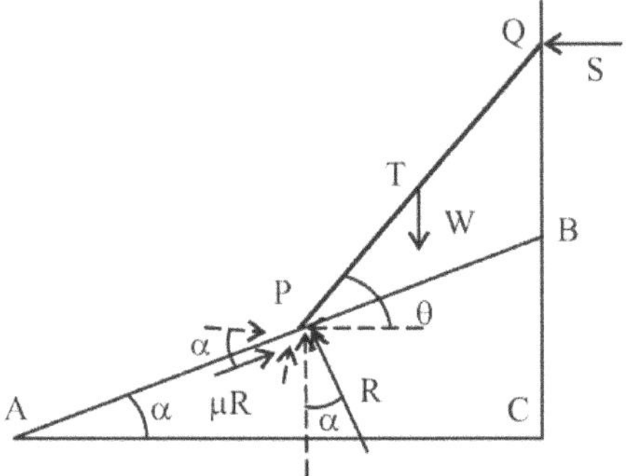

End P of the ladder is likely to slide down due to its weight. Therefore, normal reaction R and frictional resistance μR exist at P. Let 'L' be length of the ladder. Since the wall is smooth, only normal reaction S is developed at the other end, without any frictional resistance. Force equilibrium equations are

$$\sum F_Y = R \cos \alpha + \mu R \sin \alpha - W = 0$$

and $\sum F_X = \mu R \cos \alpha - R \sin \alpha - S = 0$

$\Rightarrow \quad W/S = R (\cos \alpha + \mu \sin \alpha) / [R (\mu \cos \alpha - \sin \alpha)]$

Since the forces are non-concurrent, for moment equilibrium, taking moments about P

$$\Sigma M_P = S \times (PQ \sin \theta) - W \times (PT \cos \theta) = 0$$

Since $PT = PQ/2$, $\tan \theta = W / (2 S) = (\cos \alpha + \mu \sin \alpha)/[2 (\mu \cos \alpha - \sin \alpha)]$

$$= (\cos \alpha + \tan\lambda \sin \alpha) / 2 (\tan \lambda \cos \alpha - \sin \alpha) \quad \text{with } \mu = \tan \lambda$$

$$= (\cos \lambda \cos \alpha + \sin\lambda \sin \alpha) / 2 (\sin \lambda \cos \alpha - \sin \alpha \cos \lambda)$$

$$= \cos (\lambda - \alpha) / 2 \sin(\lambda - \alpha)$$

or $\quad 2 \tan \theta = \cot (\lambda - \alpha)$

5.6 POWER SCREW OR SCREW JACK

Screw jack (one of the many applications of power screws) is commonly used for lifting weight, such as an automobile, in order to change tyre. Most of these jacks operate on the principle of screw. A tripod like stand forms the nut in which a helical screw supporting the load rotates. Each rotation of the screw causes it to move axially by one lead 'L'. In a multi-start thread, $L = n \times p$ where n is the number of starts and 'p' is the pitch or the distance between two points on successive threads. Imagine the screw unwound as a 2-D plane. If 'd' is the mean diameter of the screw, it can be considered in this view as an inclined plane of angle α

where $\tan \alpha = L / \pi d = (n \times p) / \pi d$

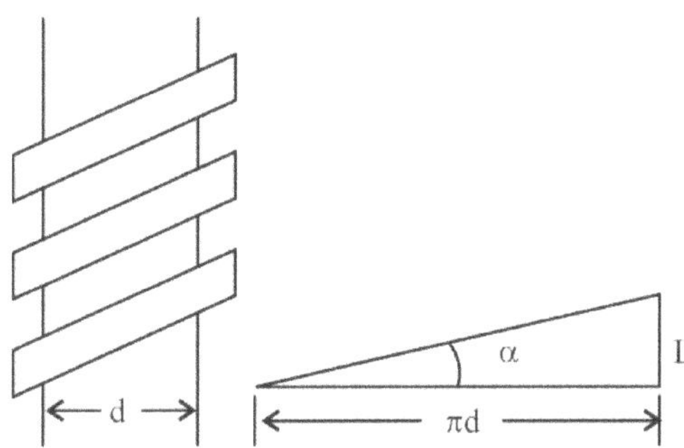

The screw is rotated in the nut by applying torque or forces tangential to the screw shaft. The screw moves up or down depending on the direction of torque. Even though load 'W' and force 'P' are spread throughout the screw thread, we assume these forces are concentrated at one point and are therefore concurrent, without any change in the result. Considering the first case of raising the load. On the average of moving up, equilibrium equations are

$$\sum F_X = P - N \sin \alpha - \mu N \cos \alpha = 0$$

and $$\sum F_Y = N \cos \alpha - \mu N \sin \alpha - W = 0$$

$\Rightarrow$ $P = (\sin \alpha + \mu \cos \alpha) N$ and $N (\cos \alpha - \mu \sin \alpha) = W$

$\Rightarrow$ $P = (\sin \alpha + \mu \cos \alpha) \times W / (\cos \alpha - \mu \sin \alpha)$

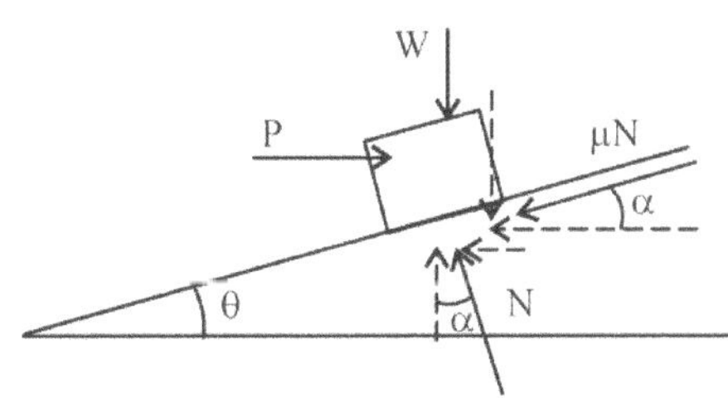

Substituting $\mu = \tan\varphi$, $P = W(\sin\alpha + \tan\varphi\cos\alpha) / (\cos\alpha - \tan\varphi\sin\alpha)$

$\Rightarrow$ $P = W(\sin\alpha\cos\varphi + \sin\varphi\cos\alpha) / (\cos\alpha\cos\varphi - \sin\varphi\sin\alpha)$

$\Rightarrow$ $P = W\sin(\alpha + \varphi) / \cos(\alpha + \varphi)$ or $\mathbf{P = W\tan(\alpha + \varphi)}$

Torque required to raise the weight W, $T = P \times (d/2)$

In most applications, a lever of length 'R' is used to lift a heavy weight with a lesser effort P′. However, torque required to raise the load remains the same. Therefore,

$$T = P \times r = P' \times R \quad \text{where} \quad r = d/2 \text{ is the radius of screw}$$

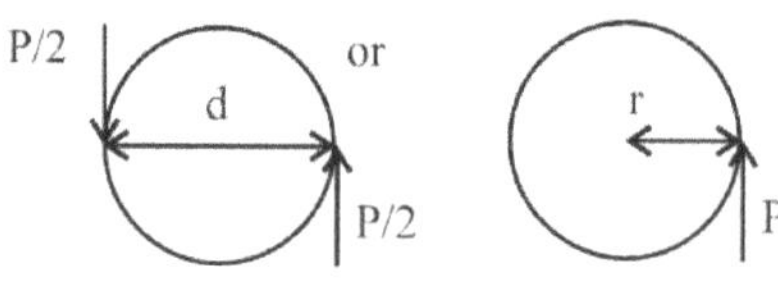

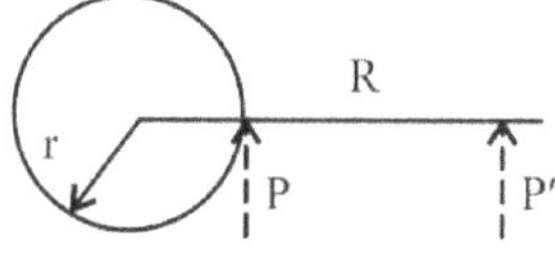

Ideal effort, when there is no friction ($\varphi = 0$), $P_0 = W\tan\alpha$

Efficiency, η = Ideal effort/Actual effort $= P_0 / P$

$= \tan\alpha / \tan(\alpha + \varphi)$

Max. efficiency is obtained

When $d\eta/d\alpha = 0 \Rightarrow \alpha = (\pi/4) - (\varphi/2)$

Then, $\eta_{max} = (1 - \sin\varphi) / (1 + \sin\varphi)$

5.6.1 LAW OF MACHINE

Effort P in a screw jack varies with load W and their relationship can be expressed by

$$P = mW + c, \quad \text{where } m \text{ and } c \text{ are constants}$$

or $W / P = (1/m) \times (1 - c/P)$

As W increases, P increases

Therefore, for large values of P, $c/P \approx 0$

and $W/P \simeq 1/m$ (for ideal machines)

Efficiency can also be defined by two other parameters of a machine.

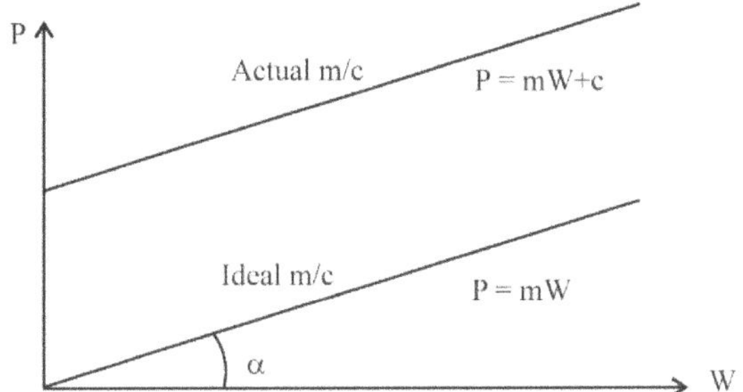

Mechanical Advantage (MA) = Load moved / Effort applied = W / P

Velocity Ratio (VR) = Distance moved by effort / Distance moved by load

$= 2\pi r / L$

Efficiency, η = MA / VR = (W/P) / (2π r / L) = W tan α / P = tan α / tan (α + φ)

From law of machines, for ideal machines, $\eta = (1/m) / VR = 1 / (m \times VR)$

A screw after raising a load W,

(a) may remain in that position without any effort (most screw jacks are designed for this), where frictional resistance is sufficient to prevent downward sliding of load. It requires effort P_1 to lower the load when desired. Such a screw is called ***self-locking screw***

(b) may slide down due to its component of weight along the screw thread overcoming frictional resistance and, hence, needs an effort P_2 to remain in its elevated position. Such a screw is called ***non self-locking screw***

5.6.2 SELF-LOCKING SCREW

In this screw, effort P_1 needs to be applied to lower the load, in the direction opposite to that for raising the load. Frictional resistance also reverses its direction, as the direction of screw movement over the nut has reversed.

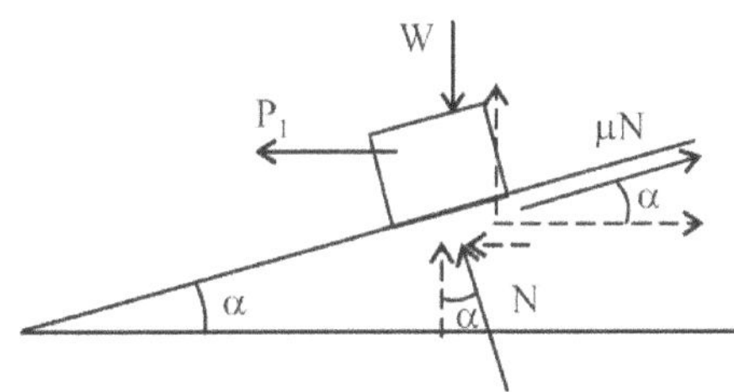

Equations of equilibrium are

$$\Sigma F_X = \mu N \cos\alpha - P_1 - N \sin\alpha = 0$$

and $$\Sigma F_Y = N \cos\alpha + \mu N \sin\alpha - W = 0$$

⇒ $$P_1 = N \times (\mu \cos\alpha - \sin\alpha) \quad \text{and} \quad N \times (\cos\alpha + \mu \sin\alpha) = W$$

⇒ $$P_1 = W \times (\mu \cos\alpha - \sin\alpha) \;/\; (\cos\alpha + \mu \sin\alpha)$$

Substituting $\mu = \tan\varphi$,

$$P_1 = W \times (\tan\varphi \cos\alpha - \sin\alpha) / (\cos\alpha + \tan\varphi \sin\alpha)$$

$\Rightarrow \quad P_1 = W \times (\sin\varphi \cos\alpha - \sin\alpha \cos\varphi) / (\cos\alpha \cos\varphi + \sin\varphi \sin\alpha)$

$\Rightarrow \quad P_1 = W \times \sin(\varphi - \alpha) / \cos(\varphi - \alpha)$

or $\quad \mathbf{P_1 = W \tan(\varphi - \alpha)}$ for $\alpha < \varphi$

Efficiency can also be defined as η = Output / Input

where $\quad$ Input = Work done = $P \times (2\pi r)$

and $\quad$ Output = Work done – Work lost in friction = $W \times L$

Therefore, Work lost in friction = Input – Output = $P \times (2\pi r) - W \times L$

Then, for a self-locking screw in which motion is irreversible,

Output < Work lost in friction or $W \times L < P \times 2\pi r - W \times L$

$\Rightarrow \quad 2\,W \times L < P \times 2\pi r$ or $W/P < (1/2) \times (2\pi r / L)$

Thus, for a self-locking screw,

$$\eta = MA/VR = (W/P) / (2\pi r / L) < 1/2 \text{ or } < 50\%$$

5.6.3 NON SELF-LOCKING SCREW

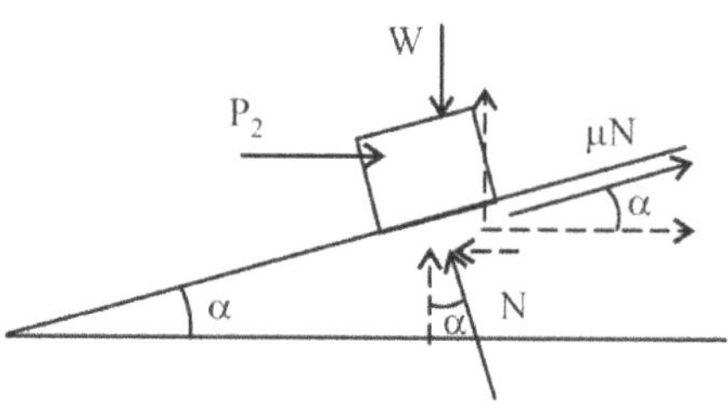

In this screw, effort P_2 needs to be applied to prevent sliding down of the weight on its own, in the same direction as that for raising the load. However, frictional resistance reverses its direction, as the direction of screw movement over the nut has reversed. Equations of equilibrium are

$\Sigma F_X = \mu N \cos\alpha + P_2 - N \sin\alpha = 0$

and $\quad \Sigma F_Y = N \cos\alpha + \mu N \sin\alpha - W = 0$

$\Rightarrow \quad P_2 = N \times (\sin\alpha - \mu \cos\alpha)$ and $N \times (\cos\alpha + \mu \sin\alpha) = W$

$\Rightarrow \quad P_2 = W \times (\sin\alpha - \mu\cos\alpha) / (\cos\alpha + \mu \sin\alpha)$

Substituting $\mu = \tan\varphi$, $P_2 = W (\sin\alpha - \tan\varphi \cos\alpha) / (\cos\alpha + \tan\varphi \sin\alpha)$

$\Rightarrow \quad P_2 = W \times (\sin\alpha \cos\varphi - \sin\varphi \cos\alpha) / (\cos\alpha \cos\varphi + \sin\varphi \sin\alpha)$

$\Rightarrow \quad P_2 = W \times \sin(\alpha - \varphi) / \cos(\alpha - \varphi)$

or $\quad \mathbf{P_2 = W \tan(\alpha - \varphi)}$ for $\alpha > \varphi$

5.6.4 DIFFERENTIAL SCREW JACK

It consists of two screws A and B, one with right handed screw and the other with left handed screw. The screw A is connected to the effort 'P' and screw B is connected to the load 'W'. One rotation of load arm will move screw A upwards by L_A while at the same time, screw B moves down by L_B. Net movement of the screw for one rotation of the load arm = $L_A - L_B$.

Therefore, Velocity Ratio (VR) = $2\pi R / (L_A - L_B)$

$$\text{Efficiency, } \eta = MA / VR = (W/P) / [2\pi R / (L_A - L_B)]$$

Example 14

A screw press is used to compress books. The screw is a double thread with a pitch of 4 mm and mean radius of 25 mm. Find effort to be applied for a load of 500 N at the end of a lever of length 500mm. Assume $\mu = 0.3$

Solution:

$$\tan\alpha = L / \pi d = n\,p / \pi d = (2 \times 4) / (\pi \times 2.5) = 0.0509$$

or $\alpha = 2^0\,55'$

and $\mu = \tan\varphi = 0.3$ or $\varphi = 16^0\,42'$

Effort, $P = W\tan(\alpha + \varphi) = 500\tan(2^0\,55' + 16^0\,42') = 178.2\text{ N}$

Torque, $T = P \times r = 178.2 \times 25 = 4455\text{ N mm}$

Load to be applied at the end of lever, $P' = T/R = 4455 / 500 = 8.91\text{ N}$

Example 15

A screw jack has square threads of 50 mm mean diameter and 10 mm pitch. The load on the jack revolves with the screw. The coefficient of friction at the screw thread is 0.05

(i) Find the tangential load required at the end of 300 mm lever to lift a load of 6000 N

(ii) State whether the jack is self locking. If not, find the torque which must be applied to keep the load from descending.

Solution:

$$\tan\alpha = L / \pi d = 10 / (\pi \times 50) = 0.06363 \quad \text{or} \quad \alpha = 3.6426^0$$

and $\mu = \tan\varphi = 0.05$ or $\varphi = 2.8624^0$

$$P = (d/2R)\,W\tan(\alpha + \varphi) = [\,50 / (2\times300)\,] \times 6000 \times \tan(\alpha + \varphi) = 57.01\text{ N}$$

Velocity Ratio = $2\pi R / L = 2\pi \times 300 / 10 = 188.496$

Mechanical Advantage = $W / P = 6000 / 57.01 = 105.245$

Efficiency, η = Mechanical Advantage / Velocity Ratio

$= 0.5583$ or $55.83\% > 50\%$

Therefore, the screw jack is ***not*** self-locking

Torque required to keep the load from descending,

$$T = (d/2)\ W \tan(\alpha - \varphi) = 25 \times 6000 \times \tan(0.7802^0)$$
$$= 2043 \text{ N mm}$$

Example 16

The following are the specifications of a differential screw jack. Pitch of smaller screw is 5 mm; pitch of larger screw is 10 mm; lever arm length from center of screw is 500 mm. The screw jack raises a load of 15 kN with an effort of 185N. Calculate efficiency. Determine law of machine if an effort of 585N can raise a load of 50 kN.

Solution:

Velocity Ratio $= 2\pi R / (L_A - L_B) = 2\pi \times 500 / (10 - 5) = 628.32$

Mechanical Advantage $= W / P = 15000 / 185 = 81.08$

Efficiency, η = Mechanical Advantage / Velocity Ratio = 0.129 or 12.9%

Law of machine is given by $P = m\ W + c$

or $185 = 15000\ m + c$ and $585 = 50000\ m + c$

Solving these two equations, $m = 4/350$ and $c = 13.57$

Therefore, $P = (4/350)\ W + 13.57$

5.7 BELT DRIVES

5.7.1 TYPES OF BELT DRIVES

In many engineering applications, mechanical power generated by one equipment (like Internal Combustion Engine, electric motor, ..) is to be used by another equipment situated elsewhere (like water pump, lathe machine, ..). Rotary power can be transferred by many ways, such as gear drives, chain drives, belt drives. Of these, belt drive is the simplest and cheapest. It can be used with a simple pulley and does not involve costly machining as in a chain drive or a gear drive. It can be used over long distances, upto 5m. It can be assembled or replaced easily at the site. It is made of rubber, nylon or fiber. Its main disadvantage is that it elongates over a period of usage due to creep of belt material and needs to be tightened to ensure transfer of power. Its cross section is usually circular, as in a sewing machine, or rectangular (also called flat belt), as in a rice or floor mill, or trapezoidal (also called V-belt), as in automobiles. V-belts and some circular section belts come in standard lengths and the equipment to be connected have to be properly located. Syllabus in this subject usually covers rectangular section (or flat) belts and so the discussion hereafter is limited to flat belt drives.

5.7.2 ARRANGEMENTS OF FLAT BELT DRIVES

There are two common types of flat belt drives – Open belt and crossed belt. Let R_B be the radius of the bigger pulley, R_S be the radius of the smaller pulley and C the distance between the centers O and O_1 of the two pulleys. The belt will pass over the pulleys tangentially. So the lines OI and O_1L are perpendicular to IL. Similarly, OJ and O_1K are perpendicular to JK. Then, the lines OJ and O_1K are parallel and are inclined at α with the vertical diameter. Lines OI and O_1L are also similarly inclined. The angle over which the belt is in contact with the pulley is called **wrap angle.**

(a) In **open belt arrangement**, both the pulleys rotate in the same direction.

Wrap angle, $\theta = \pi + 2\alpha > 180^0$ on larger pulley

and Wrap angle, $\theta = \pi - 2\alpha < 180^0$ on smaller pulley

Length of belt, $L = IJ + JK + KL + LI$

$$= (\pi + 2\alpha)\, R_B + C \cos\alpha + (\pi - 2\alpha)\, R_S + C \cos\alpha$$

From ΔILM, $\sin\alpha = IM/LM = (R_B - R_S) / C$

and $\cos\alpha = \sqrt{1-\sin^2\alpha} = \sqrt{\left[C^2 - (R_B - R_s)^2\right]} / C$

Therefore, $L \approx (\pi + 2\alpha)\, R_B + (\pi - 2\alpha)\, R_S + 2C \cos\alpha$

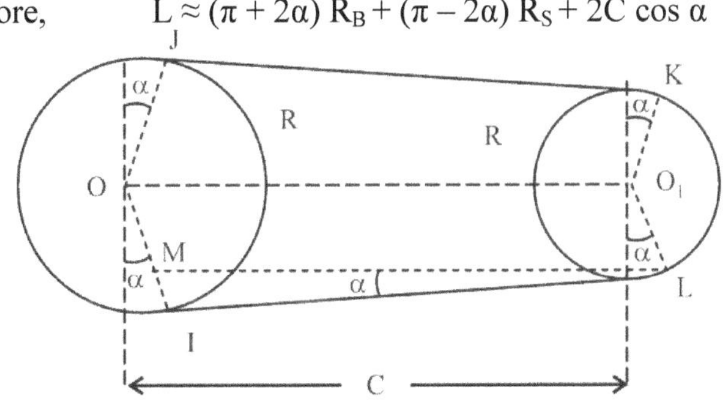

(b) In **crossed belt arrangement**, both the pulleys rotate in opposite directions and

Wrap angle, $\theta = \pi + 2\alpha > 180^0$ on both pulleys

Length of belt, $L = IJ + JK + KL + LI$

$= (\pi + 2\alpha)\, R_B + C \cos \alpha + (\pi + 2\alpha)\, R_S + C \cos \alpha$

From, $\Delta OO_1 M$, $\sin \alpha = O_1 M / OO_1 = (R_B + R_S)/C$

$$\cos \alpha = \sqrt{1 - \sin^2 \alpha} = \sqrt{\left[C^2 - (R_B + R_S)^2\right]} / C$$

Therefore, $L \approx (\pi + 2\alpha)(R_B + R_S) + 2C \cos \alpha$

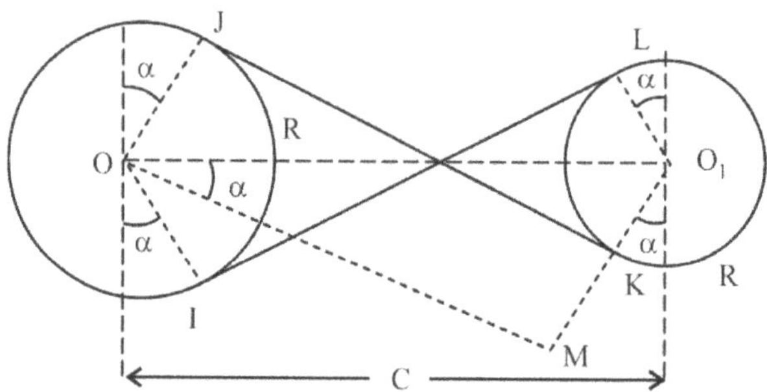

Open and crossed belt drives refer to closed loop (or endless) belt drives.

A pulley connected to a power source pulls the belt from the other (driven) pulley, thereby developing tension T_1 (tight side) in the belt. While passing over the driving pulley, some tension is lost in overcoming the friction between the pulley and the belt and the tension in the slack side reduces to T_2.

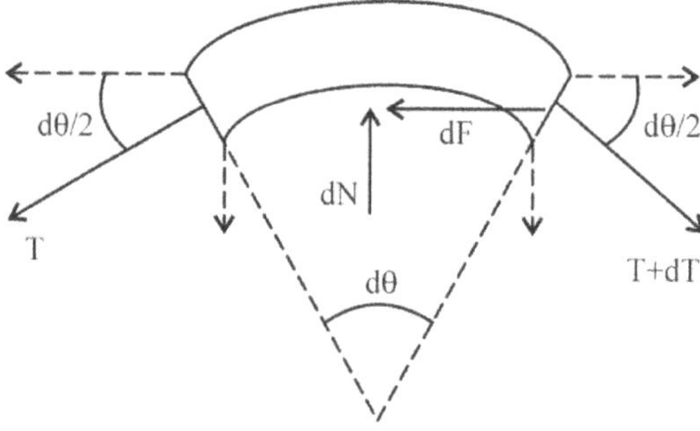

Assuming clockwise rotation of the driven pulley, tension in a small segment of the belt, covering wrap angle of $d\theta$, is assumed to vary from T on the slack side to T+dT on the tight side. These forces along with normal reaction dN and frictional resistance dF are considered concurrent. Assuming the segment of belt is symmetric about a vertical line, the tensions along the tangent to the pulley are considered inclined at an angle $d\theta/2$ from the horizontal. Relationship between them can be found by considering force equilibrium equations.

$$\Sigma F_X = (T + dT) \times \cos(d\theta/2) - T \times \cos(d\theta/2) - dF = 0$$

$\Rightarrow$ $dF = dT \times \cos(d\theta/2) \approx dT$ since $\cos(d\theta/2) \approx 1$ for small angles of $d\theta$

$$\Sigma F_Y = dN - (T + dT) \times \sin(d\theta/2) - T \times \sin(d\theta/2)$$

$\Rightarrow$ $dN \approx 2T \times \sin(d\theta/2)$ neglecting $dT \sin(d\theta/2)$, a product of small terms

$\Rightarrow$ $dN \approx 2T\,(d\theta/2) \approx T\,d\theta$ for small values of $d\theta$

Therefore, $dT = dF = \mu\,dN = \mu\,T\,d\theta$ or $dT/T = \mu\,d\theta$

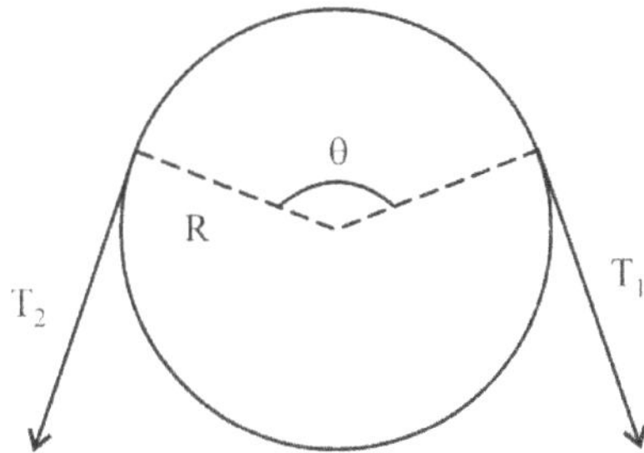

Integrating, $\log(T_1/T_2) = \mu\,\theta$ or $T_1/T_2 = e^{\mu\,\theta}$

where, θ is the wrap angle over which the belt is in contact with the pulley

A rotating mass of belt is subjected to a radially outward **centrifugal force,**

$$F_C = M \times V^2/r = (m\,rd\theta) \times V^2/r = m\,d\theta \times V^2$$

where, 'm' is the mass of the belt per unit length

and 'V' is the linear velocity of the belt

which tends to increase the belt length, thereby resulting in additional tension T_C in the belt on either side.

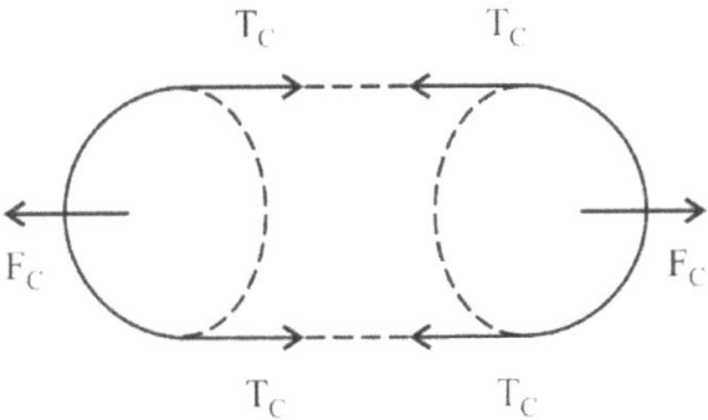

Then, resolving forces vertically,

$$F_C = T_C \times \sin(d\theta/2) + T_C \times \sin(d\theta/2)$$

$$\approx 2T_C\,(d\theta/2) = T_C\,d\theta = m \times d\theta\,V^2 \quad \text{or} \quad T_C = mV^2$$

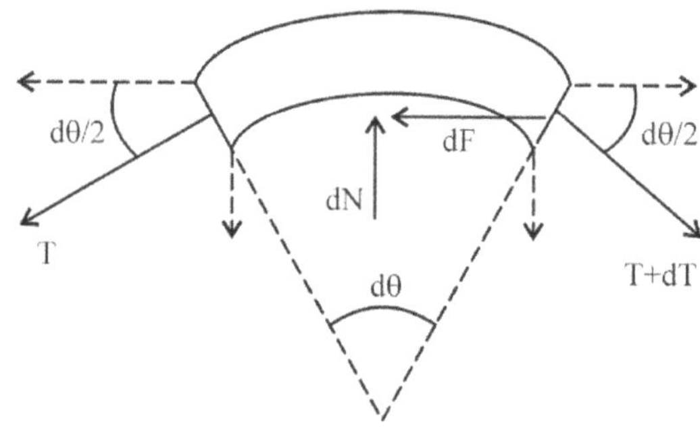

Total belt tensions are given by

$$T_1' = T_1 + T_C \quad \text{and} \quad T_2' = T_2 + T_C$$

Then, in terms of T_C, $T_1/T_2 = (T_1' - T_C) / (T_2' - T_C) = e^{\mu\theta}$

Limiting value of belt tension is given by $T_1' = T_{max} = \sigma\, b\, t$

where σ is the allowable tensile stress

b is the width and t is the thickness of belt

Power, $P = 2\pi NT / 60$ = Torque (T) × Angular velocity (ω)

$$= [(T_1 - T_2) R]\, \omega = (T_1 - T_2) V$$

where, T_1 and T_2 are the belt tensions on the tight side and slack side

T = torque on the pulley (assuming tensions T_1 and T_2 to be parallel)

$V = R\,\omega$ is Linear velocity of the pulley at the outer radius

and $\omega = 2\pi N / 60$ rad/sec, corresponding to speed of the pulley (N rpm)

$$P = (T_1 - T_2) V = (T_1 - T_1/e^{\mu\theta}) V = (1 - 1/e^{\mu\theta}) (T_1' - T_C) V$$

$$= (1 - 1/e^{\mu\theta}) (T_{max} - mV^2) V$$

For maximum power transmission,

$$dP/dV = [\, 1 - (1/e^{\mu\theta})\,] \times (T_{max} - 3mV^2) = 0$$

Since 1st term can not be zero,

$$T_{max} = 3\, m\, V^2 = 3\, T_C \quad \text{or} \quad V = \sqrt{(T_{max} / 3m)}$$

Since $V_S = V_B$ and $V = \pi DN/60$, in the absence of slip

speed ratio, $N_S/N_B = D_B/D_S$ neglecting thickness of belt

$= (D_B + t)/(D_S + t)$ if thickness of belt is also considered

$= (D_B/D_S) \times (1 - s/100)$ with a slip of s% on one pulley

$= [(D_B + t)/(D_S + t)] \times (1 - s/100)$ considering belt thickness and slip

Reaction on the bearing of the pulley,

$$R = (T_1 + T_2) \times \cos\alpha \approx T_1 + T_2 \text{ for small } \alpha$$

There may be situations where the belt or rope is used to get mechanical advantage, due to friction over the pulleys, in supporting a load and the two ends of the belt drive are not joined. All the equations derived above are applicable for these drives also. If a belt or rope is passing over a series of pulleys, similar to the figure shown, then tension at the driving end of the rope is given by

$$\frac{W}{T} = \frac{W}{T_1} \times \frac{T_1}{T_2} \times \frac{T_2}{T} = e^{(\mu\theta)1} \times e^{(\mu\theta)2} \times e^{(\mu\theta)3} = e^{\Sigma\mu\theta}$$

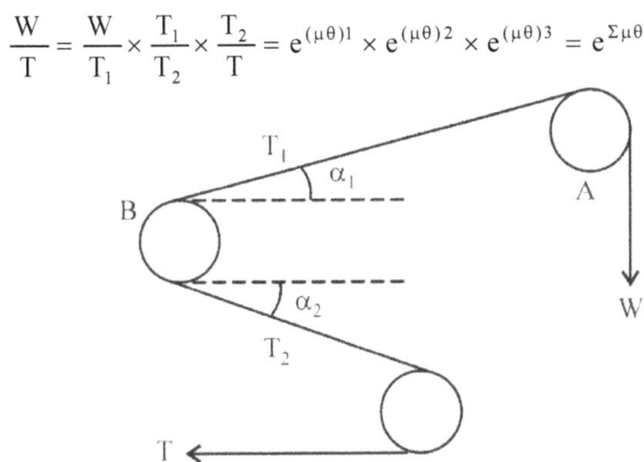

Thus, tension T required at one end of the drive to support the load W is very low

A continuous belt will have uniform speed throughout its length. If a belt passes over two pulleys of radii R_1 and R_2 and the two pulleys rotate at speeds of N_1 and N_2 rpm, then

$$\text{Velocity} = 2\pi R_1 \times (N_1 / 60) = 2\pi R_2 \times (N_2 / 60)$$

or **speed reduction**, $n_r = N_2/N_1 = R_1/R_2$

If a large speed reduction is proposed using a belt drive, radius of the slower pulley will be very large and not feasible for manufacture and operation. In such cases, **multi-stage speed reduction** is usually adopted. This is also called **compound belt drive.** Then, the final speed reduction is a product of speed reduction in each stage and, so, the radii of the pulleys in each stage are within practical limits. For example, in a 3 stage speed reduction

$$n_r = \frac{N_1}{N_6} = (n_r)_1 \times (n_r)_2 \times (n_r)_3 = \frac{N_1}{N_2} \times \frac{N_3}{N_4} \times \frac{N_5}{N_6} = \frac{R_2}{R_1} \times \frac{R_4}{R_3} \times \frac{R_6}{R_5}$$

Here, $N_2 = N_3$; $N_4 = N_5$ as the pairs of pulleys are connected to the same shaft.

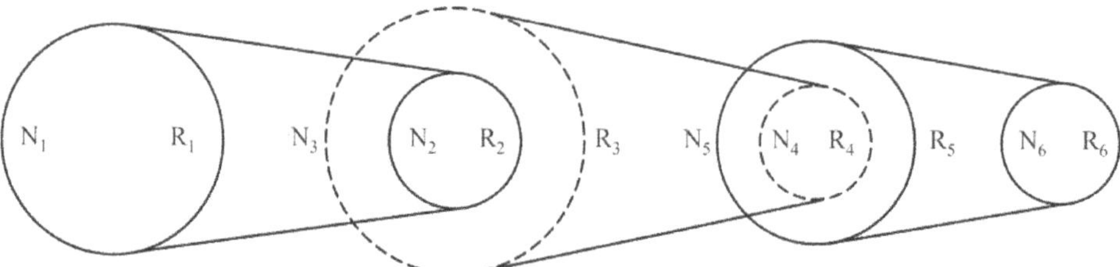

5.7.3 SPECIAL FEATURES OF BELT DRIVES

(a) V-belt: Between engine and radiator in automobiles etc, V-belt with trapezoidal cross section is used for larger torque / power transmission. It comes in closed form of fixed lengths and the shafts have to be located properly to match their length. They can only be used on pulleys with matching grooves. They have minimum slip and can not be used on conical or stepped shafts.

(b) Loose and Fast pulleys : It is often necessary to keep the driven shaft idle while the driving shaft is running continuously. This is achieved by using fast and loose pulleys. Fast pulley is one, which is keyed to the driving shaft and rotates the driven shaft. Loose pulley is one, which rotates freely without transmitting power or rotary motion to the driven shaft. Sliding bar with belt forks is used to shift the belt from fast pulley to loose pulley when the driven shaft is to be stopped and from loose pulley to fast pulley when the driven shaft is to be connected. This is not possible with V-belts.

(c) Crowning of pulleys : Slip of flat belt from pulley, due to any lateral force, can be prevented by crowned pulleys, whose rim is convex. This crowning helps in automatic adjustment of the belt position so that the belt runs centrally. Height of crown varies from 25mm to 70mm per meter width of the pulley. This is not applicable for V-belts.

(d) Initial belt tension : Every belt is subjected to initial tension (T_0) when it is stretched over a pair of pulleys at rest. To ensure transfer of motion, certain minimum initial tension, related to the tensions on tight side and slack side when it is rotating over the pulleys, need to be provided. Stretch on the tight side due to this initial tension is equal to contraction on the slack side of the belt, so that C $(T_1 - T_0)$ = C $(T_0 - T_2)$, where C is a constant depending on the type of belt material.

Thus, Minimum $T_0 = (T_1 + T_2)/2$

$= (T_1 + T_2 + 2T_C)/2$

if centrifugal tension is also considered

(e) **Creep of belt :** Length of the belt increases due to belt tension for long time and is a function of belt material and operating conditions (such as temperature). This decreases tension in the belt and increases slip of the belt over the pulley adversely affecting power transmission between the driving shaft and the driven shaft. Utilisation of the belt can be prolonged either by periodical adjustment of distance between the driving shaft and the driven shaft (as in automobiles) or by the use of an idler ***jockey pulley*** pressing the belt on the slack side near the smaller pulley.

Example 17

If the centers of three pulleys A, B and C are located at (10,8), (4,6) and (5,2), and rope passes over these pulleys to lift a load of 500 N, calculate tension (T) in the rope at the free end of the rope. Assume coefficient of friction at all the pulleys as 0.3.

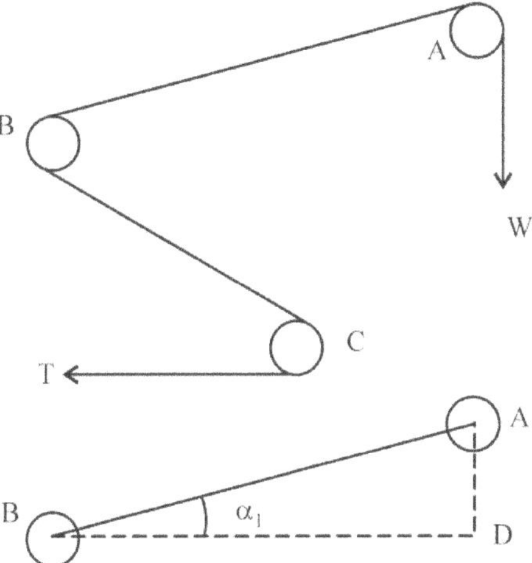

Solution:

Since the diameters of the pulleys are not given, the rope can be assumed to be passing over points A, B and C, representing pulleys of very small dimensions.

Let α_1 and α_2 be the angles made by BA and BC with the horizontal.

Then, $\tan \alpha_1 = AD / BD = (Y_A - Y_B) / (X_A - X_B) = (8 - 6)/(10 - 4)$

$= 1/3 \quad \Rightarrow \alpha_1 = 0.322$ radians (in 1st quadrant)

Similarly, $\tan \alpha_2 = EC / BE = (Y_C - Y_B) / (X_C - X_B) = (2 - 6)/(5 - 4)$

$= -4 \quad \Rightarrow \alpha_2 = 1.326$ radians (in 4th quadrant)

Since opposite angles are equal, angle of inclination of rope at A and C are also α_1 and α_2, as shown

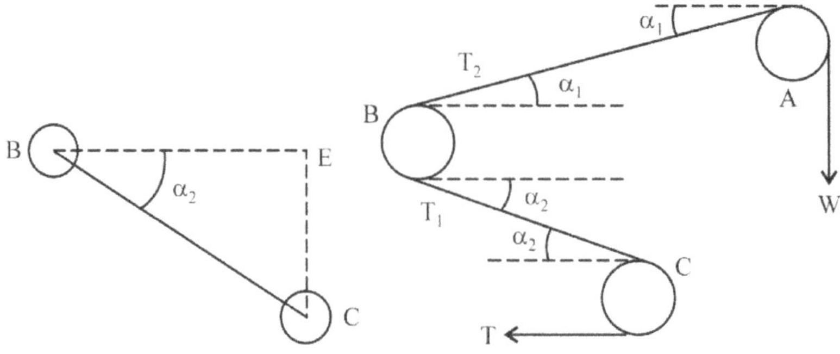

Wrap angle at A, $\theta_1 = (\pi/2) + \alpha_1 = 1.893$ radians

Since AB and BC are inclined at α_1 and α_2 to the horizontal at B,

angle of wrap at B, $\theta_2 = \pi - \alpha_1 - \alpha_2 = 1.495$ radians

Wrap angle at C, $\theta_3 = \pi - \alpha_2 = 1.817$ radians

Tension in the rope after pulley C, required to lift load W (500N) applied at the end of rope before pulley A, is given by

$$T = W\, e^{\Sigma \mu \theta} = 500 \times e^{0.3\,(1.893 + 1.495 + 1.817)} = 500 \times e^{1.5615}$$

$$= 500 \times 4.766 = 2383\text{N}$$

Note that T is very large compared to W, since it loses significant force for overcoming frictional resistance over the three pulleys.

Example 18

A vertical belt drive transmits a torque of 300 Nm and passes over a horizontal pulley of 1 m diameter. Calculate tensions in the tight side and slack side of the belt, assuming that the belt is in contact with the pulley over half perimeter of the pulley and $\mu = 0.3$. Also, calculate reaction on drum bearing.

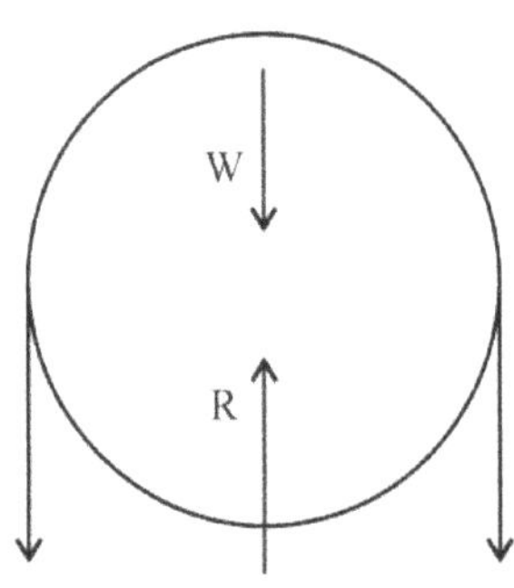

Solution:

Since the belt is in contact with the pulley over half perimeter of the pulley,

angle of wrap, $\theta = \pi$ radians

Tight side and slack side belt tensions are given by

$T_1/T_2 = e^{\mu\theta} = e^{0.3 \times \pi} = 2.57$

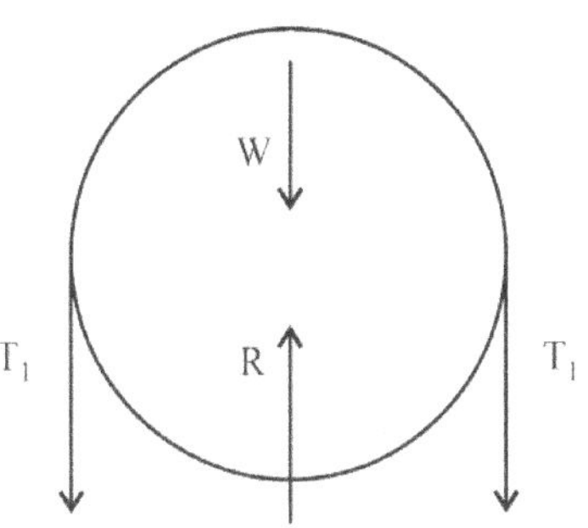

Torque = $(T_1 - T_2)\,(d/2)$

or 300 Nm $= (T_1 - T_2) \times (1/2)$

$\Rightarrow T_1 - T_2 = 300 \times 2 = 600$ N

Therefore, $T_1 = 982$ N and $T_2 = 382$ N

Reaction on drum bearing, $R = T_1 + T_2 + W = 1564$ N

Example 19

Two pulleys of 450 mm and 200 mm diameter are on parallel shafts 1.95 m apart and are connected by a belt. Calculate length of the belt and power transmitted in open belt and crossed belt arrangements, if the larger pulley rotates at 200 rpm; max. permissible belt tension is 1 kN and $\mu = 0.25$

Solution :

(a) Open belt drive:

$$\sin\alpha = (r_1 - r_2) / C = (0.225 - 0.1) / 1.95$$

$$\Rightarrow \quad \alpha = 0.0641 \text{ rad}$$

Angles of wrap, $\theta_1 = \pi + 2\alpha = 3.27$ rad on larger pulley

$\theta_2 = \pi - 2\alpha = 3.01$ rad on smaller pulley

Length of belt, $L = r_1\,\theta_1 + r_2\,\theta_2 + 2C\cos\alpha = 4.929$ m

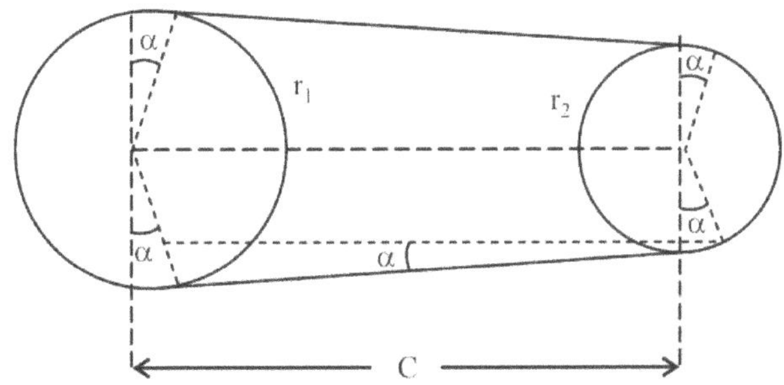

Ratio of belt tensions, $T_1/T_2 = e^{\mu\theta} = 2.265$ and $T_1 = T_{max} = 1000$ N

Therefore, $T_2 = 1000/2.265 = 441.5$ N

Power, $P = (T_1 - T_2) \times (2\pi\, r_1\, n_1 / 60)$

$= (1000 - 441.5) \times 2\pi \times 0.225 \times 200 / 60$

$= 2632\text{W}$ or 2.632kW

(b) Crossed belt drive :

$\text{Sin}\ \alpha = (r_1 + r_2) / C = (0.225+0.1) / 1.95$

$\Rightarrow$ $\alpha = 0.1675$ rad

Angles of wrap, $\theta_1 = \theta_2 = \pi + 2\alpha = 3.477$ rad

Length of belt, $L = \theta_1\,(r_1 + r_2) + 2C\cos\alpha = 4.98$ m

Ratio of belt tensions, $T_1/T_2 = e^{\mu\theta} = 2.385$ and $T_1 = T_{max} = 1000$ N

Therefore, $T_2 = 1000/2.385 = 419$ N

Power, $P = (T_1 - T_2) \times (2\pi\, r_1\, n_1 / 60)$

$= (1000 - 19) \times 2\pi \times 0.225 \times 200 / 60$

$= 2738\text{W}$ or 2.738kW

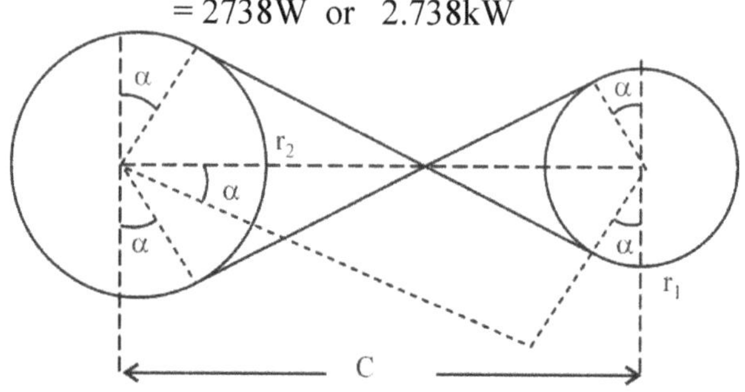

Example 20

A rotating wheel of 50cm diameter is braked by a belt AB attached to the lever ABC, which is hinged at B. The coefficient of friction between belt and wheel is 0.5. Find braking moment M exerted by a vertical weight of 100N. Take AB = 25cm and BC = 50cm. Angle of wrap, $\theta = \pi + (\pi/4)$.

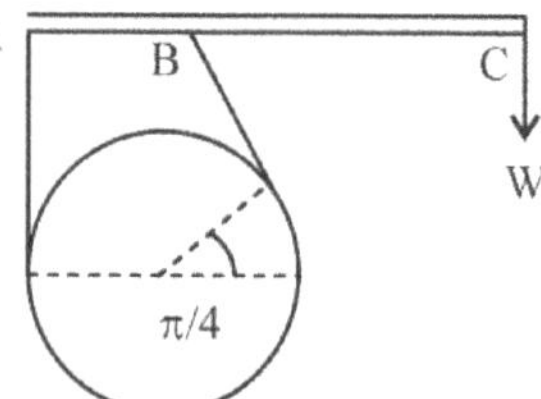

Solution:

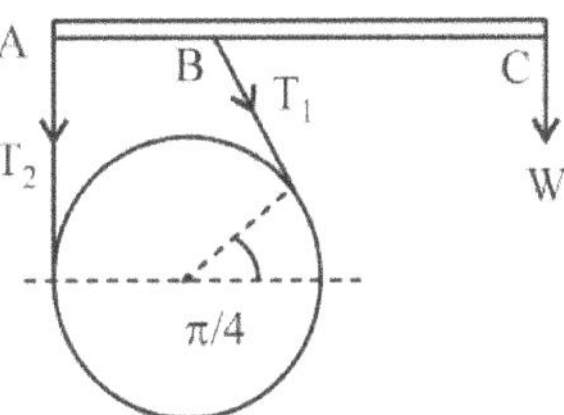

Braking moment is applied on the rotating wheel by the belt tension, which depends on the suspended weight W for moment balance of the static lever ABC.

Ratio of belt tensions, $T_2/T_1 = e^{\mu\theta} = 7.12$

or $T_1 = T_2/7.12$

For equilibrium of the lever ABC, taking moments about B,

$$AB \times T_2 = BC \times W$$

$\Rightarrow$ $25\ T_2 = 50 \times 100$ or $T_2 = 200N$

Braking moment, $M = (T_2 \times r - T_1 \times r) = T_2 \times [\ 1 - (1/7.12)\ ] \times r$

$= 200 \times (6.12/7.12) \times 25$ Ncm

$= 42.98$ Nm

SUMMARY

1. Four Laws of friction

    (a) Frictional force 'F' is the resistance to movement between two bodies in contact; Acts parallel to the surface of contact and opposite to the direction of motion

    (b) Frictional force is just sufficient to prevent the body from moving. i.e., as the load increases, frictional force increases, upto a limit

    (c) Limiting frictional resistance F_{max} bears a constant ratio with the normal reaction, 'N' (***Coulomb's law of dry friction***) and is independent of area of contact.

    $F_{max}\ \alpha\ N$ or $F_{max} = \mu_S N$

    where, $\mu = \tan \alpha_{max}$ is called ***coefficient of friction***

    and α_{max} is called the ***angle of friction***

    (d) When motion takes place, kinetic frictional resistance 'F_k' will be less than the limiting value. $F_k < F_{max}$

2. **Cone of friction** – Line of action of resultant force R for a body moving on a flat surface lies on the surface of an inverted cone, for various directions of motion, with a semi-vertex angle α.

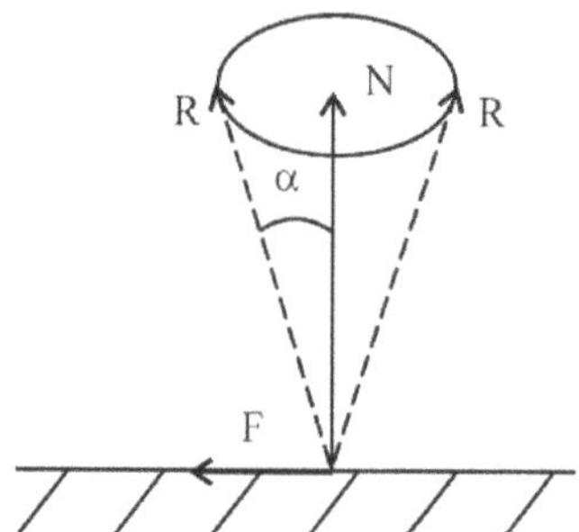

3. ***Angle of repose* ‘θ’** – It is the maximum angle of inclined plane at which a body placed on the inclined plane starts sliding down due to its own weight. $\theta = \alpha_{max}$

4. Whenever two blocks tend to move one over the other, normal reaction and frictional resistance from block A to block B will be equal and opposite to those from block B to block A, to ensure force equilibrium at the contact point between the blocks

5. A screw jack is used to raise or lower load along the axis of the screw when an effort is applied tangential to the screw. The screw can be considered as a plane inclined at an angle ‘α’, given by $\tan \alpha = L/\pi d$, where ‘d’ is the mean diameter of the screw, ‘L’ ($= n \times p$) is the lead of the screw for multi-start (n) threads and ‘p’ is the pitch of the screw threads.

6. Tangential load ‘P’ to be applied on the screw to raise a load of W,

$$\mathbf{P = W \tan (\alpha + \varphi)}$$

where φ is the angle of friction

Torque required to raise the weight W, $T = P \times (d/2)$

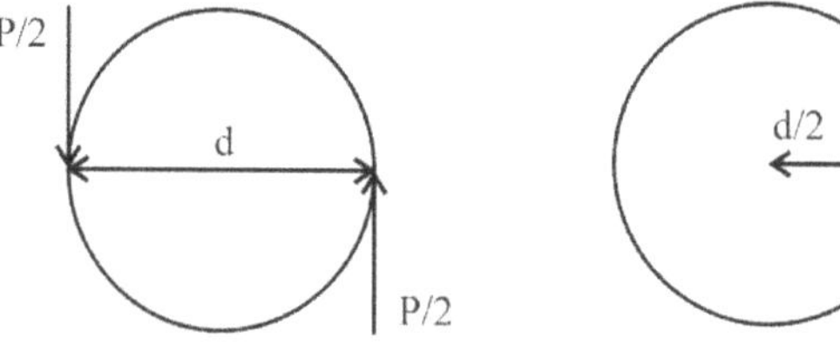

Effort required at the end of a lever of length R,

$$P' = T/R = P \times (r/R)$$

7. **Ideal effort**, when there is no friction ($\varphi = 0$),

$$P_0 = W \tan \alpha$$

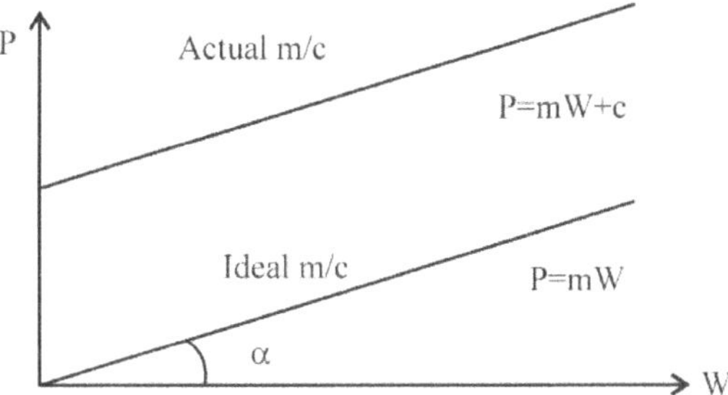

Efficiency, η = Ideal effort/Actual effort

$$= P_0 / P = \tan \alpha / \tan (\alpha + \varphi)$$

Max. efficiency (i.e., $d\eta/d\alpha = 0$) is obtained when

$$\alpha = (\pi/4) - (\varphi/2) \quad \text{Then,} \quad \eta_{max} = (1 - \sin \varphi) / (1 + \sin\varphi)$$

Law of machine $P = m W + c$, where, m and c are constants

For ideal machine, $P = P_0 = W \tan \alpha = m W$

8. **Mechanical Advantage (MA)** = Load moved / Effort applied = W / P

Velocity Ratio (VR)

= Distance moved by effort / Distance moved by load = $2\pi r / L$

Efficiency, η = MA / VR = $(W/P) / (2\pi r / L)$

$$= W \tan \alpha / P = \tan \alpha / \tan (\alpha + \varphi)$$

9. If a screw remains in the raised position without any effort (frictional resistance is sufficient to prevent downward sliding of load) is called ***self-locking screw***

10. A self-locking screw requires effort P_1 to lower the load 'W' when desired

$$P_1 = W \tan (\varphi - \alpha) \quad \text{if} \quad \alpha < \varphi$$

11. A non self-locking screw needs an effort P_2 to remain in its elevated position

$$P_2 = W \tan (\alpha - \varphi) \quad \text{if} \quad \alpha > \varphi$$

12. Input = Work done and Output = Work done – Work lost in friction

For a self-locking screw in which motion is irreversible,

Output < Work lost in friction $\Rightarrow$ Efficiency, $\eta < 1/2$ or 50%

13. **Differential screw jack** consists of two screws A and B. The screw A is connected to the effort 'P' and screw B is connected to the load 'W'. One rotation of load arm will move screw A upwards by L_A while at the same time, screw B moves down by L_B. Net movement of the screw for one rotation of the load arm = $L_A - L_B$.

Therefore, Velocity Ratio (VR) = $2\pi R / (L_A - L_B)$

Efficiency, $\eta = MA / VR = (W/P) / [2\pi R / (L_A - L_B)]$

14. Belt drive, compared to a gear drive or chain drive, is a cheaper method of transferring rotary motion or power from one shaft to another, over long distances - open ended belt drives are less common while closed loop (or endless) belt drives are most common and are discussed here. The angle over which the belt is in contact with the pulley is called wrap angle 'θ'.

15. In **open belt arrangement**, both the pulleys rotate in the same direction. Wrap angle,

$$\theta = \pi + 2\alpha > 180^0 \text{ on larger pulley}$$

and $$\theta = \pi - 2\alpha < 180^0 \text{ on smaller pulley}$$

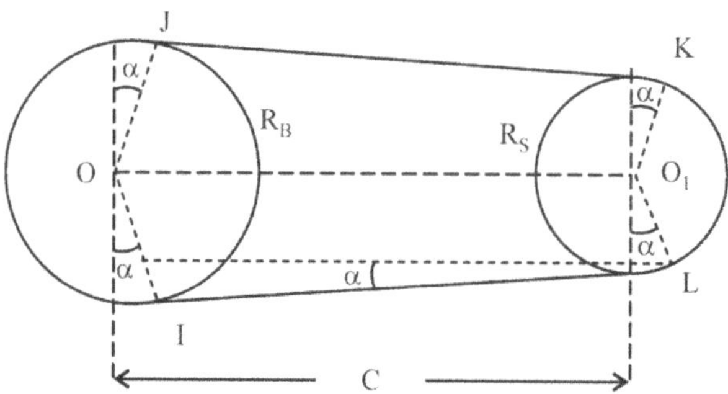

Length of belt,

$$L \approx (\pi + 2\alpha) R_B + (\pi - 2\alpha) R_S + 2\, C \cos \alpha$$

In **crossed belt arrangement**, both the pulleys rotate in opposite directions. Wrap angle,

$$\theta = \pi + 2\alpha > 180^0 \text{ on both pulleys}$$

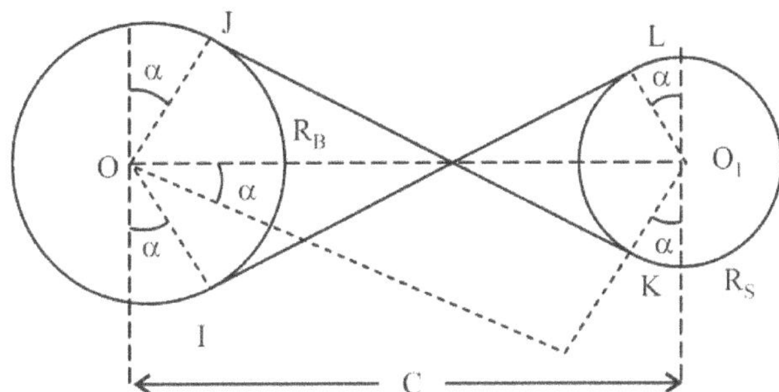

Length of belt,

$$L \approx (\pi + 2\alpha) \times (R_B + R_S) + 2\ C \cos\alpha$$

16. Due to friction between pulley & belt, tension changes from T_1 on tight side to T_2 on slack side given by $T_1/T_2 = e^{\mu\,\theta}$

    where, μ is the coefficient of friction and θ is the wrap angle

17. Additional tension on each side due to centrifugal force, $T_C = mV^2$

    where, m is the mass per unit length or mass density x area of cross section of belt

18. Total belt tensions are given by $T_1' = T_1 + T_C$ and $T_2' = T_2 + T_C$

    Limiting value of belt tension is given by $T_1' = T_{max} = \sigma\ b\ t$

    Where, σ is the allowable tensile stress of belt material

    b and t are the width and thickness of belt cross section

19. Power, P = Torque (T) × Angular velocity (ω)

    $= [(T_1 - T_2)\ R]\omega = (T_1 - T_2)\ V$ assuming T_1 and T_2 are parallel

    where, $V = R\ \omega$ is Linear velocity of the pulley at the outer radius

    and $\omega = 2\pi N / 60$ rad/sec, corresponding to speed of the pulley (N rpm)

20. For maximum power transmission, $dP/dV = 0$

    $\Rightarrow \quad T_{max} = 3\ m\ V^2 = 3\ T_C \quad$ or $\quad V = \sqrt{(T_{max} / 3m)}$

21. Since $V_S = V_B$ and $V = \pi DN/60$, in the absence of slip,

    Speed ratio, $N_S/N_B = D_B/D_S$ neglecting thickness of belt

    $= (D_B + t)/(D_S + t)$ if mean diameters (with thickness) are considered

    $= (D_B/D_S) \times (1 - s/100)$ with a slip of s% on one pulley

    $= (D_B + t)/(D_S + t) \times (1 - s/100)$ considering belt thickness and slip

22. Reaction on the bearing of the pulley, $R = (T_1 + T_2) \cos \alpha \approx T_1 + T_2$ for small α

23. In a **multi-stage** or **compound belt drive.** final speed ratio is a product of speed ratio (or ratio of radii of the pulleys) in each stage. For example, in a 3-stage drive,

$$n_x = (n_r)_1 \times (n_r)_2 \times (n_r)_3 = (N_1/N_2) \times (N_3/N_4) \times (N_5/N_6)$$

$$= (R_2 / R_1) \times (R_4/R_3) \times (R_6/R_5)$$

and $N_2 = N_3$; $N_4 = N_5$ as the pairs of pulleys are connected to the same shaft

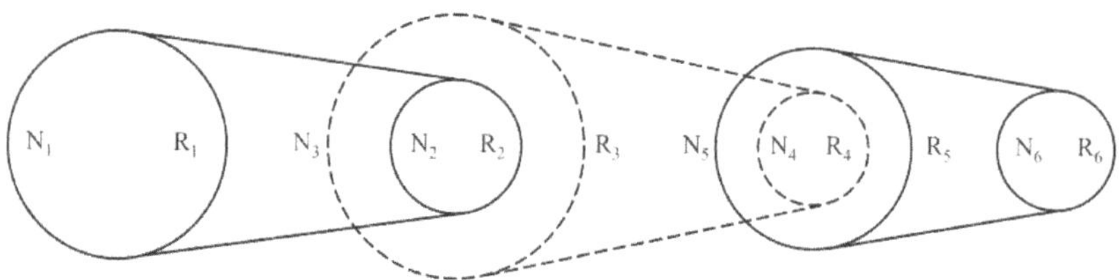

24. Belt tensions in a open ended belt drive are used as a brake to stop a rotating shaft

PROBLEMS FOR PRACTICE

1. Two equal bodies A and B of weight 'W' each, connected by a light string, are placed on a rough inclined plane. If $\mu_A = ½$ and $\mu_B = 1/3$, show that the bodies will be both on the point of motion when the plane is inclined at $\tan^{-1}(5/12)$.

2. A 15^0 wedge is used to move a block of 1000 N on a horizontal surface, as shown. Find the force P required to be applied on the wedge. Angle of friction between all contact surfaces is 14^0.

(***Ans:*** 232N)

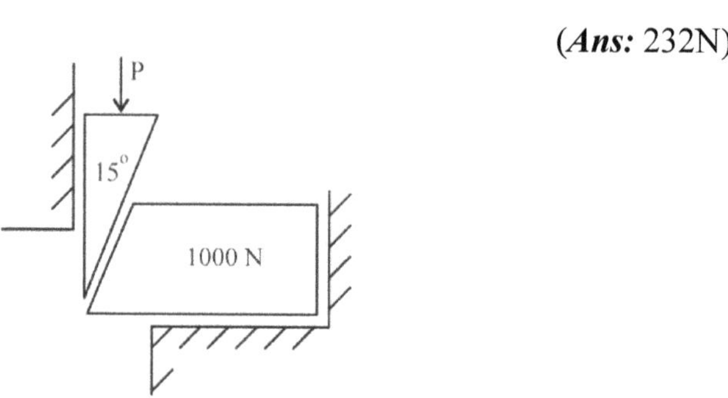

3. A load of 2500 N is to be raised by a screw jack with mean diameter of 7.5 cm and pitch of 12 mm. Find efficiency if $\mu = 0.075$

   (*Ans:* 40.3%)

4. A leather belt 9mm x 250mm is used to drive a cast iron pulley 900 mm in dia at 336 rpm. If angle of wrap is 120^0 and the stress on tight side is 2 MPa, find power transmitted by the belt. Mass density of belt material is 980 kg.m^3 and $\mu = 0.35$ (*Ans:* 32.47kW)

5. A flat belt is required to transmit 30 kW power from a pulley of 1.5 m dia running at 300 rpm. Angle of contact is spread over 11/24 of the circumference. Determine width of the belt, taking into centrifugal tension. Assume thickness of belt 9.5 mm ; mass density of belt material 1100 kg/m^3 and permissible tensile stress of belt material 2.5MPa.

   (*Ans:* 122.5mm)

6. An open belt runs between two pulleys 400 mm and 150 mm dia and their centers are 1000 mm apart. If μ for the larger pulley is 0.3, what should be μ on the smaller pulley, to avoid slipping.

   (*Ans:* 0.352)

7. A belt 100 mm wide and 10 mm thick is transmitting power at 1000 m/min. Belt tension on tight side is 1.8 times the tension on the slack side. If allowable tensile stress of belt material is 1.6 N/mm^2, calculate power transmitted at this speed. Also calculate maximum power that can be transmitted and the speed at which it is transmitted. Assume mass of belt material 0.56 kg/m.

   (*Ans:* 10.68kW, 14.49kW)

8. A braking system has a lever of 1.2m, hinged at A. It is used to stop a wheel of radius 0.6m and mass 200kg rotating at 200rpm, by a load of 100N applied at its free end C. If the point of contact of lever with the wheel is at a distance of 0.4m from the hinged end and coefficient of friction between the lever and the wheel is 0.25, calculate

   (a) braking torque and

   (b) number of revolutions made by the wheel before coming to rest.

   (*Ans:* 45Nm, 55.87 revolutions)

9. A body of weight 500 N is connected to a rope passing over semicircular cylinder. Other end of the rope makes two turns on a vertical capstan and is pulled by a force P. (a) Calculate minimum amount of force required to keep the body in equilibrium. (b) Calculate force required to lift the body upwards. Assume $\mu = 0.3$ at all contact surfaces.

(***Ans:*** 47.67N, 5275N)

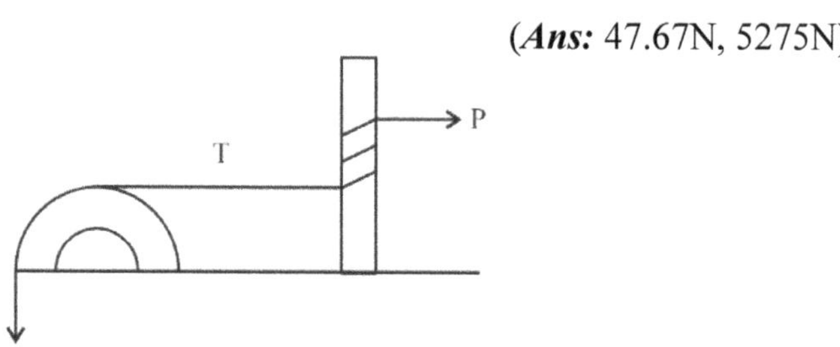

10. A 6m ladder weighing 300N is placed against a vertical wall at an angle of 60^0 with the horizontal. The ladder is about to slip when a person weighing 700N reaches a point 3.7m from the lower end of the ladder. Determine the coefficient of friction between the ladder and the floor if the coefficient of friction between the ladder and the wall is 0.23

11. If the coefficient of friction is 0.32 between the belt and the fixed drum shown in figure, find the pull P necessary to move 150N block upwards.

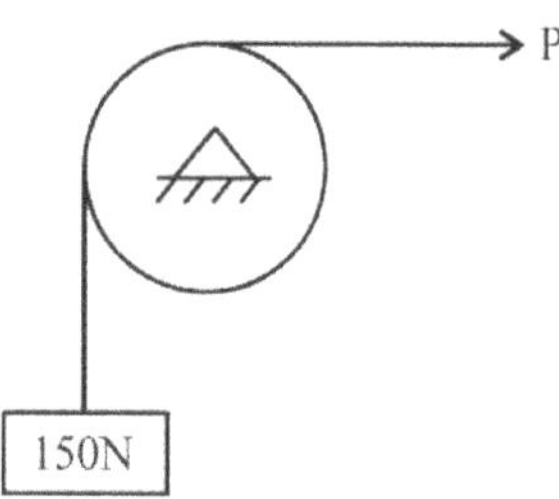

12. Two blocks A and B of weight 100N and 300N respectively are resting on a rough inclined plane as shown. Find the value of angle α when the block B is about to slide. Take coefficient of friction between the two blocks as well as between block B and the inclined plane as 0.25

(***Ans :*** 22.62^0)

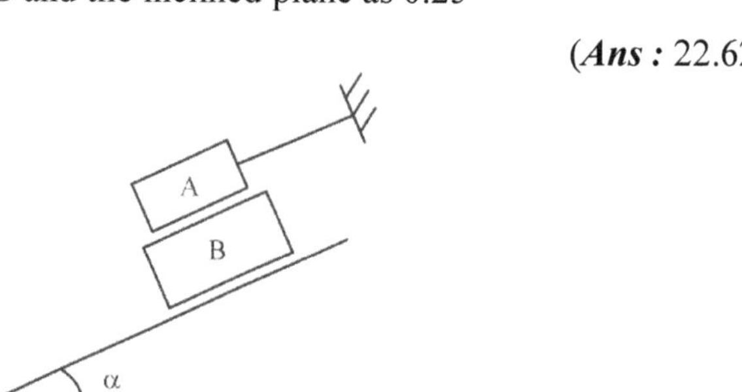

13. Two blocks A and B are connected by a horizontal rod and supported on two rough planes as shown. The coefficient of friction for block A is 0.4 while the angle of friction for block B is 20^0. Find the smallest weight of the block A for which equilibrium can exist, if weight of block B is 2250N.

(*Ans :* 4725N)

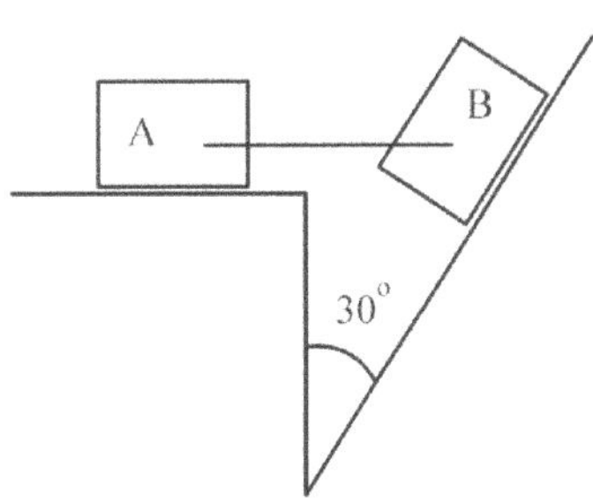

14. A spherical solid shell weighs 980N. It rests between two mutually perpendicular surfaces. The coefficient of friction between shell and surfaces is 0.3. What force applied tangentially at the top towards the vertical surface will cause the shell to slip.

(*Ans :* 434.78N)

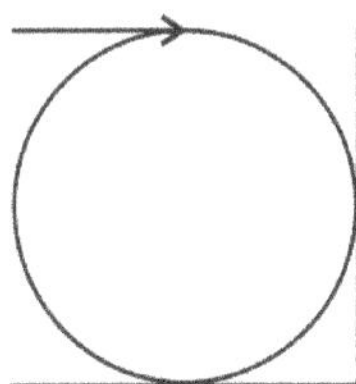

15. A homogeneous cylinder of radius 'r' and weight 'W' rests on horizontal floor in contact with vertical wall. If the coefficient of friction for all contact surfaces be 'μ', determine the couple 'M' acting on the cylinder, which will start counter-clockwise rotation.

(*Ans :* μ r W$(1 + \mu) / (1 + \mu^2)$)

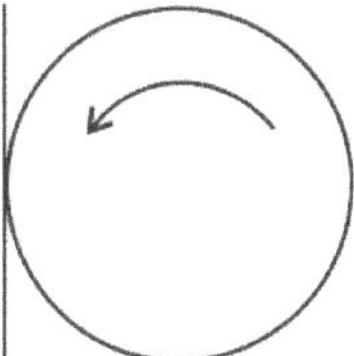

CHAPTER 6

CENTROID

6.1 CENTROID, CENTER OF MASS & CENTER OF GRAVITY

Every physical object has its properties, such as weight, distributed throughout. In any analysis involving combination of many such objects, it is very difficult to consider their distributed properties. The analysis is simplified by considering each such object as a point, where the distributed property is assumed concentrated. ***Centroid,*** usually represented by G, is defined as the point where the entire physical property is assumed to be concentrated to give the same first moment as that obtained by considering distributed property of the body. These points for length, area and volume are usually called ***centroids*** while these points for the distribution of mass and gravitational force are called ***center of mass*** and ***center of gravity*** respectively.

The distance 'r_G' of centroid 'G' of a line element is given by the moment equation as

$$L \times r_G = (\int r \times dL)$$

where, $L = \int dL$ is the length of the given line

$$\Rightarrow \quad r_G = (\int r \times dL) / L$$

It can also be expressed in terms of cartesian coordinates in the form

$$X_G = (\int X\, dL) / L ; \quad Y_G = (\int Y\, dL) / L ; \quad Z_G = (\int Z\, dL) / L$$

Similarly, the position vectors of area and volume elements are given by

$$r_G = (\int r\, dA) / A$$

or $$X_G = (\int X\, dA) / A ; \quad Y_G = (\int Y\, dA) / A ; \quad Z_G = (\int Z\, dA) / A$$

and $$r_G = (\int r\, dV) / V$$

or $$X_G = (\int X\, dV) / V ; \quad Y_G = (\int Y\, dV) / V ; \quad Z_G = (\int Z\, dV) / V$$

L, A and V for the entire element can be expressed as $\int dL$, $\int dA$ and $\int dV$ respectively

The centroid may or may not lie on the element itself. This method of locating centroid of an element is called ***moment method***. It is used here to locate centroid of some simple lines, areas and volumes.

An ***important property of centroids*** applicable to all line and area elements, is -

Distance of centrod from X-axis, $Y_G = 0$,

for elements <u>symmetric about X-axis</u>

and Distance of centrod from Y-axis, $X_G = 0$,

for elements <u>symmetric about Y-axis</u>

For volume elements,

Distance of centroid from X-axis, $Y_G = Z_G = 0$,

for elements <u>symmetric about X-axis</u>

Similarly, $Z_G = X_G = 0$ ($X_G = Y_G = 0$)

for elements <u>symmetric about Y-axis</u> (Z-axis)

It is thus obvious that for solids of revolution <u>about X-axis</u> , $Y_G = Z_G = 0$ and so on

6.1.1 CENTROIDS OF COMPOSITE BODIES

There are many situations in which a particular component can be treated as a combination of different simple elements. Centroid of such combinations can be calculated by using the centroid coordinates of the simpler elements, which together make up the desired component. The combination may include ***addition and/or deletion of simpler elements***. Integrals used in the case of continuously distributed property are replaced by the algebraic addition/deletion of the property of individual elements. A few examples will illustrate practical application of this approach.

6.2 CENTROID OF CIRCULAR ARCS

Let us consider a circular arc of radius 'R' and subtending angle $-\alpha$ to $+\alpha$

Consider a small segment of the arc at an angle 'θ' and subtending an angle '$d\theta$' at the origin. Length of this segment, $dL = R \times d\theta$

Center of this segment 'C' is located at '$R \sin \theta$' and '$R \cos \theta$' from X and Y axes

Then, $X_C = (\int x \times dL) / (\int dL) = (\int R\cos\theta \times R\, d\theta) / (\int R\, d\theta)$

$= (R^2 \sin\theta)/(R\,\theta) = R \sin\theta / \theta$ for $-\alpha \le \theta \le +\alpha$ .....(6.1)

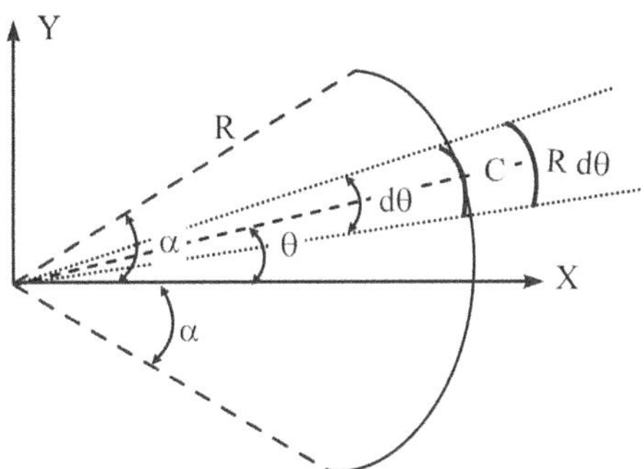

Similarly, $Y_C = (\int y \times dL) / (\int dL) = (\int R \sin\theta \times R\, d\theta) / (\int R\, d\theta)$

$= -(R^2 \cos\theta) / (R\,\theta)$

$= -R\cos\theta / \theta$ for $-\alpha \le \theta \le +\alpha$ .....(6.2)

For a semi-circular arc with limits of 'θ' equal to $-\pi/2$ and $+\pi/2$,

$X_C = [R \sin\theta / \theta] = R\,[\sin(\pi/2) - \sin(-\pi/2)] / [(\pi/2) - (-\pi/2)]$

$= 2R/\pi$

and $Y_C = [-R\cos\theta/\theta] = -R\,[\cos(\pi/2) - \cos(-\pi/2)] / [(\pi/2) - (-\pi/2)] = 0$

These values are tabulated here for a few simple cases.

Case	Line	Limits of θ	X_C	Y_C
1	Full circle (Sym about X & Y axes)	0 to $+2\pi$	0	0
2	Semi circle (Sym about X-axis)	$-\pi/2$ to $+\pi/2$	$2R/\pi$	0
3	Semi circle (Sym about Y-axis)	0 to $+\pi$	0	$2R/\pi$
4	Quarter circle (Sym about X-axis)	$-\pi/4$ to $+\pi/4$	$2\sqrt{2}R/\pi$	0
5	Quarter circle	0 to $+\pi/2$	$2R/\pi$	$2R/\pi$

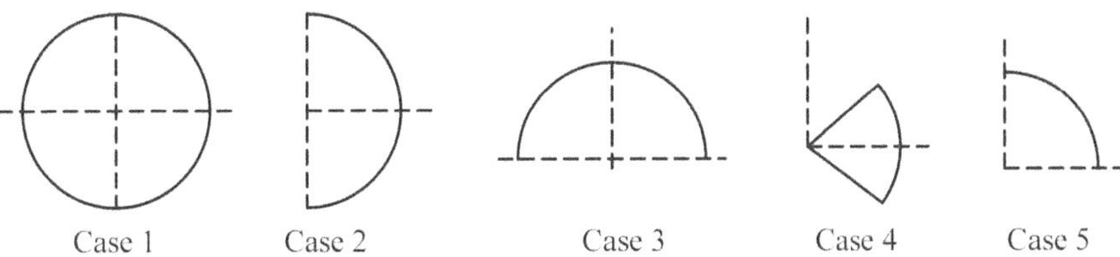

Important rule: If a body is suspended from any point on it, its _center of gravity lies on the vertical line through the point of support,_ so that weight through center of gravity and normal (upward vertical) reaction through the point of support balance each other and being collinear do not produce any unbalanced moment – thus satisfying all conditions of equilibrium

Example 1

Determine the length of a thin homogeneous wire which is bent into a semi-circular arc of radius R together with extensions on either end as shown, such that the centroid is located at the center of the arc, O.

Solution:

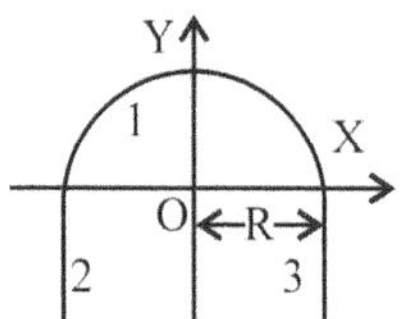

Since the geometry of the bent wire is symmetric about Y-axis, its centroid lies on the Y-axis. i.e., $X_G = 0$

Considering the bent wire as made up of three segments – semicircular ring and two straight lines, say each of length 'y' Then, length and coordinates of centroid of these segments are

$L_1 = \pi R$; $x_1 = 0$; $y_1 = 2R/\pi$

$L_2 = y$; $x_2 = -R$; $y_2 = -y/2$

$L_3 = y$; $x_3 = R$; $y_3 = -y/2$

These values may be more conveniently represented in a table

Segment	Length, L_i	X-coordinate of its centroid, X_i	Y-coordinate of its centroid, Y_i
1	πR	0	$2R/\pi$
2	y	–R	–y/2
3	y	–R	–y/2

X-coordinate of the centroid of the composite figure is given by

$X_G = [\pi R \times (0) + y \times (-R) + y \times R] / [\pi R + y + y] = 0$, as expected from symmetry

Note: Even though this result is obvious from the symmetry of the bent wire about Y-axis, such calculations help in cross checking correctness of entries of the table.

Y-coordinate of the centroid of the composite figure is given by

$Y_G = [L_1 \times Y_1 + L_2 \times Y_2 + L_3 \times Y_3]/[L_1 + L_2 + L_3]$

$= [\pi R \times (2R/\pi) + y \times (-y/2) + y \times (-y/2)] / [\pi R + y + y]$

$= [2R^2 - y^2/2 - y^2/2] / [\pi R + 2y] = [2R^2 - y^2] / [\pi R + 2y]$

$= 0$, since it should lie at the origin

Therefore, $2R^2 - y^2 = 0$ or $y = \sqrt{2}\ R$

Total length of the bent wire, $L = \pi R + y + y = (\pi + 2\sqrt{2})R = \boldsymbol{5.97\ R}$

Example 2

A homogenous wire ABC is bent and attached to the hinge at A. Find the length of the portion BC, for which BC is horizontal. AB = 25cm & inclined at 30^0 to BC.

Solution:

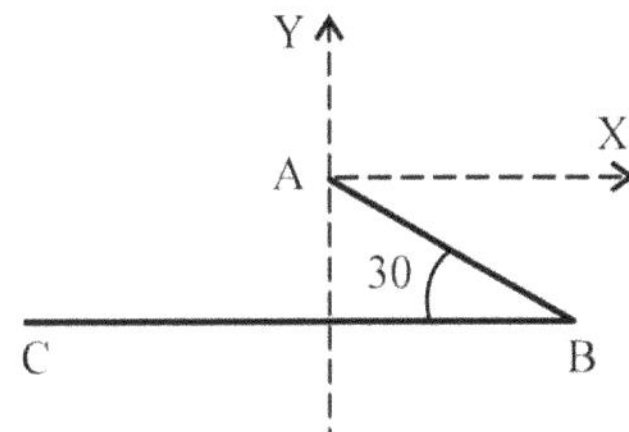

If the wire BC remains horizontal, X-coordinate of the centroid of the bent wire must lie on Y-axis. i.e., $X_G = 0$. Since the wire consists of two segments AB and BC of lengths L_{AB} and L_{BC},

$X_G = (L_{AB} \times X_{AB} + L_{BC} \times X_{BC}) / (L_{AB} + L_{BC})$

$= [25 \times \{(AB/2) \cos 30\} + L_{BC} \times \{AB \cos 30 - (BC/2)\}] / (25 + L_{BC})$

$= [25 \times (25/2) \times \left(\sqrt{3}/2\right) + L_{BC} \times \{25 \times \left(\sqrt{3}/2\right) - (L_{BC}/2)\}] / (25 + L_{BC})$

$= 0$

$\Rightarrow \quad 25 \times 25 \times \sqrt{3}/4 + 25 \times \sqrt{3}\ L_{BC}/2 - L_{BC}^2/2 = 0$

or $\quad 2L_{BC}^2 - 50\sqrt{3}\ L_{BC} - 625\sqrt{3} = 0$

$$L = 50\sqrt{3} \pm \sqrt{[(50\sqrt{3})^2 - 4 \times 2 \times 625\sqrt{3}]} / (2 \times 2) = 53.43$$

Example 3:

A wire is bent into a semicircular form of radius 'R' and suspended from one of its ends. Find the angle that the line joining the two ends of the wire makes with the vertical.

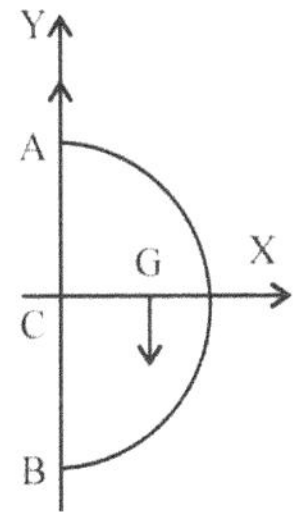

Solution:

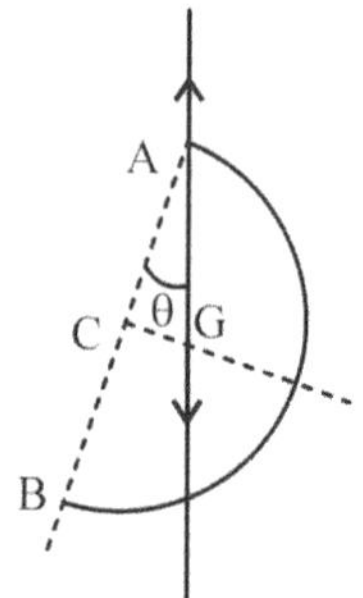

Let the bent wire be suspended from A. Then, weight of the wire acting downwards through the centroid 'G' and the vertical reaction (to balance weight of wire) through the hinge at A are not collinear. Hence, these two forces produce a clockwise moment. Coordinates of centroid of the semi-circular wire are given by

$$X_G = 2R/\pi \text{ and } Y_G = 0 \text{ (due to symmetry about X-axis)}$$

Equilibrium position is reached when the moment is zero i.e., the two forces are collinear, possible only by the rotation of the wire by an angle 'θ' such that centriod lies on Y-axis

Then, $\tan\theta = CG/AC = X_G/R = (2R/\pi)/R = 2/\pi = 0.6363$

or $\boldsymbol{\theta = 32.5^0}$

Example 4

A thin wire is bent in the form A-O-B-C, as shown, such that AO = OB = BC = a. Calculate coordinates of the centroid of the wire.

Solution:

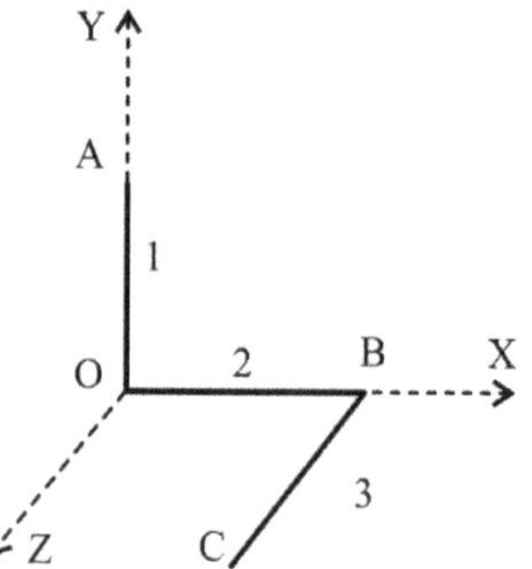

The wire consists of three segments, AO, OB and BC, each of length 'a' and parallel to one of the three coordinate axes X, Y or Z.

Length and centroid coordinates of these segments are tabulated as

Segment	Length, L_i	X-coordinate of its centroid, X_i	Y-coordinate of its centroid, Y_i	Z-coordinate of its centroid, Z_i
1 A-O	a	0	a/2	0
2 O-B	a	a/2	0	0
3 B-C	a	a	0	a/2

Then, $X_G = (L_1 \times x_1 + L_2 \times x_2 + L_3 \times x_3) / (L_1 + L_2 + L_3)$
$= (0 + a^2/2 + a^2) / 3a = \mathbf{a / 2}$

$Y_G = (L_1 \times y_1 + L_2 \times y_2 + L_3 \times y_3) / (L_1 + L_2 + L_3)$
$= (a^2/2 + 0 + 0) / 3a = \mathbf{a / 6}$

$Z_G = (L_1 \times z_1 + L_2 \times z_2 + L_3 \times z_3) / (L_1 + L_2 + L_3)$
$= (0 + 0 + a^2/2) / 3a = \mathbf{a / 6}$

Example 5

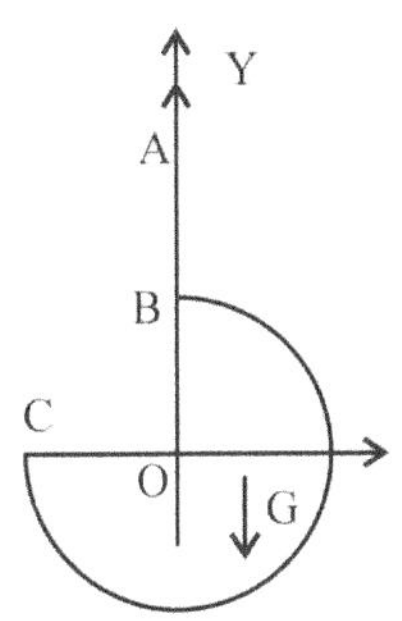

A wire is bent into the form shown below, such that AB = r and radius of the bent portion BC = r. Find the angle that the line AB makes with the vertical, when the bent wire is suspended from A.

Solution:

Weight of the wire acting downwards through the centroid 'G' and the vertical reaction (to balance weight of wire) through the hinge at A are not collinear. Hence, these two forces produce a clockwise moment. Equilibrium position is reached when the moment is zero i.e., the two forces are collinear, possible only by the rotation of the wire by an angle 'θ' such that centriod lies on Y-axis.

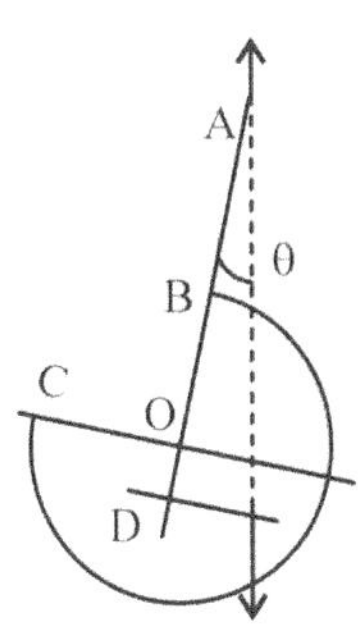

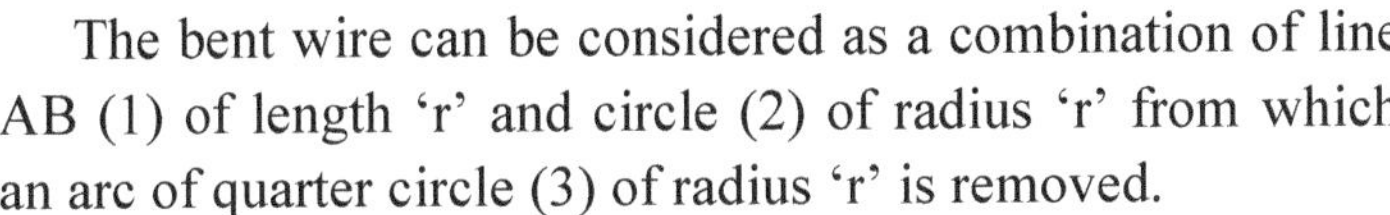

The bent wire can be considered as a combination of line AB (1) of length 'r' and circle (2) of radius 'r' from which an arc of quarter circle (3) of radius 'r' is removed.

Length and centroid coordinates of the three segments are given in the table as

Segment	Length, L_i	X-coordinate of its centroid, X_i	Y-coordinate of its centroid, Y_i
1 Line	r	0	OB + AB/2 = 3r/2
2 Circle	$2\pi r$	0	0
3 Arc	$\pi r/2$	$-2r/\pi$	$2r/\pi$

Then, coordinates of the centroid of the bent wire w.r.t. the origin O are given by

$$X_G = (L_1 \times x_1 + L_2 \times x_2 - L_3 \times x_3) / (L_1 + L_2 - L_3)$$

$$= [0 + 0 - (\pi r/2) \times (-2r/\pi)] / [r + 2\pi r - (\pi r/2)] = r^2 / [r \times (2 + 3\pi)/2]$$

$$= 2r / (2 + 3\pi) = 0.175\ r$$

$$Y_G = (L_1 \times y_1 + L_2 \times y_2 - L_3 \times y_3) / (L_1 + L_2 - L_3)$$

$$= [r \times (3r/2) + 0 - (\pi r/2) \times (2r/\pi)] / [r + 2\pi r - (\pi r/2)]$$

$$= [(3r^2/2) - r^2] / [r(2 + 3\pi)/2] = r / (2 + 3\pi) = 0.0875\ r$$

Inclination of line AB with the Y-axis is given by,

$\tan\theta = DG/DA = X_G / (DO + OB + BA) = X_G / (Y_G + r + r)$

$= (0.175\ r) / (2.0875\ r) = 0.0915$

$\Rightarrow$ $\boldsymbol{\theta = 5.23^0}$

6.3 CENTROID OF PLANE AREAS (ASSUMING UNIFORM THICKNESS AND DENSITY)

6.3.1 CENTROID OF A SECTOR OF A CIRCULAR PLATE

Consider a circular plate of radius 'R' and subtending angle $-\alpha$ to $+\alpha$

Let us consider a small segment of the plate at mean radius 'r', width 'dr', mean angle 'θ' and subtending an angle 'dθ' at the origin.

Area of this segment, $dA = (r\ d\theta) \times dr$

Center of this segment 'C' is located at '$r \sin\theta$' and '$r\cos\theta$' from X and Y axes

Then, $X_C = (\int x \times dA) / (\int dA) = (\int\int r\cos\theta \times r\,d\theta \times dr) / (\int\int r\,d\theta \times dr)$

for $0 \le r \le R$ and $-\alpha \le \theta \le +\alpha$

$= (R^3 \sin\theta / 3) / (R^2\,\theta / 2)$

$= 2R\sin\theta / 3\theta$ for $-\alpha \le \theta \le +\alpha$ .....(6.3)

Similarly, $Y_C = (\int y \times dA) / (\int dA)$

$= (\int r\sin\theta \times r\,d\theta \times dr) / (\int r\,d\theta \times dr)$

$= -(R^3 \cos\theta/3) / (R^2\,\theta/2)$

$= \boldsymbol{-2R\cos\theta / 3\theta}$ .....(6.4)

For a semi-circular plate with limits of 'θ' equal to $-\pi/2$ and $+\pi/2$,

$X_C = [2R\sin\theta / 3\theta]$ for $-\pi/2 \le \theta \le +\pi/2$

$= 2R[\sin(\pi/2) - \sin(-\pi/2)] / 3[(\pi/2) - (-\pi/2)] = 4R/3\pi$

and $Y_C = [-2R\cos\theta / 3\theta]$ for $-\pi/2 \le \theta \le +\pi/2$

$= -2R[\cos(\pi/2) - \cos(-\pi/2)] / 3[(\pi/2) - (-\pi/2)] = 0$

These values are tabulated here for a few simple cases.

Case	Plate	Limits of θ	X_C	Y_C
1	Full circle (Sym about X & Y axes)	0 to $+2\pi$	0	0
2	Semi circle (Sym about X-axis)	$-\pi/2$ to $+\pi/2$	$4R/3\pi$	0
3	Semi circle (Sym about Y-axis)	0 to $+\pi$	0	$4R/3\pi$
4	Quarter circle (Sym about X-axis)	$-\pi/4$ to $+\pi/4$	$4\sqrt{2}\,R/3\pi$	0
5	Quarter circle	0 to $+\pi/2$	$4R/3\pi$	$4R/3\pi$

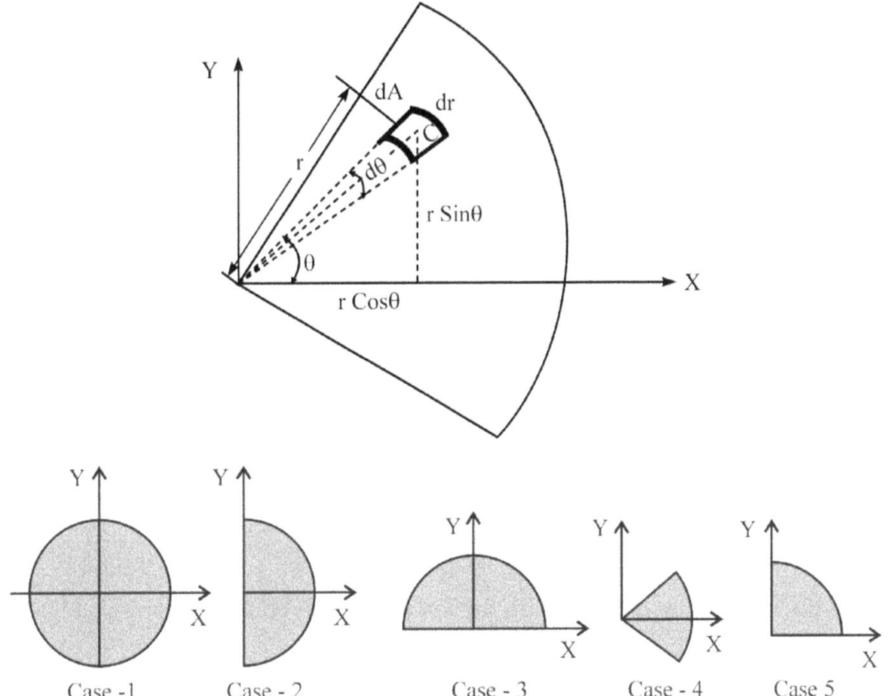

6.3.2 CENTROID OF OTHER PLANE AREAS

(i) **Triangle:** Let us consider a triangle of base 'b' and height 'h'. Its centroid can be calculated by two approaches.

Method-1

Let us consider a small elemental area of 'dA', identified by EF at a distance of 'y' from the base. Let BC = b; AD = h; GD = y and EF = x_1, which is function of y

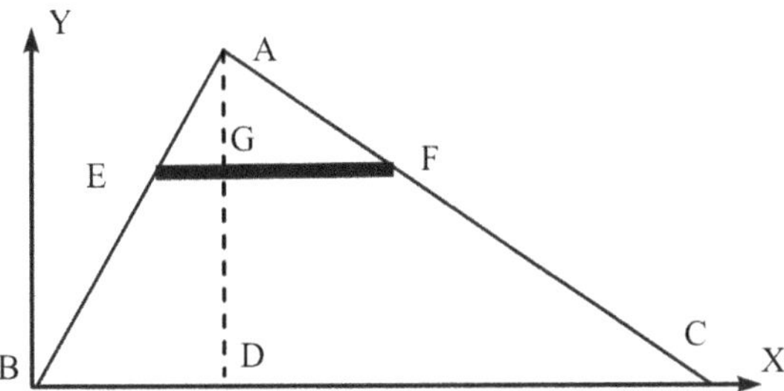

From similar Δs AEF and ABC, EF/BC = AG/AD or $x_1 / b = (h - y)/h$

and Area of the segment,

$dA = x_1 \times dy = [\, b \times (h - y)/h \,]\, dy$

Therefore $Y_C = (\int y \times dA) / (\int dA)$

$= (\int y \times [\, b \times (h - y)/h \,]\, dy) / (\int [\, b \times (h - y)/h \,]\, dy)$

$= [\, hy^2/2 - y^3/3] / [\, hy - y^2/2 \,]$ for $0 \leq y \leq h$

$= (h^3/6) / (h^2/2) = \boldsymbol{h/3}$ .....(6.5)

Method-2

Let us consider a small rectangular area dA, identified by its sides dx and dy and located at a distance of 'y' and 'x' from X and Y axes. Here, limits of x are functions of y

$$Y_C = (\int y \times dA) / (\int dA)$$

$$= (\int \int y \times dx \times dy) / (\int \int dx \times dy)$$

$$= [\int (\int dx) \times y\, dy] / [\int (\int dx)\, dy]$$

$$= [\int (x) \times y\, dy] / [\int (x)\, dy]$$

within the limits of x, identified by the lines AB and AC as

$$y = m_1 x + c_1 \quad \text{and} \quad y = m_2 x + c_2$$

and is more involved for a general case.

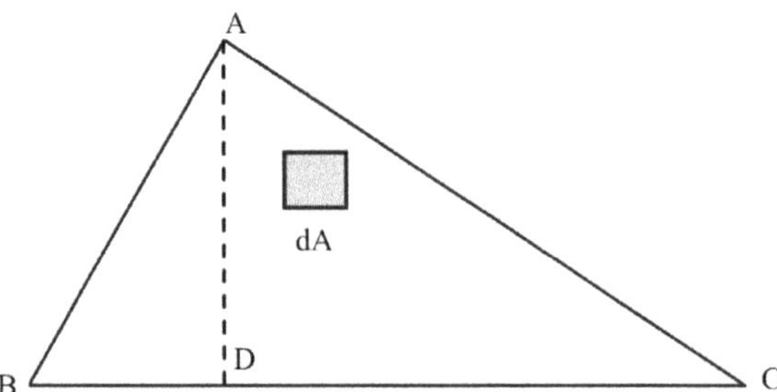

For the particular case of a right angled triangle with AB perpendicular to BC, the limits of x are 0 and $b \times (h - y)/h$

$$Y_C = [\int \{ (b (h - y) / h) - 0\} \times y\, dy] / [\int \{(b (h - y) / h) - 0\}\, dy]$$

$$= [hy^2/2 - y^3/3] / [hy - y^2/2] \quad \text{for } 0 \leq y \leq h$$

$$= (h^3/6) / (h^2/2) = \boldsymbol{h/3}$$

(ii) ***Parabola***: Let us consider a parabola, defined by $y^2 = k \times x$, and bounded by the X-axis and x = b line (parallel to Y-axis). Let us consider a small elemental area of 'dA', identified by EF at a distance of 'y' from the base.

Let BC = b ; AC = h ; FC = y and EF = x_1

Then, at point A on the parabola, $b = h^2/k$

and $x_1 = b - x_E = b - (y^2/k) = (h^2 - y^2) / k$

Area of the segment, $dA = x_1 \times dy$

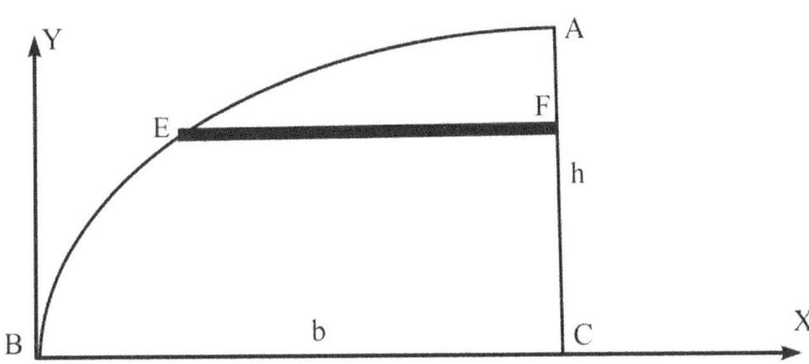

$$Y_C = (\int y \times dA) / (\int dA) = (\int y \times x_1 \times dy) / (\int x_1 \times dy)$$
$$= (\int y \times [(h^2 - y^2) / k] dy) / (\int [(h^2 - y^2) / k] dy)$$
$$= [h^2 y^2 / 2 - y^4/4] / [h^2 y - y^3/3] \quad \text{for } 0 \le y \le h$$
$$= (h^4 / 4) / (2h^3 / 3) = \boldsymbol{3h/8}$$

$$X_C = (\int x_2 \times dA) / (\int dA) \quad \text{where, } x_2 = (y^2/k) + x_1 / 2 = (h^2 + y^2) / 2k$$
$$= (\int [(h^2 + y^2) / 2k] \times [(h^2 - y^2) / k] dy) / (\int [(h^2 - y^2) / k] dy)$$
$$= [(h^4 y - y^5/5)/2k^2] / [(h^2 y - y^3/3)/k] \quad \text{for } 0 \le y \le h$$
$$= (4h^5 / 10 k^2) / (2h^3 / 3k) = 3h^2 / 5k$$
$$= \boldsymbol{3b/5}$$

$$A = \int x_1 dy = \int \left[\left(h^2 y - y^3 / 3\right) / K\right]_0^h = 2h^3 / 3K = 2bh / 3$$

(iii) Parabolic *spandrel*: Let us consider a parabolic spandrel, defined by $y = k \times x^n$, and bounded by the X-axis and '$x = b$' line (parallel to Y-axis).

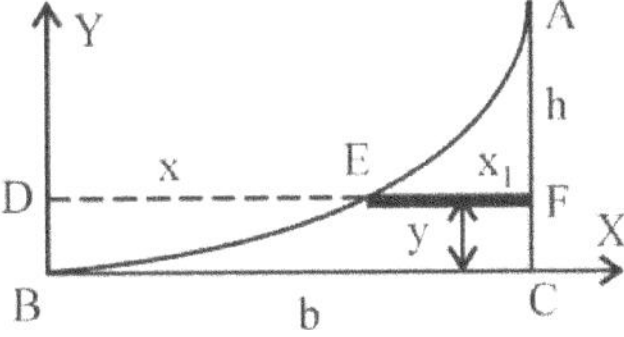

Let us consider a small elemental area of 'dA', identified by EF at a distance of 'y' from the base.

Let BC = b; AC = h; FC = y and EF = $x_1 = b - x$

Then, $h = k\,b^n$; $dA = x_1\,dy$ and $dy = n\,k\,x^{n-1}\,dx$

$\mathbf{X_G} = (\int x_2 \times dA) / (\int dA)$

where,

x_2 = distance of the center of EF from Y-axis = DE + (EF/2)

$= x + (x_1/2) = x + (b - x)/2 = (b + x)/2$

and $dA = x_1 \times dy = (b - x) \times n\,k\,x^{n-1}\,dx = nk \times (b\,x^{n-1} - x^n)\,dx$

$= (\int [(b + x)/2] \times [nk \times (b\,x^{n-1} - x^n)\,dx]) / (\int nk \times (b\,x^{n-1} - x^n)\,dx)$

$= (\int [(b^2\,x^{n-1} - b\,x^n + b\,x^n - x^{n+1}) / 2]\,dx) / (\int [(b\,x^{n-1} - x^n]\,dx)$

$= (\int [(b^2\,x^{n-1} - x^{n+1}) / 2]\,dx) / (\int [(b\,x^{n-1} - x^n]\,dx)$

$= [(b^2\,x^n / 2n) - x^{n+2} / 2(n+2)] / [(b\,x^n / n) - x^{n+1}/(n + 1)]$

for $0 \le x \le b$

$= [(b^{n+2} / 2n) - b^{n+2} / 2(n + 2)] / [(b^{n+1} / n) - b^{n+1}/(n+1)]$

$= b^{n+2} \times [(n + 2) - n] / [2n(n + 2)] / [b^{n+1}\{(n + 1) - n\}/\{n(n + 1)\}]$

$= b\,[1/\{n \times (n + 2)\}] / [1/\{n \times (n + 1)\}]$

$\mathbf{= (n + 1) \times b / (n + 2)}$

Similarly, $\mathbf{Y_G} = (\int y\,dA) / \int dA = \int y \times (x_1\,dy) / \int x_1\,dy$

$\mathbf{= (n + 1) \times h / (4n + 2)}$

Same results can be obtained by taking a vertical strip of area $y \times dx$ at a distance x from B

For $y = k \times x^2$, $\boldsymbol{X_G = 3\,b / 4}$ and $\boldsymbol{Y_G = 3\,h / 10}$

Example 6

Calculate centroid of the area bounded by $y^2 = a \times x$ and $y = x / 2$

Solution :

Centroid location is calculated by two approaches, for highlighting the convenience of combination (addition / deletion) of simple shapes approach.

Method-1:

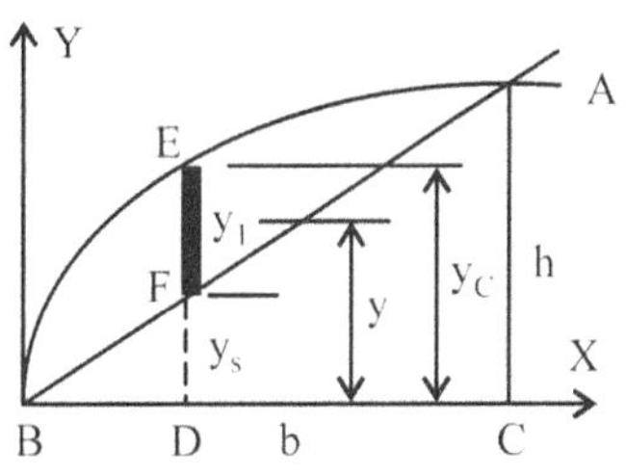

Let us consider the area B-E-A-F-B bounded by a parabola, defined by $y^2 = a \times x$, and the line $y = x/2$, starting from origin to the point of their intersection at A.

Let us consider a small elemental area 'dA', of height EF and thickness 'dx' at a distance of 'x' from Y-axis.

Let BC = b; AC = h; and EF = y_1

Substituting coordinates b and h in the equations of the curve and the straight line at the point of intersection A, $h^2 = a \times b$ and $h = b/2$

$$\Rightarrow \quad a = h^2/b = (b/2)^2 / b = b/4$$

Height of the elemental area, $y_1 = DE - DF = y_C - y_S = \sqrt{(ax)} - (x/2)$

Also, $y = y_1/2 + y_S = [\sqrt{(ax)} - (x/2)]/2 + (x/2) = [\sqrt{(ax)} + (x/2)]/2$

and Area of the segment, $dA = y_1 \times dx = [\sqrt{(ax)} - (x/2)] \times dx$

$$X_G = (\int x \times dA) / (\int dA) = (\int x \times y_1 \times dx) / (\int y_1 \times dx)$$

$$= (\int x \times [\sqrt{(ax)} - (x/2)]\, dx) / (\int [\sqrt{(ax)} - (x/2)]\, dx)$$

$$= [\sqrt{a}\, x^{5/2} / (5/2) - x^3 / 6] / [\sqrt{a}\, x^{3/2} / (3/2) - x^2/4]$$

within the limits $0 \le x \le b$

$$= [\sqrt{a}\, b^{5/2} / (5/2) - b^3 / 6] / [\sqrt{a}\, b^{3/2} / (3/2) - b^2/4]$$

$$= [b^3/5 - b^3/6] / [(2/3)\, b^2/2 - b^2/4] \quad \text{after substituting } a = b/4$$

$$= (b^3 / 30)[6-5] / (b^2 / 12)[4-3] = \boldsymbol{2b/5}$$

$$Y_G = (\int y \times dA) / (\int dA) = (\int y \times y_1 \times dx) / (\int y_1 \times dx)$$

$$= \left(\int\left[\sqrt{(ax)} + (x/2)\right] \times \left[\sqrt{(ax)} - (x/2)\right]dx\right) \Big/ \left(\int\left[\sqrt{(ax)} - (x/2)\right]dx\right)$$

$$= (1/2)\int\left[(ax) - (x/2)^2\right]dx \Big/ \left(\int\left[\sqrt{(ax)} - (x/2)\right]dx\right)$$

$$= (1/2)[ax^2/2 - x^3/12] / [\sqrt{a}\, x^{3/2} / (3/2) - x^2/4]$$

within the limits $0 \le x \le b$

$$= (1/2)[ab^2/2 - b^3/12] / \left[\sqrt{a} b^{3/2} / (3/2) - b^2/4\right]$$

$= (1/2)\ [\ b^3/8 - b^3/12\ ]\ /\ [\ (2/3)\ b^2/2 - b^2/4\ ]$ after substituting $a = b/4$

$= (1/2)\ [b^3\ (3-2)\ /\ 24]\ /\ [b^2\ (4-3)\ /\ 12] = b\ /\ 4 = \boldsymbol{h/2}$

Method-2:

Let us consider the same area B-E-A-F-B as a combination of area-1 under the curve $y^2 = a \times x$ from $x = 0$ to $x = b$ or 2h from which area-2 under the line $y = x/2$ within the same limits of x is subtracted. Centroid of this area is now obtained using the known positions of centroids of these two basic elements.

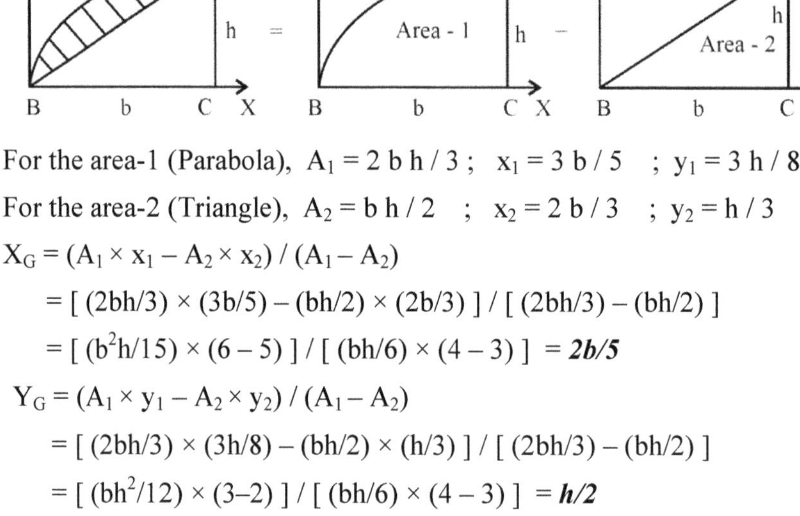

For the area-1 (Parabola), $A_1 = 2\ b\ h\ /\ 3$; $x_1 = 3\ b\ /\ 5$; $y_1 = 3\ h\ /\ 8$

For the area-2 (Triangle), $A_2 = b\ h\ /\ 2$; $x_2 = 2\ b\ /\ 3$; $y_2 = h\ /\ 3$

$X_G = (A_1 \times x_1 - A_2 \times x_2)\ /\ (A_1 - A_2)$

$= [\ (2bh/3) \times (3b/5) - (bh/2) \times (2b/3)\ ]\ /\ [\ (2bh/3) - (bh/2)\ ]$

$= [\ (b^2h/15) \times (6 - 5)\ ]\ /\ [\ (bh/6) \times (4 - 3)\ ] = \boldsymbol{2b/5}$

$Y_G = (A_1 \times y_1 - A_2 \times y_2)\ /\ (A_1 - A_2)$

$= [\ (2bh/3) \times (3h/8) - (bh/2) \times (h/3)\ ]\ /\ [\ (2bh/3) - (bh/2)\ ]$

$= [\ (bh^2/12) \times (3-2)\ ]\ /\ [\ (bh/6) \times (4 - 3)\ ] = \boldsymbol{h/2}$

Method-2 is more convenient, provided centroid positions of the basic elements are known. Method-1, by integration of elemental area, is applicable in all cases.

Example 7

Calculate centroid of a trapezium of height ‘h’ and parallel sides ‘a’ and ‘b’

Solution :

Let AB = a ; DC = b ; EF = h

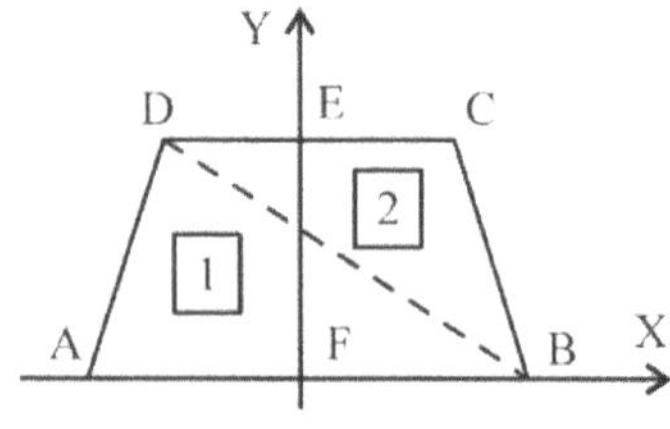

The trapezium can be considered as a combination of two triangles ABD and BCD. Since the trapezium is symmetric about Y-axis, $X_G = 0$

For ΔABD, $A_1 = ah/2$; $y_1 = h/3$

For ΔBCD, $A_2 = bh/2$;

$y_1 = 2h/3$ from AB (h/3 from side CD)

Then, $Y_G = [A_1 \times y_1 + A_2 \times y_2] / (A_1 + A_2)$

$= [(ah/2) \times (h/3) + (bh/2) \times (2h/3)] / [(ah/2) + (bh/2)]$

$= (a + 2b) \times (h^2/6) / [(a + b) \times (h/2)] = \boldsymbol{h(a + 2b) / 3(a + b)}$

Distance of centroid from CD = $h - Y_G$ = $\boldsymbol{h(2a + b) / 3(a + b)}$

Example 8

A rectangle of side 1.5m is kept on the smaller side of a trapezium of sides 3m and 1.5m and height 2m, as shown. Find height of the rectangle 'd', if centroid of the combined area lies on the joining plane.

Solution :

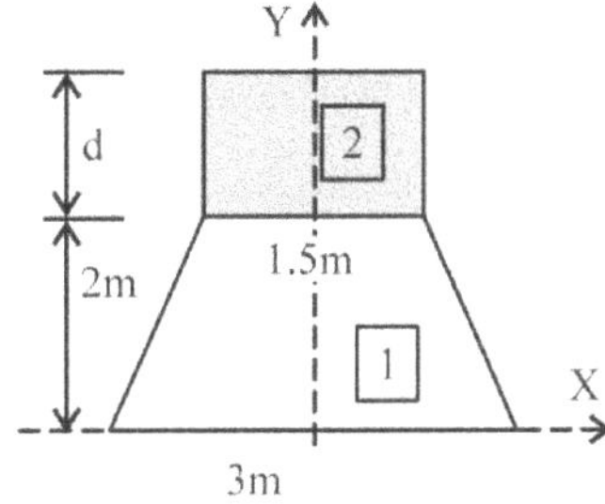

Given, a = 3m ; b = 1.5m and h = 2m

For the trapezium, $A_1 = (a + b)h/2$

$= (3 + 1.5) \times 2/2 = 4.5m^2$

$y_1 = h(a + 2b)/3(a + b)$

$= [2 \times (3 + 2 \times 1.5)] / [3 \times (3 + 1.5)] = 4/4.5m$

For the rectangle, $A_2 = 1.5d$; $y_2 = 2 + d/2$ from X-axis

Because of symmetry of the composite area about Y-axis, $X_G = 0$

When the centroid lies on the joining plane, $Y_G = 2m$

$Y_G = [A_1 \times y_1 + A_2 \times y_2] / (A_1 + A_2)$

$= [4.5 \times 4/4.5 + 1.5d \times (2 + d/2)] / [4.5 + 1.5d] = 2$

$\Rightarrow$ $4 + 3d + 0.75d^2 = 9 + 3d$ or $0.75d^2 = 5$

Therefore, $d = \sqrt{(5 / 0.75)} = 2.58m$

Alternatively, by considering the X-axis along the joining plane, areas of the two elements remain same while distances of element centroids from X-axis will change to

$y_1 = -[h(2a + b)] / [3(a + b)] = -5/4.5m$ (along –ve Y direction)

and $y_2 = d/2$ From this X-axis, $Y_G = 0$

$\Rightarrow$ $Y_G = [A_1 \times y_1 + A_2 \times y_2] / (A_1 + A_2) = 0 \Rightarrow [A_1 \times y_1 + A_2 \times y_2] = 0$

$\Rightarrow \quad 4.5\,(-5/4.5) + 1.5d\,(d/2) = 0 \quad \text{or} \quad 0.75d^2 = 5$

$\Rightarrow \quad d = \sqrt{(5/0.75)} = 2.58\text{m}$

Example 9

Calculate the position of the centroid of an L-section of base width 80 mm, height 100 mm and thickness of flange and web 20 mm.

Solution:

The L-section can be considered as a combination of two rectangles, as shown

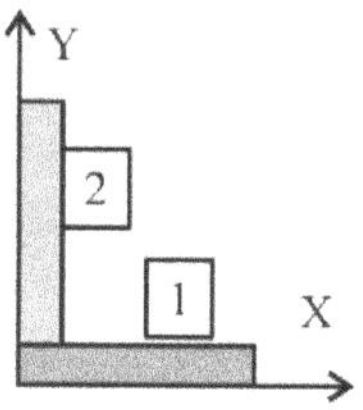

$A_1 = 80 \times 20 = 1600\ \text{mm}^2$

$x_1 = 80/2 = 40\text{mm}\ ;\ y_1 = 20/2 = 10\ \text{mm}$

$A_2 = 20 \times 80 = 1600\ \text{mm}^2$

$x_2 = 20/2 = 10\text{mm}\ ;\ y_2 = 20 + (80/2) = 60\text{mm}$

$X_G = [A_1 \times x_1 + A_2 \times x_2] / (A_1 + A_2) = (1600 \times 40 + 1600 \times 10) / (1600 + 1600)$

$= (1600 \times 50) / (1600 \times 2) = 25\text{mm}$

$Y_G = [A_1 \times y_1 + A_2 \times y_2] / (A_1 + A_2) = (1600\times10 + 1600\times60) / (1600+1600)$

$= (1600\times70) / (1600\times2) = 35\text{mm}$

Alternatively, the two rectangles can also be considered as shown here.

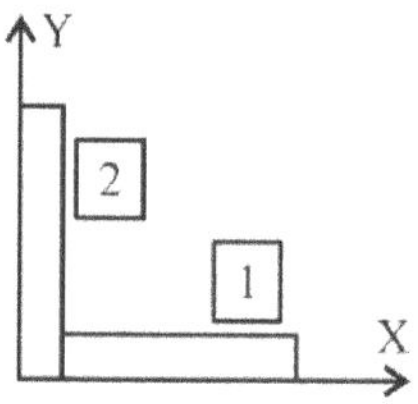

Correspondingly element areas and centroid coordinates change.

$A_1 = 60 \times 20 = 1200\ \text{mm}^2$

$x_1 = 20 + (60/2) = 50\text{mm};\ y_1 = 20/2 = 10\ \text{mm}$

$A_2 = 20 \times 100 = 2000\ \text{mm}^2;\ x_2 = 20/2 = 10\text{mm};\ y_2 = 100/2 = 50\text{mm}$

$X_G = [A_1 \times x_1 + A_2 \times x_2] / (A_1 + A_2) = (1200 \times 50 + 2000 \times 10) / (1200 + 2000)$

$= 80000 / 3200 = 25\text{mm}$

$Y_G = [A_1 \times y_1 + A_2 \times y_2] / (A_1 + A_2) = (1200\times10 + 2000\times50) / (1200+2000)$

$= 112000 / 3200 = 35\text{mm}$

Example 10

A quarter circle of radius 40mm and a right angled triangle of sides 120mm × 40mm are cut from a semicircle of radius 80mm as shown. Determine the angle AB makes with the vertical, when the remaining area is suspended from A.

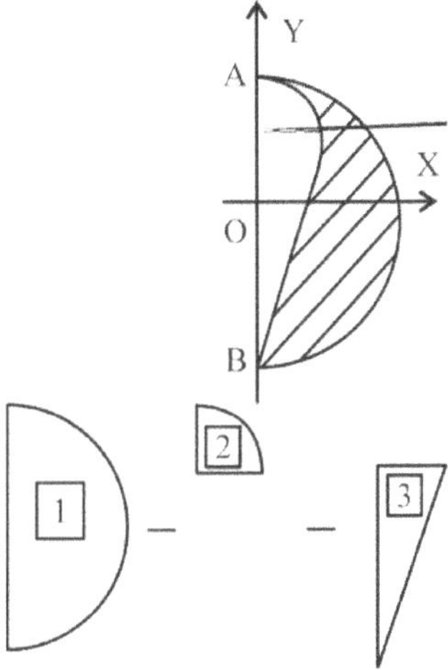

Solution :

The composite body consists of three areas – (1) a semicircle of radius R, (2) a quarter circle of radius r and (3) a triangle,

where, $R = 80mm$; $r = 40mm$; $b = 40mm$ and $h = 120mm$

For the area-1, $A_1 = \pi R^2/2$; $x_1 = 4R/3\pi$; $y_1 = 0$

For the area-2, $A_2 = \pi r^2/4$; $x_2 = 4r/3\pi$; $y_2 = (R - r) + 4r/3\pi = 40 + 4r/3\pi$

For the area-3, $A_3 = b\,h/2$; $x_3 = b/3$; $y_3 = 2h/3 - 80$

$$X_G = [A_1 \times x_1 - A_2 \times x_2 - A_3 \times x_3] / (A_1 - A_2 - A_3)$$

$$= [(\pi R^2/2) \times (4R/3\pi) - (\pi r^2/4) \times (4r/3\pi) - (bh/2) \times (b/3)] / [(\pi R^2/2) - (\pi r^2/4) - (bh/2)]$$

$= 45.0$ mm (represented by DG in the figure)

$$Y_G = [A_1 \times y_1 - A_2 \times y_2 - A_3 \times y_3] / (A_1 - A_2 - A_3)$$

$$= [(\pi R^2/2) \times 0 - (\pi r^2/4) \times (40 + 4r/3\pi) - (bh/2) \times (2h/3 - 80)] / [(\pi R^2/2) - (\pi r^2/4) - (bh/2)]$$

$= -11.2$ mm ($|Y_G| = 11.2$mm is represented by OD in the figure)

When the remaining area is suspended from A, AB tilts by an angle 'θ' with the vertical, in such a way that the centroid lies vertically below A. Then, weight of the plate acting through its centroid and reaction at A form a system of two equal, opposite and collinear forces, so as to keep the plate in equilibrium.

Then, $\tan\theta = DG/AD = DG/(AO + OD) = X_C/(R + |Y_G|)$

$$= 45/(80 + 11.2) = 0.493 \quad \text{or} \quad \boldsymbol{\theta = 26^0}$$

6.4 CENTROIDS OF SOLIDS OF REVOLUTION

An arc rotated about an axis generates a hollow body while an area rotated about an axis generates a solid body. While arcs and areas have two coordinates (X_C and Y_C) for their centroids, a solid body has three coordinates (X_C, Y_C and Z_C). *A body generated by rotating an area about Y-axis will have its centroid on the axis of rotation* i.e., $X_C = 0$ and $Z_C = 0$. Centroid of a solid body of revolution is calculated by integration of an elemental volume property, similar to the method used for calculating centroids of arcs or areas.

Example 11

Find centroid of a right circular cone of radius 'R' and height 'h'.

Solution:

The cone is generated by rotating a right angled triangle of sides 'R' and 'h' about the edge of length 'h'. Since the cone is symmetric about Y-axis, its centroid lies along the axis of revolution. Hence, X and Z coordinates of centroid are zero. Considering an elemental volume dV of a circular disc of radius r and thickness 'dy' at 'y' from the origin,

$$dV = \pi r^2\, dy \quad \text{and from similar } \Delta\text{s ABC \& ADE,} \quad r/R = (h - y)/h$$

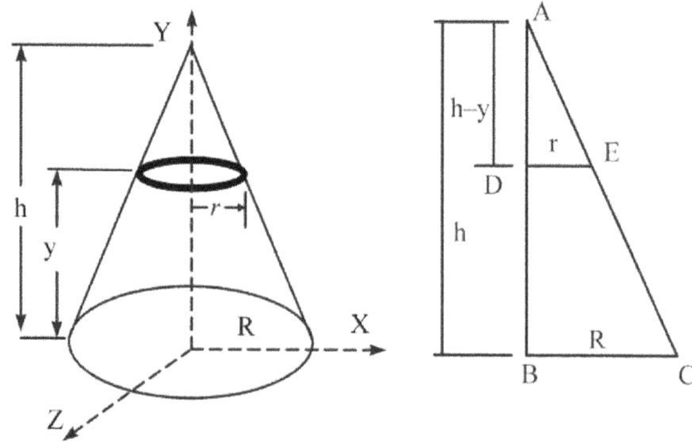

Then, $Y_C = \int y \times dV / \int dV = [\int y \times (\pi r^2 dy)] / [\int \pi r^2 dy]$

$= [\int y \times (h-y)^2 \times (R/h)^2 dy] / [\int (h-y)^2 \times (R/h)^2 dy]$ for $0 \le y \le h$

$= [h^2y^2/2 - 2hy^3/3 + y^4/4] / [h^2y - 2hy^2/2 + y^3/3]$

$= \boldsymbol{h/4}$

Volume of cone,

$V = \int \pi r^2 dy = \int \pi\{R(h-y)/h\}^2 dy$

$= (\pi R^2/h^2) \int (h^2 - 2hy + y^2) dy$

$= (\pi R^2/h^2) [h^2y - 2hy^2/2 + y^3/3]$ within the limits $0 \le y \le h$

$= (\pi R^2/h^2) [h^3/3] = \boldsymbol{\pi R^2 h/3}$

Example 12

Find centroid and volume of a **rectangular pyramid** from its base, if its length, width and height are L, B and H respectively.

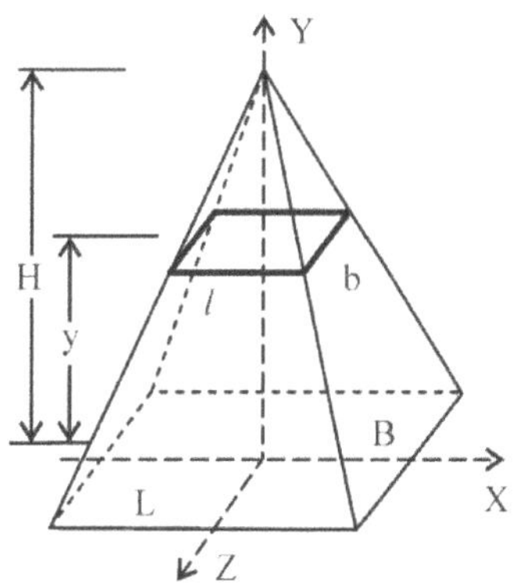

Solution:

Since the pyramid is symmetric about y-axis, its centroid lies along Y-axis. Hence, X and Z coordinates of centroid are zero. Considering an elemental volume dV of rectangle of sides 'ℓ' × 'b' and thickness 'dy' at a distance y from the origin, $dV = \ell \times b \times dy$

From similar Δs in X–Y plane, $\ell / L = (H-y) / H$

and From similar Δs in Z–Y plane, $b / B = (H-y) / H$

Then, $Y_G = \int y \times dV / \int dV$

$= [\int \ell \times b \times dy \times y] / [\int \ell \times b \times dy]$

$= [\int (L/H) \times (H-y) \times (B/H) \times (H-y) \times y \times dy] / [\int (L/H) \times (H-y) \times (B/H) \times (H-y) dy]$

$= [\int (H-y)^2 y \times dy] / [\int (H-y)^2 dy]$

$= [\int (H^2y - 2Hy^2 + y^3) \times dy] / [\int (H^2 - 2Hy + y^2) dy]$

$= [H^2y^2/2 - 2Hy^3/3 + y^4/4] / [H^2y - 2Hy^2/2 + y^3/3]$

within the limits $0 \le y \le H$

$= [H^4(6 - 8 + 3)/12] / [H^3(3 - 3 + 1)/3] = \boldsymbol{H/4}$

Volume of pyramid,

$$\mathbf{V} = \int \ell \times b \times dy = \int (L/H) \times (H - y) \times (B/H) \times (H - y)\, dy$$
$$= (L \times B/H^2) \times \int (H - y)^2\, dy = (L \times B/H^2) \times \int (H^2 - 2Hy + y^2)\, dy$$
$$= (L \times B/H^2) \times [H^2y - 2Hy^2/2 + y^3/3] \text{ within the limits } 0 \le y \le H$$
$$= (L \times B/H^2) \times [H^3(3 - 3 + 1)/3] = \mathbf{L\,B\,H / 3}$$

Example 13

Find centroid of a hemisphere of radius 'R'.

Solution:

The hemisphere is generated by rotating a quarter circle of radius 'R' about its vertical edge. Since the hemisphere is symmetric about Y-axis, its centroid lies along the axis of revolution. Hence, X and Z coordinates of centroid are zero. Considering an elemental volume dV of radius 'r' and thickness 'dy' at a distance of 'y' from the origin, $dV = \pi r^2 dy$. Here, r and y are related by $r^2 = R^2 - y^2$ from the triangle shown, since any point A on the periphery is at a distance of 'R' from the center of base or origin 'O'.

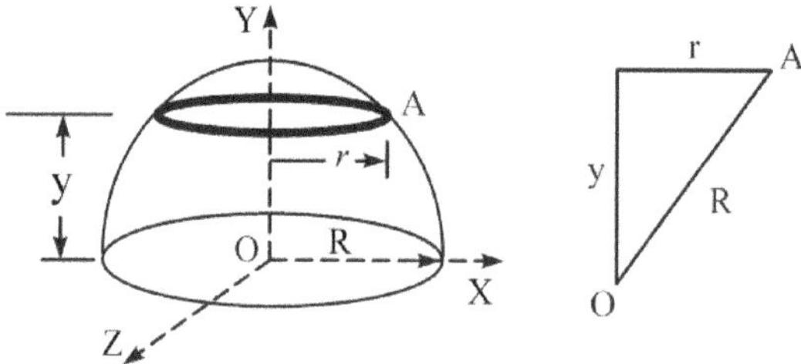

Then, $$Y_C = \int y \times dV \,/ \int dV = [\int y \times (\pi r^2 dy)] / [\int \pi r^2 dy]$$
$$= [\int y \times (R^2 - y^2) \times dy] / [\int (R^2 - y^2)\, dy] \quad \text{for} \quad 0 \le y \le R$$
$$= [R^2y^2/2 - y^4/4] / [R^2y - y^3/3] = (R^4/4)/(2R^3/3) = \boldsymbol{3R/8}$$

6.5 CENTROIDS OF COMPOSITE SOLIDS OF REVOLUTION

Centroids of composite bodies of revolution are calculated following the same procedure used for composite areas, by treating the composite body as a combination of different simple bodies and identifying their individual centroids.

Example 14

A homogeneous body consists of a circular cylindrical body of radius 'R' and height 'h', attached to a hemispherical body of the same radius. Find 'h' such that centroid of the composite body lies at the center of the circular plane of the hemisphere.

Solution:

Since the composite body is symmetric about Y-axis, its centroid lies along the axis of revolution. Hence, $X_C = Z_C = 0$. Considering hemisphere as part-1 and cylinder as part-2, volume and distance of centroid from the origin are :

$$V_1 = 2\pi R^3/3 \; ; \quad Y_1 = h + 3R/8 \qquad \text{and} \qquad V_2 = \pi R^2 h \; ; \quad Y_2 = h/2$$

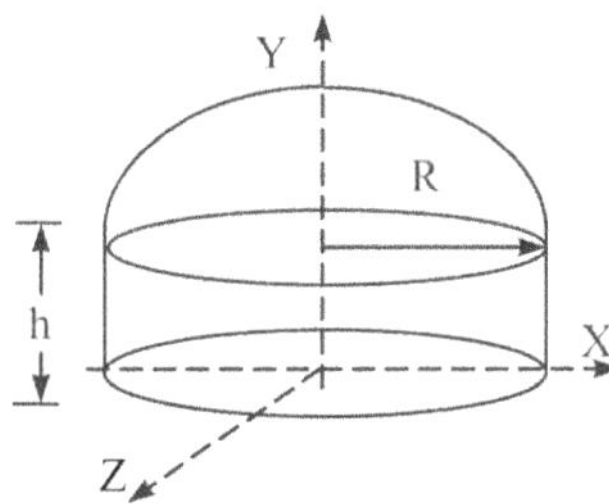

Then, $Y_C = (V_1 \times Y_1 + V_2 \times Y_2) / (V_1 + V_2)$

$$h = [\,(2\pi R^3/3) \times (h + 3R/8) + (\pi R^2 h) \times (h/2)] \;/\; [\,(2\pi R^3/3) + (\pi R^2 h)]$$

or $\quad 4h(2R + 3h)\pi R^2 = (8Rh + 3R^2 + 6h^2)/\pi R^2$

$\Rightarrow \quad 3R^2 = 6h^2$ or $\mathbf{h = R/\sqrt{2}}$

6.5.1 PAPPU'S THEOREM

The area of surface generated by revolving a plane curve of length 'L' about a non-intersecting axis in its plane is equal to the product of length of the curve and the distance traveled by the centroid of the curve during the generation of the surface.

If a surface is generated by revolving a curve about X-axis, then

$$\text{Surface area} = L \times (2\pi\, y_G)$$

Similarly, If a surface is generated by revolving a curve about Y-axis, then

$$\text{Surface area} = L \times (2\pi\, x_G)$$

6.5.2 GULDINUS THEOREM (ALSO CALLED PAPPUS THEOREM-2)

The volume of a body generated by revolving a plane of area 'A' about a non-intersecting axis in its plane is equal to the product of area of the plane and the distance traveled by the centroid of the area during the generation of the volume.

If a volume is generated by revolving an area about X-axis, then

$$\text{Body volume} = A \times (2\pi\, y_G)$$

Similarly, If a volume is generated by revolving an area about Y-axis, then

$$\text{Body volume} = A \times (2\pi\, x_G)$$

Example 15

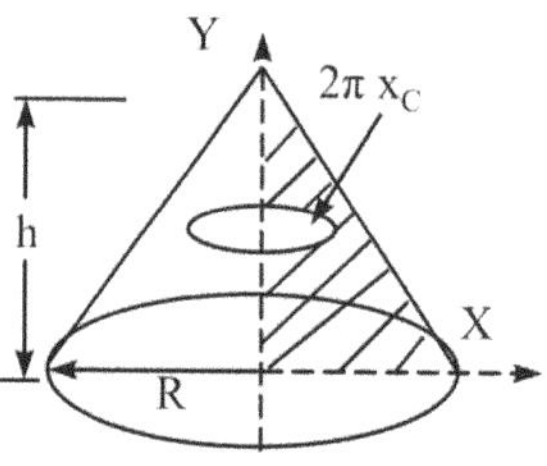

Volume of a cone of base radius 'R' and height 'h',

V = (Area of the triangle of base 'R' and height 'h') × ($2\pi X_C$)

where, X_C is the distance of centroid of the triangle from Y-axis

$= (R \times h/2) \times [2\,\pi \times (R/3)] = \boldsymbol{\pi R^2 h/3}$

Example 16

Volume of a hemisphere of radius 'R',

V = (Area of the quarter circle of radius 'R') × ($2\pi X_C$)

where, X_C is the distance of centroid of the quarter circle from Y-axis

$= (\pi R^2/4) \times [2\,\pi \times (4R/3\pi)\,] = \boldsymbol{(2/3)\,\pi R^3}$

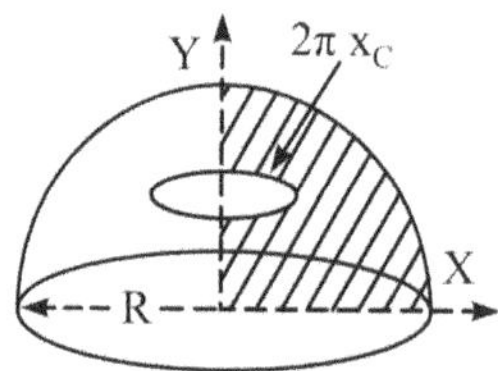

Similarly, Volume of a sphere, $V = (\pi R^2/2) \times [2\pi \times (4R/3\pi)\,] = (4/3)\,\pi R^3$

Example 17

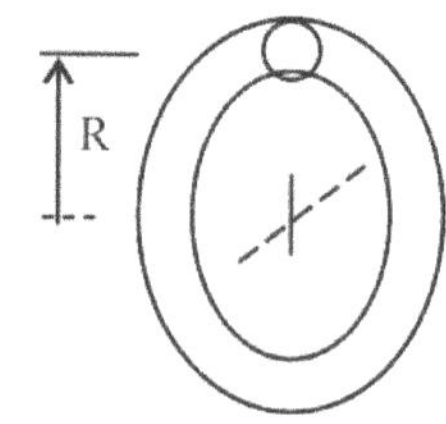

Surface area of annular torus of mean radius 'R', with hollow circular cross section of radius 'r' is obtained by rotating a circular line of radius 'r' about a center located at a distance 'R'. Then,

A = Length of circular line (cross section) × distance covered by centroid ($2\pi\, y_G$)

where

y_G is the distance of centroid of the cross section from the axis of revolution

$$= (2\pi r) \times (2\pi R) = 4\pi^2 r R$$

Example 18

Volume of fluid in an annular torus of mean radius 'R', with circular cross section of radius 'r' is obtained by rotating a circular area of radius 'r' about a center located at a distance 'R'. Then,

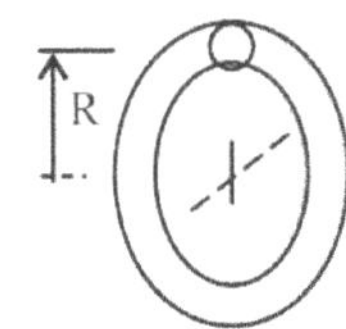

$$V = \text{Area of circular section} \times \text{distance covered by centroid}$$

$$= (\pi r^2) \times (2\pi R) = 2\pi^2 r^2 R$$

Pappu's theorem can also be used to calculate centroid of an area / volume of a solid of revolution, if the generated area / volume is known.

Example 19

Calculate centroid of a semi-circular plate from the volume of the sphere generated by rotating it about its diameter

Volume of sphere = Area of semi-circular plate × distance covered by its centroid

$$(4/3)\, \pi r^3 = (\pi r^2/2) \times (2\pi x_G)$$

Therefore, $\boldsymbol{x_G = 4r/3\pi}$

SUMMARY

1. ***Centroid*** is the point where the entire physical property such as length, area or volume of a body is assumed to be concentrated, to give the same first moment as that obtained by considering distributed property of the body.
2. These imaginary points where distributed mass and gravitational force in a body assumed concentrated are called ***center of mass*** and ***center of gravity*** respectively
3. For a line, distance of centroid from a chosen origin,

$r_G = (\int r \times dL) / \int dL$

It can also be expressed in terms of cartesian coordinates in the form

$X_G = \int X \times dL / \int dL$; $Y_G = \int Y \times dL / \int dL$; $Z_G = \int Z \times dL / \int dL$

Similarly, for area and volume elements,

$r_G = \int r\, dA / \int dA$

or $X_G = \int X\, dA / \int dA$; $Y_G = \int Y\, dA / \int dA$; $Z_G = \int Z\, dA / \int dA$

and $r_G = \int r\, dV / \int dV$

or $X_G = \int X\, dV / \int dV$; $Y_G = \int Y\, dV / \int dV$; $Z_G = \int Z\, dV / \int dV$

4. The centroid may or may not lie on the element itself
5. Distance of centroid from X-axis, $\boldsymbol{Y_G = Z_G = 0}$, for elements *symmetric about X-axis*

   Similarly, $\boldsymbol{Z_G = X_G = 0}$ ($\boldsymbol{X_G = Y_G = 0}$) for elements *symmetric about Y-axis* (Z-axis)
6. For composite bodies which can be considered as a combination (addition and / or deletion) of simple bodies, coordinates of the centroid can be obtained from the coordinates of the simple bodies. Thus, for line elements,

   $X_G = \sum X \times dL / \sum dL$; $Y_G = \sum Y \times dL / \sum dL$; $Z_G = \sum Z \times dL / \sum dL$

   Similarly for area and volume elements,

   $X_G = \sum X \times dA / \sum dA$; $Y_G = \sum Y \times dA / \sum dA$; $Z_G = \sum Z \times dA / \sum dA$

   $X_G = \sum X \times dV / \sum dV$; $Y_G = \sum Y \times dV / \sum dV$; $Z_G = \sum Z \times dV / \sum dV$
7. If a body is suspended from any point on it, its ***center of gravity lies on the vertical line through the point of support***, so that weight through center of gravity and normal (upward vertical) reaction through the point of support form a system of two equal, collinear and opposite forces, without producing any unbalanced moment, and keep the body in equilibrium.
8. Pappu's theorem - 1

   If a surface is generated by revolving a curve about X-axis, then

   Area of generated surface = $L \times (2\pi\, y_G)$

   Similarly, If a surface is generated by revolving a curve about Y-axis, then

   Area of generated surface = $L \times (2\pi\, x_G)$

Ex : Surface area of annular torus (car tube) of mean radius 'R' and radius of circular section 'r' = $(2\pi r) \times (2\pi R) = 4\pi^2 r R$

9. Pappu's theorem – 2 or Guldinus theorem

If a volume is generated by revolving an area about X-axis, then

Volume of generated body = $A \times (2\pi y_G)$

Similarly, If a volume is generated by revolving an area about Y-axis, then

Volume of generated body = $A \times (2\pi x_G)$

Ex: Volume of fluid in an annular torus (car tube) of mean radius 'R' and radius of circular section 'r' = $(\pi r^2) \times (2\pi R) = 2\pi^2 r^2 R$

SOME USEFUL RESULTS

Coordinates of centroid of

1. Quarter circular arc of radius R,

$$X_G = Y_G = 2R/\pi$$

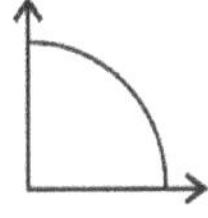

2. Quarter circular plate of radius R,

$$X_G = Y_G = 4R/3\pi$$

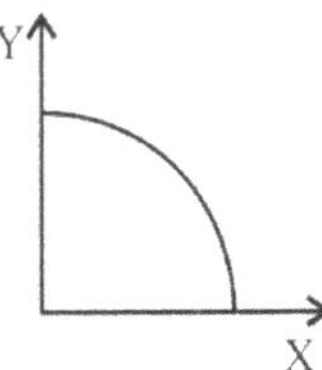

3. Rectangle or cylinder of height 'h', $Y_G = h/2$
4. Triangular plate of base 'b' and height 'h' about its base,

$$Y_G = h/3$$

5. Trapezium of base 'a', top 'b' and height 'h', $Y_G = h(a + 2b) / 3(a + b)$
6. Area below a parabola defined by $y^2 = k x$ in the limit y = 0 to h,

$$Y_G = 3h/8$$

7. Cone or pyramid of height 'h', $Y_G = h/4$
8. Hemisphere of radius 'R', $Y_G = 3R/8$

PROBLEMS FOR PRACTICE

1. Locate centroid of the area bounded by parabolic spandrel $x^2 = a \times y$ and $y = kx$

   (***Ans:*** $x_G = a/2$, $y_G = 2a/5$)

2. A circular area of radius 'r' is removed from another circular area of radius '2r', such that the two areas have a common tangent parallel to Y. Locate centroid of the remaining area.

   (***Ans:*** $x_G = -r/3$, $y_G = 0$)

3. A square area of diagonal 'r' is removed from a circular area of radius 'r', both having a common center. Locate centroid of the remaining area.

   (***Ans:*** $x_G = -0.0946$ r, $y_G = 0$)

4. A circular area of radius 'r' is removed from a semicircular area of radius '2r', such that the two areas have a common tangent parallel to X. Locate centroid of the remaining area.

   (***Ans:*** $x_G = 0$, $y_G = 0.69765$ r)

5. A quarter circular area of radius 'a' is removed from a square area of side 'a'. Locate centroid of the remaining area.

   (***Ans:*** $x_G = y_G = 0.7766$ a)

6. A semi-circular area of radius 'r' is removed from a quarter circular area of radius '2r'. Locate centroid of the remaining area.

   (***Ans:*** $x_G = 0.69765$ r, $y_G = 1.2732$ r)

7. A rectangular area of 30mm×40mm is cut from a rectangle of 100mm × 120mm, such that top left corner of the inner rectangle is at (10,0) w.r.t. the center of the bigger rectangle. Find the position of the centroid of the remaining area.

   (***Ans:*** –2.8, 2.2)

8. One quarter of a rectangle (top right portion of 'b' × 'd') is cut from a rectangle of '2b' × '2d'. Find the position of the centroid of the remaining area, with origin at the center of the bigger rectangle.

   (***Ans:*** –b/6, –d/6)

9. A triangular area of width 'b' and height 'y' is cut from a rectangle of width 'b' and height 'd' such that the centroid of the remaining area is at the vertex A of the triangle. Find the height 'h' of the triangle.

   (***Ans:*** $y = 0.634d$)

10. Calculate the position of the centroid of an I-section of base width 200 mm, top width 100 mm, total height 140 mm and thickness of flanges and web 20 mm.

(***Ans:*** 0, 55)

11. Find the location of the centroid of the shaded area w.r.t. the given axes, if the shaded area is within a square of side 80 cm

(***Ans:*** $X_G = -4.776$ cm, $Y_G = -2.269$ cm)

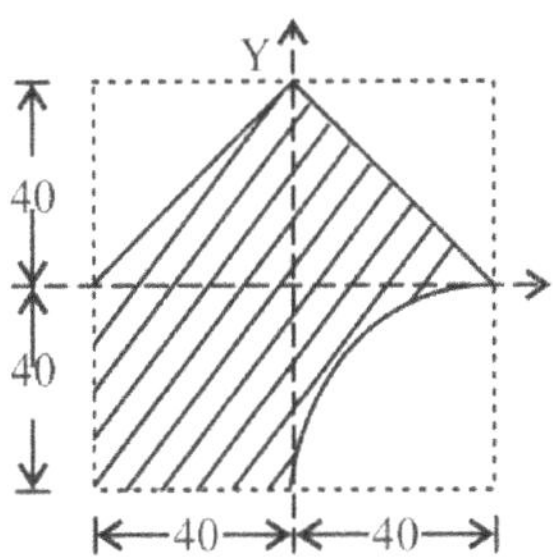

12. Find centroid of a **paraboloid** of base radius 'R' and height h, if the parabola is defined by $h - y = k\,x^2$ (since the origin is the center of base of paraboloid)

(***Ans:*** h/3)

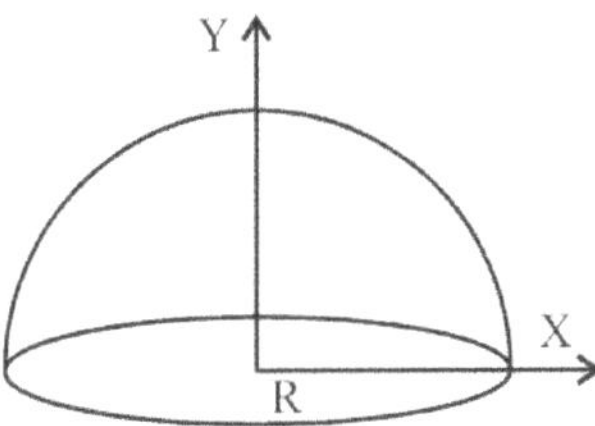

13. A homogeneous body consists of a circular cone of radius 'R' and height 'h', resting on a hemisphere of the same radius. Find 'h' such that centroid of the composite body lies at the center of the circular plane of the hemisphere.

(***An :*** $\sqrt{3}R$)

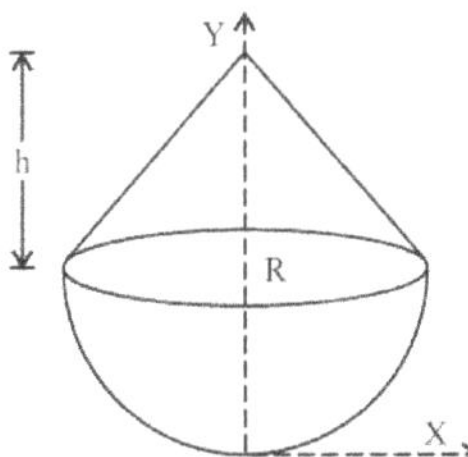

14. Find the centroid of the areas shown. All dimensions are in cm.

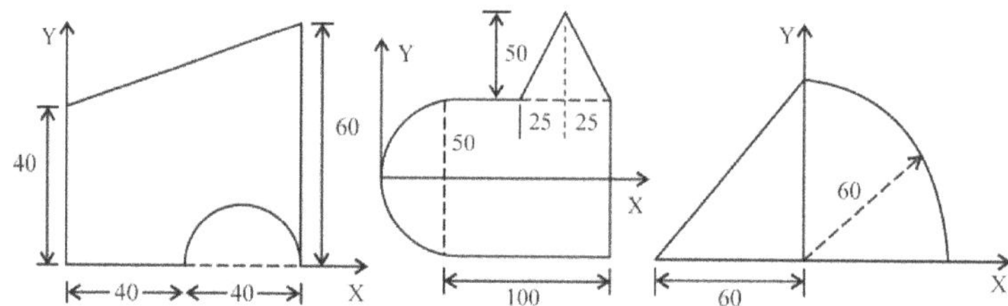

(*Ans:* $X_G = 39.44$, $Y_G = 28.47$; $X_G = 71.1$, $Y_G = 7.2$; $X_G = 7.778$, $Y_G = 23.33$)

15. Find the centroid of the wires bent as shown. All dimensions are in cm.

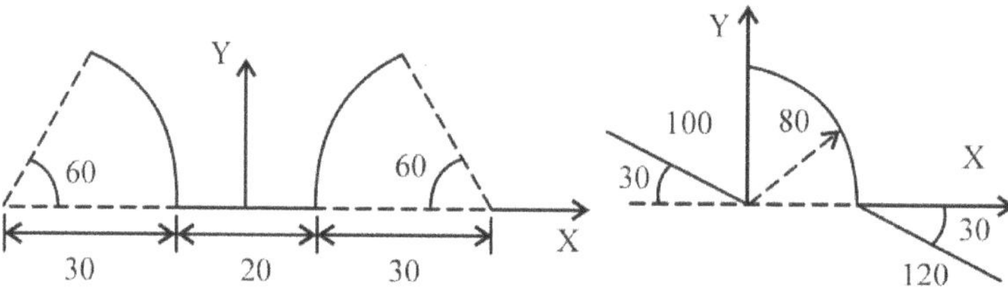

(*Ans:* $X_G = 0$, $Y_G = 10.86$; $X_G = 42.06$, $Y_G = 19.97$)

16. Find the volume generated by revolving a semi circular area about X and Y axes, as shown in the figure.

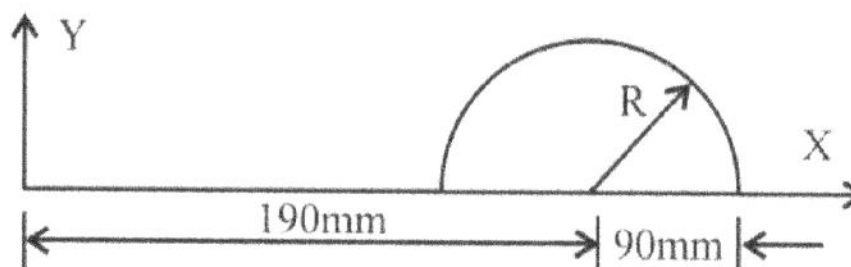

(*Ans:* $V_X = 3054.857 cm^3$; $V_Y = 7200.735 cm^3$)

CHAPTER 7

AREA MOMENT OF INERTIA

Moment of inertia (also called 2nd moment of area) is a property of cross sectional area of a member, used in the calculation of bending stresses in beams subjected to loads perpendicular to its axis. Moment of inertia about any axis is defined as the area of the cross section multiplied by square of the distance of the area from the axis. Hence, for the same area, moment of inertia varies with the particular axis about which it is calculated. Since area is always positive and distance square is also positive, moment of inertia of any area is always positive about any axis.

For a small area 'A' at a distance of x from Y-axis and y from X-axis,

$$I_{XX} = A \times y^2 \quad ; \quad I_{YY} = A \times x^2$$

Moment of inertia of a large area can be obtained by integration as

$$I_{XX} = \int (dA \times y^2) \quad ; \quad I_{YY} = \int (dA \times x^2) \qquad .....(7.1)$$

Moment of inertia about Z-axis (also called **Polar moment of inertia**), perpendicular to the plane of cross section,

$$I_{ZZ} = \int (dA \times r^2) = \int (dA \times y^2) + \int (dA \times x^2)$$

or $$I_{ZZ} = I_{XX} + I_{YY} \qquad .....(7.2)$$

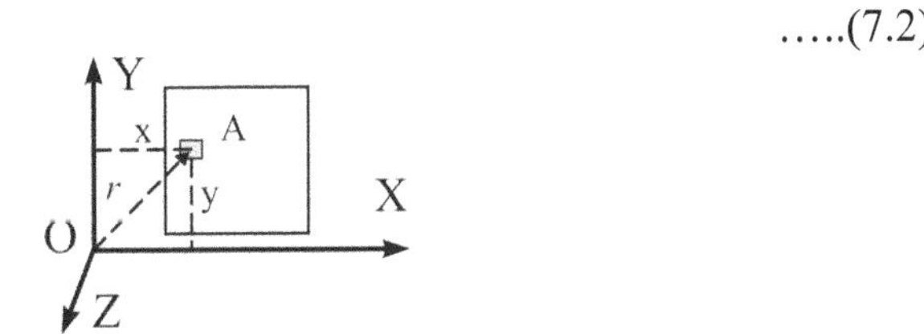

This is also called **perpendicular axis theorem**. I_{ZZ} is a property used in members subjected to torsional moment or moment about the axis (Z-direction) of the beam or shaft.

7.1 MOMENT OF INERTIA OF SOME SIMPLE SHAPES

(a) ***Rectangular section of width 'b' and depth 'd'***

Considering a small area dA of sides 'ds' and 'dt' at a distance of 't' from Y-axis and 's' from X-axis,

Moment of inertia about X-axis, passing through O, is

$$\mathbf{I_{XX}} = \int dA \times t^2 = \int\int ds \times dt\, t^2 \quad \text{for } 0 \leq s \leq b \text{ and } 0 \leq t \leq d$$

$$= [b] \int t^2\, dt = b\,[t^3 / 3] = \mathbf{b\, d^3 / 3} \qquad \text{.....(7.3)}$$

Similarly, $I_{YY} = b^3\, d / 3$

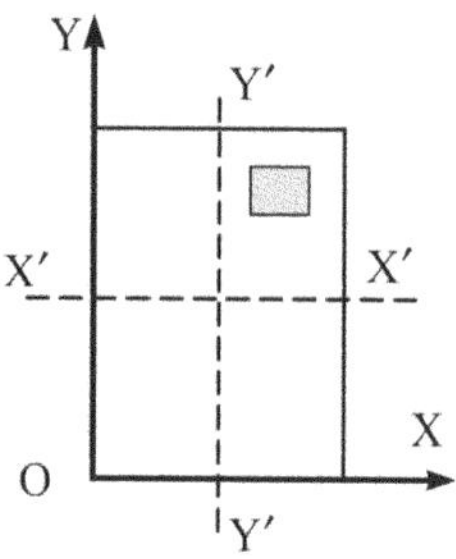

Moment of inertia about $(X' - X')$ axis passing through the centroid, can be obtained in the same way, by using appropriate limits for 't', as

$$\mathbf{I_{X'X'}} = \int dA \times t^2 = \int\int (ds \times dt)\, t^2 \quad \text{for} \quad 0 \leq s \leq b \quad \text{and} \quad -d/2 \leq t \leq d/2$$

$$= [b] \int t^2\, dt = b\,[\,t3 / 3] = \mathbf{b\, d^3 / 12} \qquad \text{.....(7.4)}$$

Similarly, $I_{Y'Y'} = b^3\, d / 12$

This method of integration w.r.t. y alone is possible because width is constant for different values of y (unlike in the case of a triangle, circle, etc.), throughout the depth.

Alternative Method

Considering a small area 'dA' of width 'b' and thickness 'dy' at a height of 'y' from the axis A – A, moment of inertia,

$$IAA = \int dA \times y^2 = \int (b \times dy) \times y^2 = b[y^3/3] \qquad \text{for} \quad 0 \leq y \leq d$$

$$= b\, d^3 / 3$$

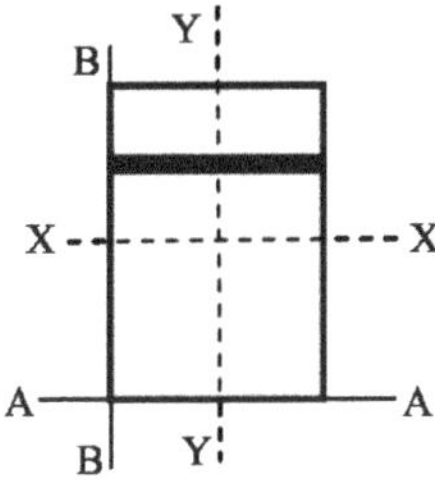

Similarly, moment of inertia about BB, $I_{BB} = b^3d / 3$

(b) ***Triangular section of width 'b' and depth 'd'***

Considering a small area dA of width 's' and thickness 'dt' at a distance of 't' from X-axis,

Moment of inertia about X-axis, passing through O, is

$$I_{XX} = \int dA \times t^2 = \int (s \times dt) \times t^2$$

with $s/b = (d - t)/d$ and $0 \le t \le d$

$$= \int [b \times (d - t) / d]\, t^2\, dt = (b/d) \times [(dt^3/3) - (t^4/4)]$$

$$= \mathbf{b\, d^3 / 12} \qquad \text{.....(7.5)}$$

Similarly, $I_{YY} = b^3\, d / 12$

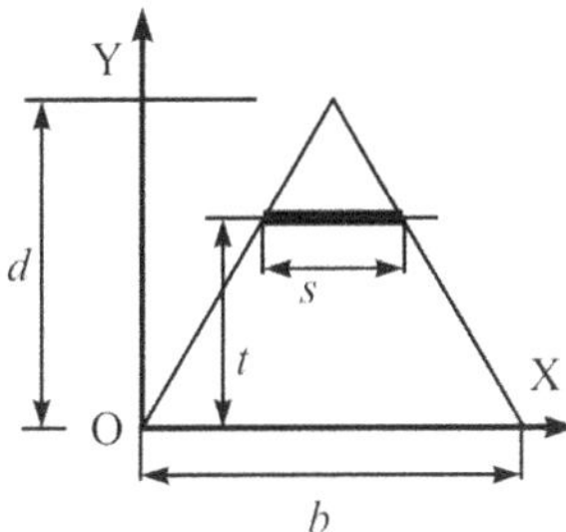

(c) ***Circular section of radius 'r'***

Consider a small elemental area (r dθ) × dr at a distance of 'y' (= r sin θ) from X-axis through its centroid,

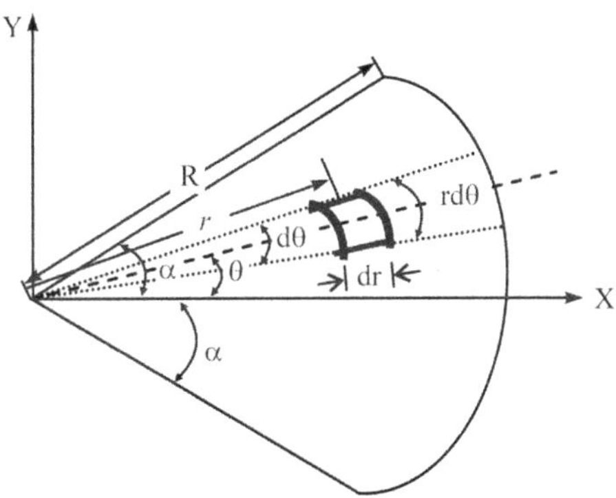

$$I_{XX} = \int dA \times y^2 = \int\int [(r\, d\theta) \times dr] \times (r \sin \theta)2$$

$$\text{for} \quad 0 \le r \le R \quad \text{and} \quad -\alpha \le \theta \le +\alpha$$

$$= [r^4/4] \int \sin^2\theta\, d\theta = (R^4/4) \int [(1 - \cos^2\theta)/2]\, d\theta$$

$$= (R^4/8) \times [\theta + \cos^2\theta] = (R^4/8) \times [2\alpha + 2 \sin 2\alpha]$$

For a circular plate, $\alpha = \pi$ and, therefore,

$$\mathbf{I_{XX} = \pi R^4 / 4 \quad \text{or} \quad \pi D^4 / 64}$$

Since circular plate is symmetric about any diametral axis,

$$\mathbf{I_{YY} = I_{XX} = \pi R^4 / 4 \quad \text{or} \quad \pi D^4 / 64} \quad \text{.....(7.6)}$$

Polar moment of inertia of a circular plate,

$$I_{ZZ} = I_{XX} + I_{YY} = \mathbf{\pi R^4 / 2} \quad \text{or} \quad \mathbf{\pi D^4 / 32} \quad \text{.....(7.7)}$$

Following the same procedure, with appropriate change of limits of θ,

Moment of inertia of a semi-circular plate (symmetrical about Y-axis),

$$\mathbf{I_{XX} = \pi R^4 / 8}$$

and Moment of inertia of a quarter circular plate,

$$\mathbf{I_{XX} = I_{YY} = \pi R^4 / 16}$$

(d) ***Thin circular ring of mean radius 'r' and thickness 't'***

Consider a small elemental area $(R\, d\theta) \times t$ at a distance of 'R' from the center of the ring.

$$I_{ZZ} = \int dA \times R^2 = \int [(R\, d\theta) \times t] \times R^2 \quad \text{for} \quad 0 \le \theta \le 2\pi$$

$$= \mathbf{2\pi R^3 t} \quad \text{or} \quad \mathbf{\pi D^3 t / 4}$$

Since the ring is symmetric about any diametral axis,

$$\mathbf{I_{XX} = I_{YY} = I_{ZZ} / 2 = \pi R^3 t} \quad \text{.....(7.8)}$$

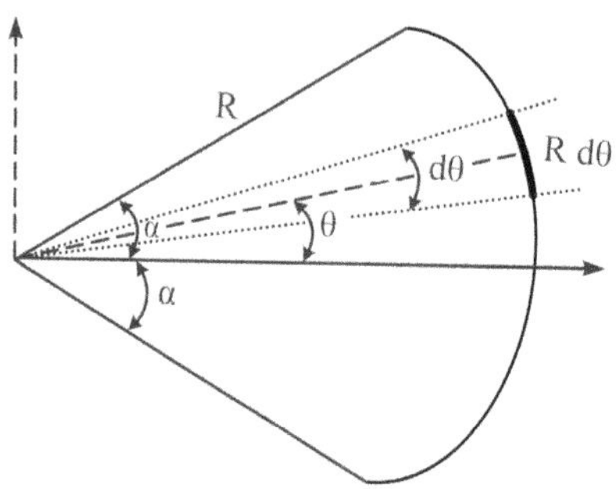

In the same way, polar moment of inertia of a **semi-circular ring**,

$$I_{ZZ} = \pi R^3 t$$

and polar moment of inertia of a ***quarter*** **circular ring**,

$$I_{ZZ} = \pi R^3 t / 2$$

(e) ***Ellipse:*** Calculate moment of inertia of an ellipse of semi-major axis 'a' and semi-minor axis 'b'.

Let us consider a small elemental area 'dA' of height '2y' and width 'dx', at a distance 'x' from theY-axis. Equation of ellipse is

$$x^2/a^2 + y^2/b^2 = 1 \quad \text{or} \quad y = b\sqrt{1-(x^2/a^2)}$$

$$I_{YY} = \int x^2 \times dA = \int x^2 \times (2y \times dx)$$

$$= \int x^2 \times 2b\sqrt{1-(x^2/a^2)} \times dx \quad \text{for} \quad -a \le x \le +a$$

This integral can be evaluated by substituting, $x = a \sin\theta$

Then, $dx = a\cos\theta \times d\theta$

with $\theta = -\pi/2$ to $+\pi/2$ corresponding to $x = -a$ to $+a$

and $\sqrt{\left[1-(x^2/a^2)\right]} = \cos\theta$

Therefore, $I_{YY} = \int (a \sin\theta)^2 \times 2b \times \cos\theta \times a\cos\theta \times d\theta$

$$= 2a^3b \int \sin^2\theta \cos^2\theta \, d\theta \qquad .....(7.9)$$

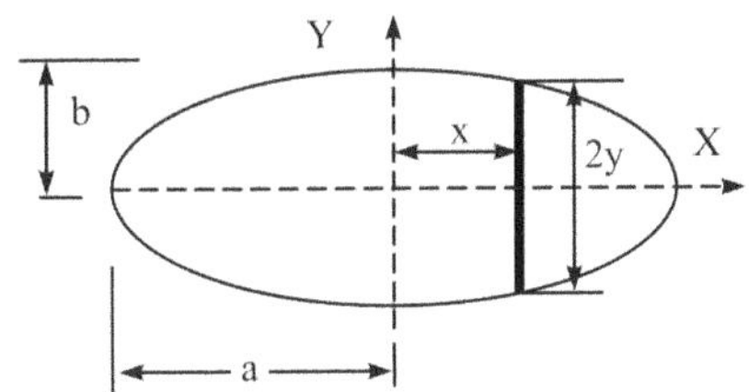

$$= 2a^3b\int[(\sin^2 2\theta)/4]\,d\theta = 2a^3b\int[(1-\cos 4\theta)/8]\,d\theta$$

$$= (2a^3b/8)\,[\theta - \sin 4\theta/4] \quad \text{for } -\pi/2 \le \theta \le +\pi/2$$

$$= (a^3b/4)[\{(\pi/2) - (-\pi/2)\} - \{\sin 2\pi - \sin(-2\pi)\}/4$$

$$= (a^3b/4)\,[\pi(0-0)/4] = \boldsymbol{\pi a^3 b/4}$$

Similarly, $\mathbf{I_{XX} = \pi\, a\, b^3/4}$ and $\mathbf{I_{ZZ} = I_{XX} + I_{YY} = \pi\, a\, b\,(a^2 + b^2)/4}$

7.1.1 RECTANGULAR SECTION

Rectangular section of width 'b' and depth 'd'. Considering a small area dA of sides 'ds' and 'dt' at a distance of 't' from Y-axis and 's' from X-axis,

Moment of inertia about X-axis, passing through O, is

$$I_{XX} = \int dA \times t^2 = \int (ds \times dt) \times t^2) \quad \text{for} \quad 0 \le s \le b \quad \text{and} \quad 0 \le t \le d$$

$$= (\int ds) \times (\int t^2\, dt) = b\,[t^3/3] = \boldsymbol{b\, d^3/3} \qquad .....(7.10)$$

Similarly, $I_{YY} = b^3 d / 3$

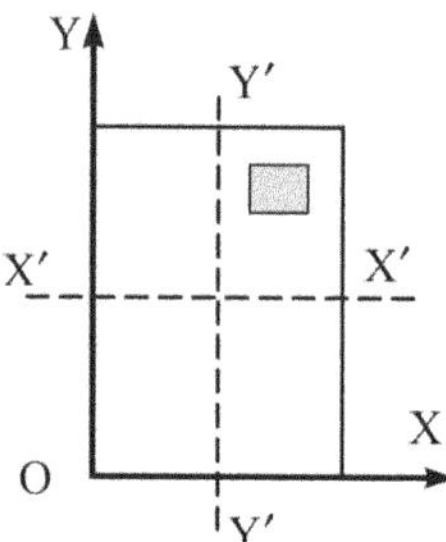

Moment of inertia about $(X' - X')$ axis passing through the centroid, can be obtained in the same way, by using appropriate limits for 't', as

$$\mathbf{I}_{X'X'} = \int dA \times t^2 = \int\int (ds \times dt)\, t^2$$

for $0 \le s \le b$ and $-d/2 \le t \le d/2$

$$= \left(\int ds\right) \times \left(\int t^2 dt\right) = b\,[\,t^3 / 3] = \boldsymbol{b\,d^3 / 12} \qquad \text{.....(7.11)}$$

Similarly, $\mathbf{I_{Y'Y'} = b^3\, d / 12}$

This method of integration w.r.t. y alone is possible because width is constant for different values of y (unlike in the case of a triangle, circle, etc.), throughout the depth.

Alternative Method

Considering a small area 'dA' of width 'b' and thickness 'dy' at a height of 'y' from the axis A – A, moment of inertia,

$$I_{AA} = \int dA \times y^2 = \int (b \times dy) \times y^2 \qquad \text{for} \quad 0 \le y \le d$$

$$= b\, d^3 / 3$$

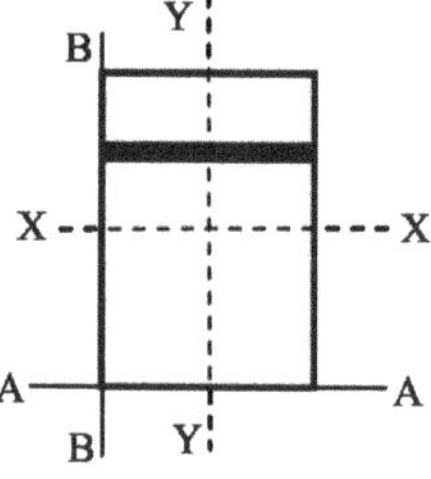

Similarly, moment of inertia about BB, $\mathbf{I_{BB} = b^3 d / 3}$

7.1.2 TRIANGULAR SECTION

Triangular section of width ‘b’ and depth ‘d’

Considering a small area dA of width ‘s’ and thickness ‘dt’ at a distance of ‘t’ from Y-axis,

Moment of inertia about X-axis, passing through O, is

$$I_{XX} = \int dA \times t^2 = \int (s \times dt) \times t^2$$

with $s/b = (d - t)/d$ and $0 \leq t \leq d$

$$= \int [b \times (d - t) / d]\, t^2\, dt = (b/d) \times [(dt^3/3) - (t^4/4)]$$

$$= \boldsymbol{b\, d^3 / 12} \qquad .....(7.12)$$

Similarly, $\mathbf{I_{YY} = b^3\, d / 12}$

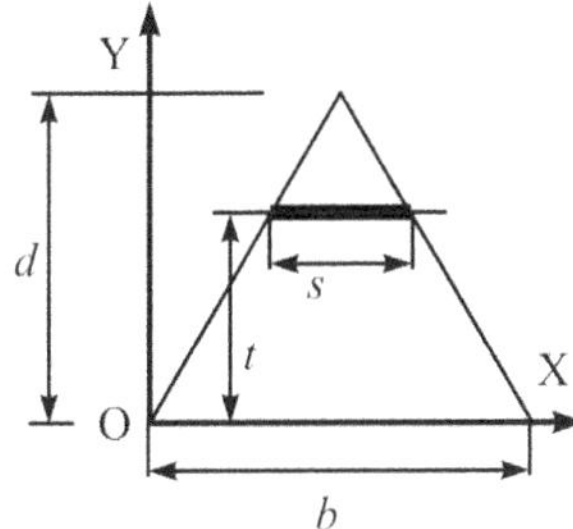

7.1.3 CIRCULAR SECTION OF RADIUS ‘R’

Consider a small elemental area (r dθ) × dr at a distance of ‘y’ (= r sin θ) from X-axis through its centroid C,

$$I_{XX} = \int dA \times y^2 = \int\int [(r\, d\theta) \times dr] \times (r \sin\theta)^2$$

for $0 \leq r \leq R$ and $-\alpha \leq \theta \leq +\alpha$

$$= \int [r^4/4] \sin^2\theta\, d\theta = (R^4/4) \int [(1 - \cos 2\theta)/2]\, d\theta$$

$$= (R^4/8) \times [\theta + \sin 2\theta] = (R^4/8) \times [2\alpha + 2 \sin 2\alpha]$$

For a circular plate, $\alpha = \pi$ and, therefore,

$$\mathbf{I_{XX} = \pi R^4 / 4 \quad \text{or} \quad \pi D^4 / 64}$$

Since circular plate is symmetric about any diametral axis,

$$\mathbf{I_{YY} = I_{XX} = \pi R^4 / 4 \quad \text{or} \quad \pi D^4 / 64} \qquad .....(7.13)$$

Polar moment of inertia of a circular plate,

$$\mathbf{I_{ZZ}} = I_{XX} + I_{YY} = \boldsymbol{\pi R^4 / 2} \quad \text{or} \quad \mathbf{\pi D^4 / 32} \qquad .....(7.14)$$

Following the same procedure, with appropriate change of limits of θ,

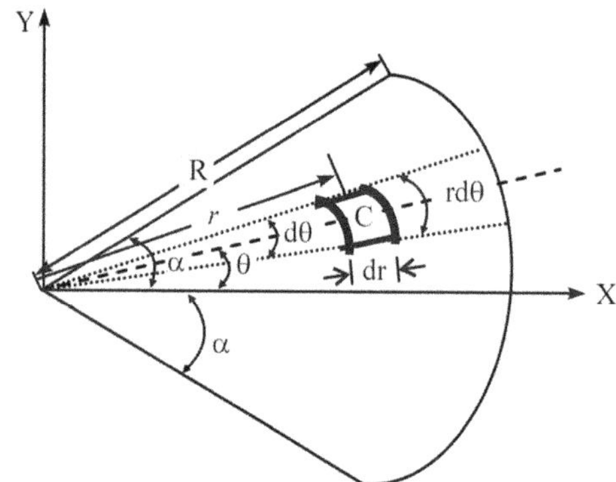

Moment of inertia of a semi-circular plate (symmetrical about Y-axis),

$$I_{XX} = \pi R^4 / 8$$

and Moment of inertia of a quarter circular plate,

$$I_{XX} = I_{YY} = \pi R^4 / 16$$

7.1.4 THIN CIRCULAR RING OF MEAN RADIUS 'R' AND THICKNESS 't'

Consider a small elemental area (R dθ) × t at a distance of 'R' from the center of the ring.

$$I_{ZZ} = \int dA \times R^2 = \int [(R\, d\theta) \times t] \times R^2 \quad \text{for} \quad 0 \le \theta \le 2\pi$$

$$= 2\pi R^3 t \quad \text{or} \quad \pi D^3 t / 4$$

Since the ring is symmetric about any diametral axis,

$$I_{XX} = I_{YY} = I_{ZZ} / 2 = \pi R^3 t \qquad \text{.....(7.15)}$$

In the same way, polar moment of inertia of a **semi-circular ring**,

$$I_{ZZ} = \pi R^3 t$$

and polar moment of inertia of a ***quarter*** **circular ring**,

$$I_{ZZ} = \pi R^3 t / 2$$

7.1.5 ELLIPSE OF SEMI-MAJOR AXIS 'a' AND SEMI-MINOR AXIS 'b'

Let us consider a small elemental area 'dA' of height '2y' and width 'dx', at a distance 'x' from the origin of the coordinate system. Equation of ellipse is

$$x^2/a^2 + y^2/b^2 = 1 \qquad \text{or} \qquad y = b\sqrt{[1-(x^2/a^2)]}$$

$$\mathbf{I_{YY}} = \int x^2 \times dA = \int x^2 \times (2y \times dx)$$

$$= \int x^2 \times 2b\sqrt{1-(x^2/a^2)} \times dx \quad \text{for} \quad -a \le x \le +a$$

This integral can be evaluated by substituting, $x = a \sin \theta$

Then, $dx = a \cos \theta \times d\theta$

with $\theta = -\pi/2$ to $+\pi/2$ corresponding to $x = -a$ to $+a$

and $\sqrt{[1-(x^2/a^2)]} = \cos\theta$

Therefore, $I_{YY} = \int (a \sin \theta)^2 \times 2b \times \cos \theta \times a \cos \theta \times d\theta$

$$= a^3 b \int [(\sin^2 2\theta)/2]\, d\theta$$

$$= 2\pi a^3 b \times \{ (\pi/8) - [0 - 0]/32 \}$$

$$= \boldsymbol{\pi a^3 b / 4} \qquad \text{.....(7.16)}$$

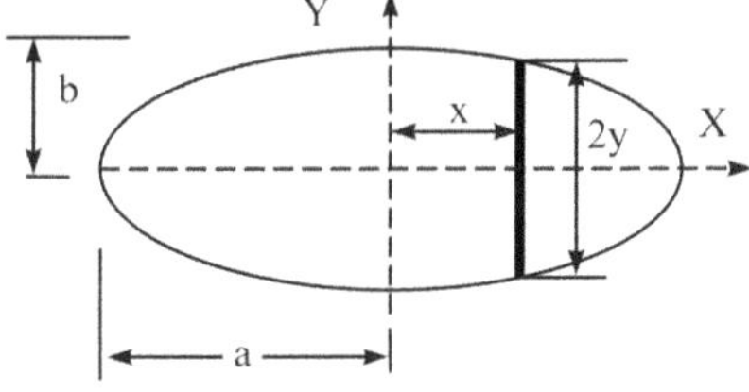

Similarly, $\mathbf{I_{XX} = \pi a b^3 / 4}$

and $\mathbf{I_{ZZ} = I_{XX} + I_{YY} = \pi a b (a^2 + b^2)/4}$

7.2 PARALLEL AXES THEOREM

Moment of inertia calculated about an axis (X-X) through its centroid can be transferred to a parallel axis using the relation,

$$I_{AA} = I_{XX} + A \times y^2 \qquad \text{.....(7.17)}$$

where, A is the area of the plane and 'y' is the distance between the two parallel axes X-X and A – A. This is also called ***Transfer of axis theorem.***

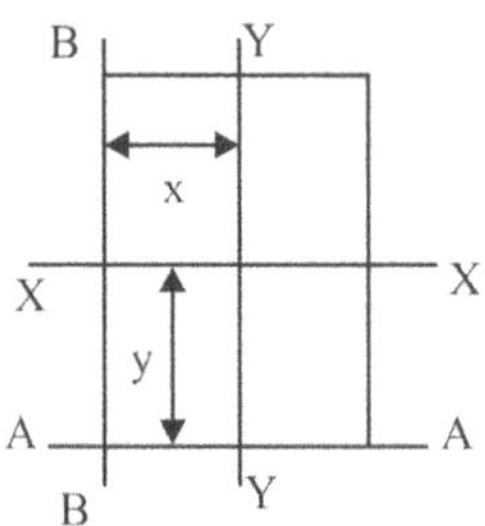

Similarly, $I_{BB} = I_{YY} + A \times x^2$ ....(7.18)

Example 1

Moment of inertia of a rectangle through its centroidal axes can be obtained from the values about one of their edges, using parallel axes theorem, as shown here.

$$I_{XX} = I_{AA} - A \times y^2 = (bd^3/3) - (b \times d) \times (d/2)2 = b\,d^3 / 12$$

Similarly, $I_{YY} = I_{BB} - A\times x^2 = (b^3d/3) - (b \times d) \times (b/2)2 = b^3d / 12$

.....(7.19)

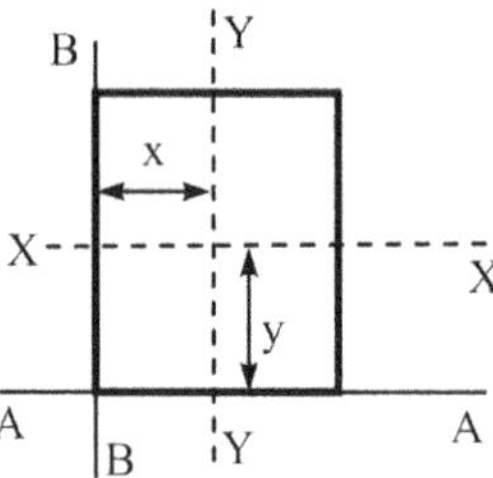

Example 2

Moment of inertia of a triangle through its centroidal axis X-X can be obtained from the moment of inertia value about its edge A-A, using parallel axes theorem, as

$$I_{XX} = I_{AA} - A \times y^2 = (bd^3/12) - (b \times d/2) \times (d/3)2$$

$$= b\,d^3 / 36$$ ....(7.20)

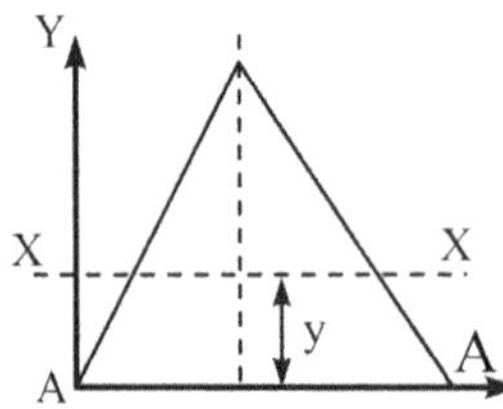

Example 3

Moment of of a semi-circular plate through its centroidal axis X-X can be obtained from the moment of inertia value about its edge A-A, using parallel axes theorem, as

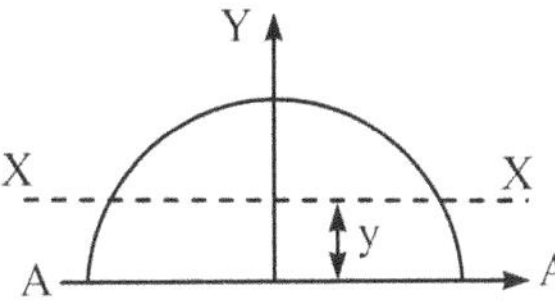

$I_{XX} = I_{AA} - A \times y^2 = (\pi\ r^4/8) - (\pi\ r^2/2) \times (4\ r/3\pi)2 = 0.11\ r^4$

Due to symmetry, centroid lies on Y-axis. So, $I_{YY} = \pi\ r^4/8$

Example 4

Moment of inertia of a quarter circle through its centroidal axis X-X can be obtained from the moment of inertia value about its edge A-A, using parallel axes theorem, as

$I_{XX} = I_{YY} = I_{AA} - A \times y^2 = (\pi\ r^4/16) - (\pi\ r^2/4) \times (4r/3\pi)^2$

$= \boldsymbol{0.055\ r^4}$ .....(7.21)

7.3 MOMENT OF INERTIA OF COMPOSITE SECTIONS

Moment of inertia of a composite section is the algebraic sum of moments of inertia of its simpler components.

$I_{XX} = \sum [\ I_{XX}\ ]i$; $I_{YY} = \sum [\ I_{yy}\ ]i$ for i = 1, n ....(7.22)

Moments of inertia of all individual sections should be added only after transferring all of them to a common axis.

7.3.1 HOLLOW RECTANGULAR SECTION

For a hollow rectangular section of outer dimensions ‘B’ and ‘D’ and inner dimensions ‘b’ and ‘d’

$I_{XX} = [I_{XX}]_1 - [I_{XX}]_2 - BD^3/12 \quad bd^3/12 = [\ BD^3 - bd^3\ ]\ /\ 12$

and $I_{YY} = [I_{YY}]_1 - [I_{YY}]_2 = B^3D/12 - b^3d/12 = [\ B^3D - b^3d\ ]\ /\ 12$

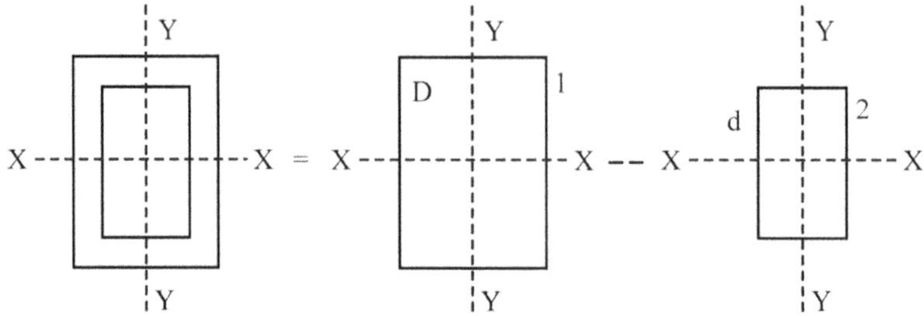

7.3.2 HOLLOW CIRCULAR SECTION

For a hollow circular section of outer diameter 'D' and inner diameter 'd'

$$I_{XX} = [I_{XX}]_1 - [I_{XX}]_2 = \pi D_4/64 - \pi d_4/64 = \pi(D_4 - d_4) / 64$$

Due to symmetry of section, $I_{YY} = I_{XX}$

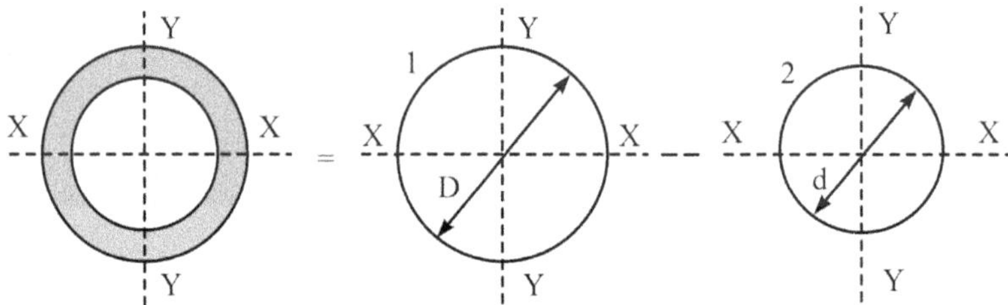

A general formula for the other cases is not convenient. Specific cases are given here.

7.3.3 T-SECTION

For a T-section of Width 20cm ; Height 30 cm ; Thickness of web and flange 4 cm

This section can be divided into two rectangles. Moment of inertia of the section through its centroidal X-axis is then obtained as the sum of the moments of inertia of the two rectangles about the same axis. It is, therefore, essential to find the location of the centroidal X-axis of the section. Let 'O' be the origin of the coordinate system and X and Y be the centroidal axes of the section

$$A_1 = b_1 d_1 = 20 \times 4 = 80 \quad ; \quad x_1 = 0; y_1 = 30 - 2 = 28$$

$$A_2 = b_2 d_2 = 4 \times (30 - 4) = 104 \quad ; \quad x_2 = 0; y_2 = (30 - 4)/2 = 13$$

Section is symmetric about Y-axis. Therefore, $X_G = 0$

It can also be found from

$$X_G = [A_1 \times x_1 + A_2 \times x_2] / [A_1 + A_2] = 0 \quad \text{since } x_1 = x_2 = 0$$

$Y_G = [A_1 \times y_1 + A_2 \times y_2] / [A_1 + A_2]$

$= [80 \times 28 + 104 \times 13] / [80 + 104]$

$= (2240 + 1352)/184 = 3592/184 = 19.52$ cm

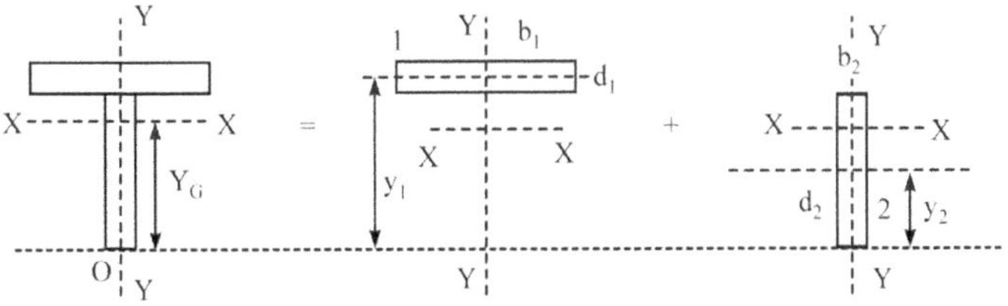

$I_{XX} = [I_{XX}]_1 + [I_{XX}]_2$

$= [(b_1 d_1^3/12) + A_1 \times (y_1 - Y_G)^2] + [(b_2 d_2^3/12) + A_2 \times (y_2 - Y_G)^2]$

$= [(20 \times 43/12) + 80 \times (28 - 19.52)2]$

$+ [(263 \times 4/12) + 104 \times (13 - 19.52)2]$

$= 16139.3$ cm^4

Since Y-axes of the two rectangles coincide with the Y-axis of the section,

$I_{YY} = [I_{YY}]_1 + [I_{YY}]_2 = (b_1^3 d_1/12) + (b_2^3 d_2/12)$

$= (203 \times 4/12) + (263 \times 4/12)$

$= 8525.4$ cm^4

7.3.4 I-SECTION

For an I-section of Width 20cm at top & 30cm at bottom, Height 40 cm and Thickness of web and flanges 4 cm

This section can be divided into three rectangles. Moment of inertia of the section through its centroidal X-axis is then obtained as the sum of the moments of inertia of the three rectangles about the same axis. It is, therefore, essential to find the location of the centroidal X-axis of the section. Let O be the origin of the coordinate system and X and Y be the centroidal axes of the section.

$A_1 = b_1 \times d_1 = 20 \times 4 = 80$; $x_1 = 0; y_1 = 40 - 2 = 38$

$A_2 = b_2 \times d_2 = 4 \times (40 - 4 - 4) = 128$; $x_2 = 0; y_2 = 40/2 = 20$

$A_3 = b_3 \times d_3 = 30 \times 4 = 120$; $x_3 = 0; y_3 = 4/2 = 2$

Section is symmetric about y-axis. Therefore, $X_G = 0$

It can also be found from

$X_G = [A_1 \times x_1 + A_2 \times x_2 + A_3 \times x_3] / [A_1 + A_2 + A_3] = 0$

since $x_1 = x_2 = x_3 = 0$

$Y_G = [A_1 \times y_1 + A_2 \times y_2 + A_3 \times y_3] / [A_1 + A_2 + A_3]$

$= [80 \times 38 + 128 \times 20 + 120 \times 2] / [80 + 128 + 120] = 17.8$ cm

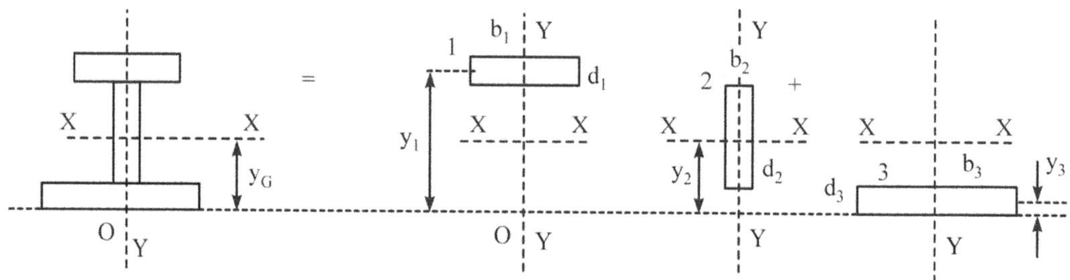

$I_{XX} = [I_{XX}]_1 + [I_{XX}]_2 + [I_{XX}]_3$

$= [(b_1 \times d_1^3/12) + (b_1 \times d_1) \times (y_1 - Y_G)^2]$

$+ [(b_2 \times d_2^3/12) + (b_2 \times d_2) \times (y_2 - Y_G)^2] + [(b_2 \times d_2^3/12)$

$+ (b_2 \times d_2) \times (y_3 - Y_G)^2]$

$= [(20 \times 4^3/12) + 80 \times (38 - 17.8)^2] + [(4 \times 32^3/12) + 128 \times (20 - 17.8)^2]$

$+ [(30 \times 4^3/12) + 120 \times (2 - 17.8)^2]$

$= 74408 \text{ cm}^4$

Since Y-axes of the three rectangles coincide with the Y-axis of the section,

$I_{YY} = [I_{YY}]_1 + [I_{YY}]_2 + [I_{YY}]_3$

$= (b_1^3 \times d_1/12) + (b_2^3 \times d_2/12) + (b_3^3 \times d_3/12)$

$= (20^3 \times 4/12) + (4^3 \times 32/12) + (30^3 \times 4/12)$

$= 11837.4 \text{ cm}^4$

7.3.5 L – SECTION

For a L-section of Width 20cm, Height 30 cm and Thickness of web and flange 4 cm

This section can be divided into two rectangles. Moment of inertia of the section through its centroidal X-axis / Y-axis is then obtained as the sum of the moments of inertia of the two rectangles about the same axis. It is, therefore, essential to find the coordinates of the centroid of the section. Let 'O' be the origin of the coordinate system and X and Y be the centroidal axes of the section

$A_1 = b_1 d_1 = 20 \times 4 = 80$; $x_1 = 20/2 = 10$; $y_1 = 4/2 = 2$

$A_2 = b_2 d_2 = 4 \times (30 - 4) = 104$; $x_2 = 4/2 = 2$; $y_2 = (30 - 4)/2 + 4 = 17$

$X_G = [A_1 \times x_1 + A_2 \times x_2] / [A_1 + A_2] = [80 \times 10 + 104 \times 2] / [80 + 104]$

$= 5.48$ cm

$Y_G = [A_1 \times y_1 + A_2 \times y_2] / [A_1 + A_2] = [80 \times 2 + 104 \times 17] / [80 + 104]$

$= 10.48$ cm

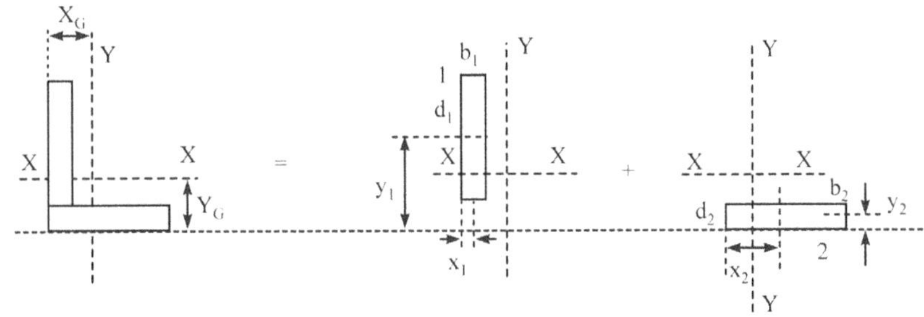

$I_{XX} = [I_{XX}]_1 + [I_{XX}]_2$

$= [(b_1 \times d_1^3/12) + (b_1 \times d_1) \times (y_1 - Y_G)^2]$

$+ [(b_2 d_2^3/12) + (b_2 d_2) \times (y_2 - Y_G)^2]$

$= [(4 \times 26^3/12) + 104 \times (17 - 10.48)^2] + [(20 \times 4^3/12) + 80 \times (2 - 10.48)^2]$

$= 16139.3\ \text{cm}^4$

$I_{YY} = [I_{YY}]_1 + [I_{YY}]_2 = [(b_1^3 \times d_1/12) + (b_1 \times d_1) \times (X_1 - X_G)^2]$

$+ [(b_2^3 \times d_2/12) + (b_2 \times d_2) \times (X_2 - X_G)^2]$

$= [(4^3 \times 26/12) + (4 \times 26 \times 3.48^2)] + [(20^3 \times 4/12) + (4 \times 20 \times 4.52^2)]$

$= 11419.3\ \text{cm}^4$

7.3.6 C – SECTION

For a C-section of Width 20cm, Height 30 cm and Thickness of web and flanges 4 cm

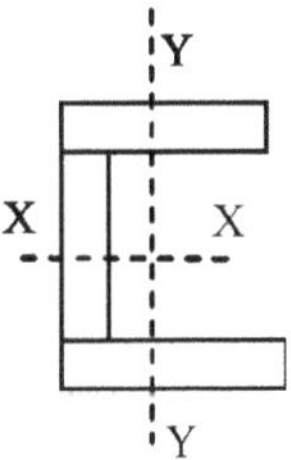

Section is symmetric about X-axis. Therefore, $Y_C = 0$

Taking top flange as area-1, vertical web as area-2, bottom flange as area-3 and taking all measurements w.r.t. the left edge,

$A_1 = A_3 = 20 \times 4$; $A_2 = (30 - 2 \times 4) \times 4$

$X_1 = X_3 = 20/2 = 10$; $X_2 = 4/2 = 2$; $Y_1 = Y_3 = (30 - 4)/2$; $Y_2 = 0$

$X_C = [A_1 \times X_1 + A_2 \times X_2 + A_3 \times X_3] / [A_1 + A_2 + A_3]$

$= [2 \times (20 \times 4) \times 10 + (4 \times 22) \times 2] / [2 \times (20 \times 4) + (4 \times 22)] = 7.16$ cm

$I_{XX} = 2\,[(b_1 \times d_1^3/12) + (b_1 \times d_1) \times (Y_C - Y_1)^2]$

$+ [(b_2 \times d_2^3/12) + (b_2 \times d_2) \times (Y_C - Y_2)^2]$

$= 2\,[(20 \times 4^3 / 12) + (20 \times 4) \times (0 - 13)^2)] + [(4 \times 22^3 / 12) + (4 \times 22) \times (0 - 0)^2]$

$= 30802$ cm^4

$I_{YY} = 2\,[(b_1^3 \times d_1/12) + (b_1 \times d_1) \times (X_C - X_1)^2]$

$+ [(b_2^3 \times d_2/12) + (b_2 \times d_2) \times (X_C - X_2)^2]$

$= 2\,[(20^3 \times 4 / 12) + (20 \times 4) \times (7.16 - 10)^2]$

$+ [(4^3 \times 22 / 12) + (4 \times 22) \times (7.16 - 2)^2]$

$= 9084.4$ cm^4

Example 5

A quarter circular area of radius 'a' is removed from a square area of side 'a'. Calculate moment of inertia of the remaining section about X and Y axes.

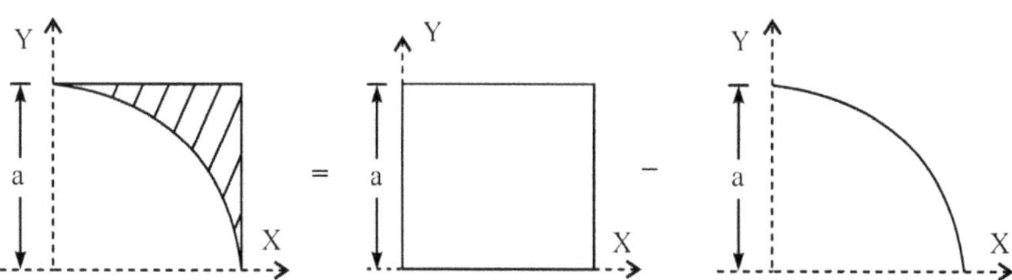

Solution :

$I_{XX} = (I_{XX})_{Square} - (I_{XX})_{Quadrant}$

$= (a^4/3) - (\pi\, a^4/16)$

$= a^4(16 - 3\pi)/48 = 0.1369\, a^4$

Similarly, $I_{YY} = (I_{YY})_{Square} - (I_{YY})_{Quadrant} = (a^4/3) - (\pi\, a^4/16) = 0.1369\, a^4$

Example 6

A circular hole of diameter 15 mm is removed from a rectangular area of width 20 mm and height 40 mm, as shown. Calculate moment of inertia of the remaining section about horizontal and vertical axes through its centroid.

Solution:

Moment of inertia of this section can be obtained from those of a rectangle (1) and a circular plate (2), transferred to a common axis. Let 'O' be the origin of the coordinate system and X and Y be the centroidal axes of the section

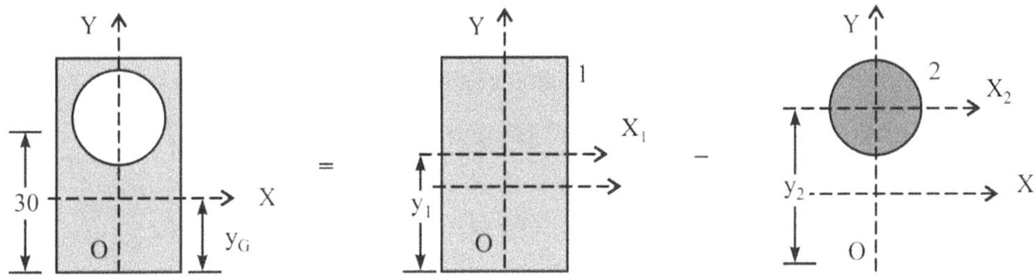

Since the section is symmetric about Y-axis, $X_G = 0$

$y_1 = 40/2$ and $y_2 = 30$

$Y_G = (A_1 \times y_1 - A_2 \times y_2) / (A_1 - A_2)$

$= [20 \times 40 \times 20 - (\pi \times 15^2/4) \times 30] / [20 \times 40 - (\pi \times 15^2/4)] = 17.16$ mm

$I_{XX} = [\ (I_{XX})_1 + A_1 \times (y_1 - Y_G)^2\ ] - [\ (I_{XX})_2 + A_2 \times (y_2 - Y_G)^2\ ]$

$= [(20 \times 40^3/12) + 20 \times 40 \times (20 - 17.16)^2]$

$- [(\pi \times 15^4/64) + (\pi \times 15^2/4) \times (30 - 17.16)^2]$

$= 81485\ mm^4$

and $I_{YY} = (I_{YY})_1 - (I_{YY})_2 = (20^3 \times 40/12) - (\pi \times 15^4/64) = 24180\ mm^4$

7.4 PRODUCT OF INERTIA

This property does not have much of practical use, except for finding out principal planes. It is defined as $I_{XY} = \int dA \times (x \times y)$. Depending on the nature of cross section and coordinate axes, IXY can take +ve or -ve values. As the coordinate axes through centroid are rotated, the product of inertia value changes from +ve to –ve or vice versa. Therefore, with reference to some particular axes, the product of inertia value becomes zero. These axes are called ***Principal axes***. These axes have significance in deciding optimum bending plane in a cross section. *Obviously, the product of inertia is zero, if any one axis or both the axes divide the section into symmetric parts.*

Thus, product of inertia for circlular, semi-circular, elliptic, parabolic, rectangular, etc.. sections about any line of symmetry is zero. Also, *the product of inertia of a given area w.r.t. the principal axes at the centroid is zero* .i.e., ***line of symmetry is a principal axis.***

Parallel axes theorem is also applicable to product of inertia with appropriate change.

i.e., $I_{X'Y'} = I_{XY} + A \times (a \times b)$,

where a and b are the distances between the two sets of axes.

(a) Rectangular section of width 'B' and height 'D', with axes coinciding adjacent sides

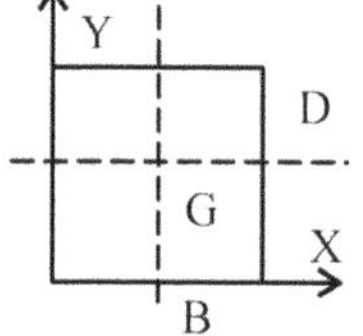

$$I_{XY} = (I_{XY})_G + (B \times D) \times (B/2) \times (D/2)$$
$$= 0 + B^2 D^2 / 4 = B^2 D^2 / 4$$

(b) Quadrant of a circle of radius 'R'

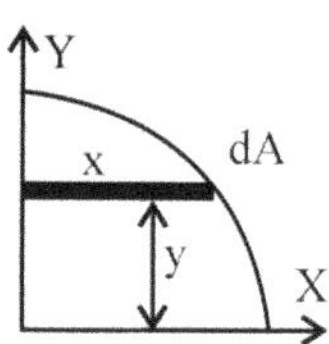

$$I_{XY} = \int (x/2) \times y \times (x \times dy) = (1/2) \int x^2 \times y \times dy$$
$$= (1/2) \int (R^2 - y^2) \times y \times dy$$
$$= (1/2) [R^2y^2/2 - y^4/4] \quad \text{with} \quad 0 \leq y \leq R$$
$$= R^4/8$$

7.5 ROTATION OF AXES

Let us consider a small area dA. With respect to axes X-Y,

$$I_{XX} = \int y^2 \times (dA); \quad I_{YY} = \int x^2 \times (dA); \quad I_{XY} = \int x \times y \times (dA)$$

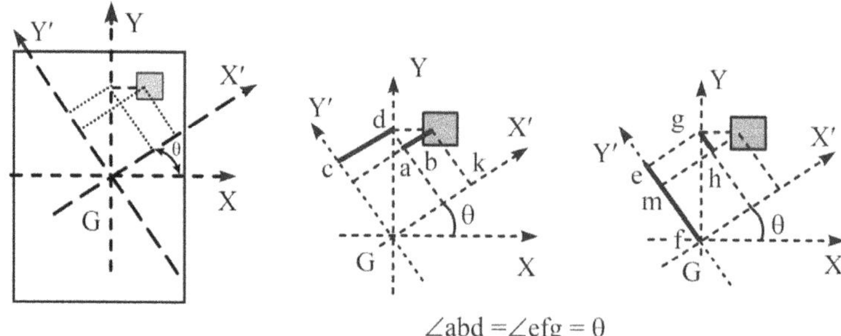

If X-Y axes are rotated through an angle θ to X′–Y′ axes, from the above fig,

$x' = Gk = ab + cd = x \cos\theta + y \sin\theta$; $y' = fm = ef - gh = y \cos\theta - x \sin\theta$

then, $I_{X'X'} = \int (y')^2\, dA = \int (y \cos\theta - x \sin\theta)^2\, dA$

$= \int y^2 \cos^2\theta\, dA + \int x^2 \sin^2\theta\, dA - \int 2xy \sin\theta \cos\theta\, dA$

$= I_{XX} \cos^2\theta + I_{YY} \sin^2\theta - 2\, I_{XY} \sin\theta \cos\theta$

$= I_{XX}(1+\cos 2\theta)/2 + I_{YY}(1-\cos 2\theta)/2 - 2I_{XY} \sin\theta\cos\theta$

$= (I_{XX} + I_{YY}) / 2 + (I_{XX} - I_{YY}) \times \cos 2\theta / 2 - I_{XY} \times \sin 2\theta$ .....(7.23)

and $I_{Y'Y'} = \int (x')^2\, dA = \int (x \cos\theta + y \sin\theta)^2\, dA$

$= \int x^2 \cos^2\theta\, dA + \int y^2 \sin^2\theta\, dA + \int 2xy \sin\theta \cos\theta\, dA$

$= I_{XX} \sin^2\theta + I_{YY} \cos^2\theta + 2\, I_{XY} \sin\theta \cos\theta$

$= I_{XX}(1-\cos 2\theta)/2 + I_{YY}(1+\cos 2\theta)/2 + 2I_{XY} \sin\theta\cos\theta$

$= (I_{XX} + I_{YY}) / 2 - (I_{XX} - I_{YY}) \times \cos 2\theta / 2 + I_{XY} \times \sin 2\theta$ .....(7.24)

Adding and subtracting these two equations, we get,

$I_{X'X'} + I_{Y'Y'} = I_{XX} + I_{YY} = I_{ZZ}$

and $I_{X'X'} - I_{Y'Y'} = I_{XX} - I_{YY} \times \cos 2\theta - 2\, I_{XY} \times \sin 2\theta$

Product of inertia w.r.t. the new axes,

$I_{X'Y'} = I_{XX} - I_{YY} \times \sin 2\theta / 2 + I_{XY} \times \cos 2\theta$

If X′-Y′ are principal axes, $I_{X'Y'} = 0$ or $\tan 2\theta = 2\, I_{XY} / (I_{YY} - I_{XX})$ .....(7.25)

Maximum and minimum moments of inertia of any section are referred to the principal axes and are given by

$$I_{X'X'},\ I_{Y'Y'} = \left(\frac{I_{XX} + I_{YY}}{2}\right) \pm \sqrt{\left(\frac{I_{XX} - I_{YY}}{2}\right)^2} \quad .....(7.26)$$

7.6 MOHR'S CIRCLE OF INERTIA

This is used to find the principal planes geometrically, by plotting I_{XX} and I_{YY} along X-axis and I_{XY} along Y-axis. Mark E and G such that $OE = I_{XX}$; $OG = I_{YY}$

Mark D and H such that $ED = -HG = I_{XY}$

Then, draw a circle with C, mid-point of GE, as the center and CD or CH as the radius.

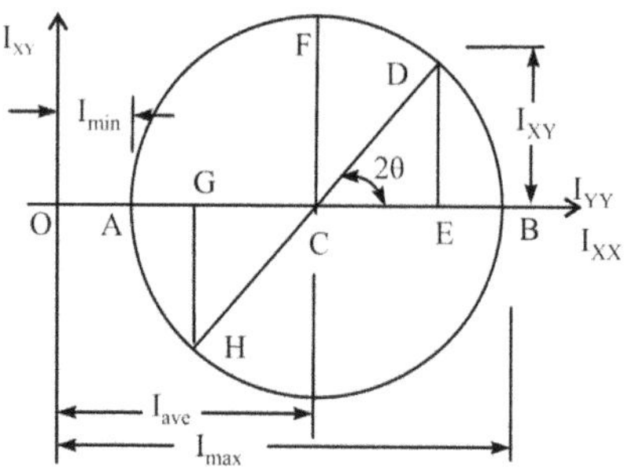

Corresponding to $I_{XY} = 0,\ I_{max} = OB = OC + CB$

and $I_{min} = OA = OC - CA$

where, $OC = I_{Ave} = (I_{XX} + I_{YY}) / 2$

and $CB = CA = CF = (I_{XY})_{max} = \sqrt{[(I_{XX} - I_{YY})/2]^2 + I_{XY}^2}$

Inclination θ of the principal axes w.r.t X-Y axes is obtained from the angle 2θ between HD and AB in the Mohr's circle.

Since the moments of inertia I_{XX} and I_{YY} are always positive, Mohr's circle of inertia will always be in the first and fourth quadrants.

Example 7

A square hole of side 'r' is cut centrally from a circular cross section of radius 'r'. Calculate moment of inertia about a diagonal of the square hole.

Solution:

Because of symmetry about X and Y axes, centroid lies at the center of the square. Moment of inertia about X-Y can be obtained by rotation of the geometry from $X' - Y'$ axes by an angle of $\theta = 45°$.

$$I_{X'X'} = I_{Y'Y'} = [I_{X'X'}]_{Circle} - [I_{X'X'}]_{SqHole} = (\pi\, r^4/4) - (r^4/12)$$

$$= r^4\, (3\pi - 1) / 12 = 0.702\, r^4$$

and $I_{X'Y'} = 0$

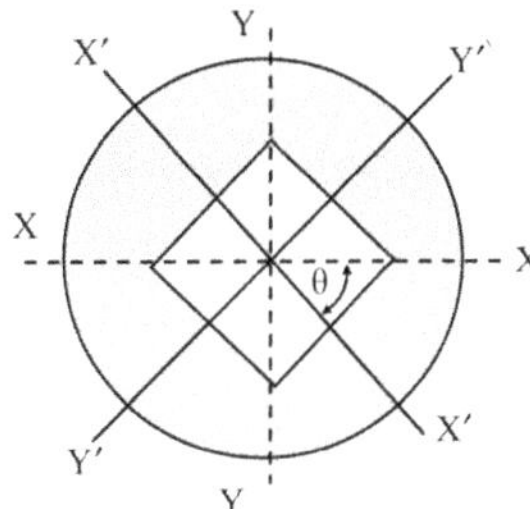

Therefore, rotating coordinate axes by θ = 450,

$$I_{XX} = I_{YY} = [I_{X'X'} + I_{Y'Y'}]/2 + [I_{X'X'} - I_{Y'Y'}] \cos 2\theta / 2 - I_{X'Y'} \sin 2\theta$$

$$= [I_{X'X'} + I_{Y'Y'}]/2 + [I_{X'X'} - I_{Y'Y'}] \times (0)/2 - (0) \times (1)$$

$$= [0.702\ r^4 + 0.702\ r^4]/2 = 0.702\ r^4$$

Example 8

A circular hole of radius 'r' is cut centrally from a square cross section of side '4r'. Calculate moment of inertia about a diagonal of the square.

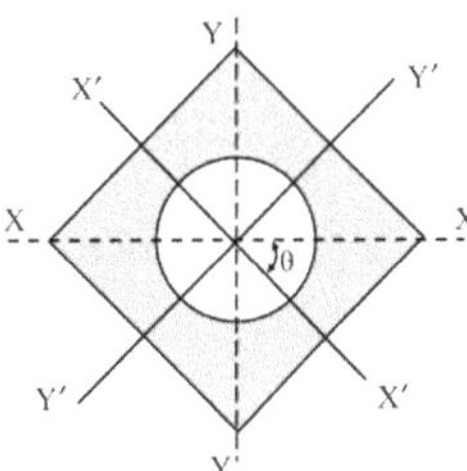

Solution:

Because of symmetry about X and Y axes, centroid lies at the center of the square. Moment of inertia about X-Y axes can be obtained by rotation of the geometry from X' - Y' axes by an angle of θ = 450.

$$I_{X'X'} = I_{Y'Y'} = [I_{X'X'}]_{Square} - [I_{X'X'}]_{Circle} = [(4r)4/12] - (\pi\ r^4/4)$$

$$= r^4\ (256 - 3\pi)/12 = 20.548\ r^4$$

and $I_{X'Y'} = 0$

Therefore, $I_{XX} = I_{YY} = [I_{X'X'} + I_{Y'Y'}]/2 + [I_{X'X'} - I_{Y'Y'}] \cos 2\theta/2 - I_{X'Y'} \sin 2\theta$

$$= [I_{X'X'} + I_{Y'Y'}]/2 + [I_{X'X'} - I_{Y'Y'}] \times (0)/2 - (0) \times (1)$$

$$= [20.548\ r^4 + 20.548\ r^4]/2$$

$$= 20.548\ r^4$$

Example 9

Compute moment of inertia of 10cm × 15cm rectangle about X-X axis, passing through corner A of the rectangle and inclined at α = sin-1(4/5) to the longer edge.

Solution:

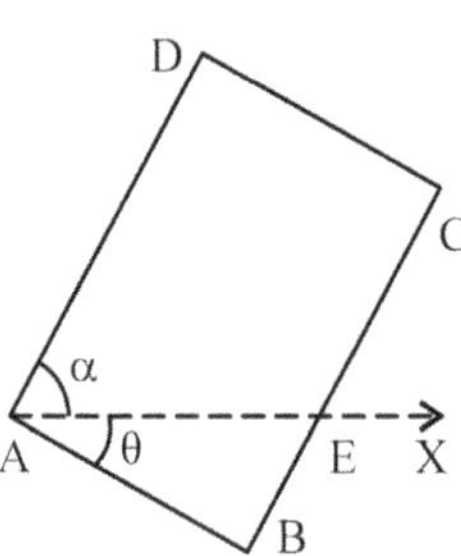

Given $\sin\theta = \cos\alpha = 3/5$; $\cos\theta = \sin\alpha = 4/5$

Method-1: The rectangle is divided into ***four triangles***, as shown, such that one side of each is parallel to X-axis

In order to find their areas, heights of these triangles have to be calculated from the given data. ∠A= 900 and $\theta = 90 - \alpha$

In ΔABE, BM =AB sin θ = 6cm ; AE =AB/cos θ = 12.5cm

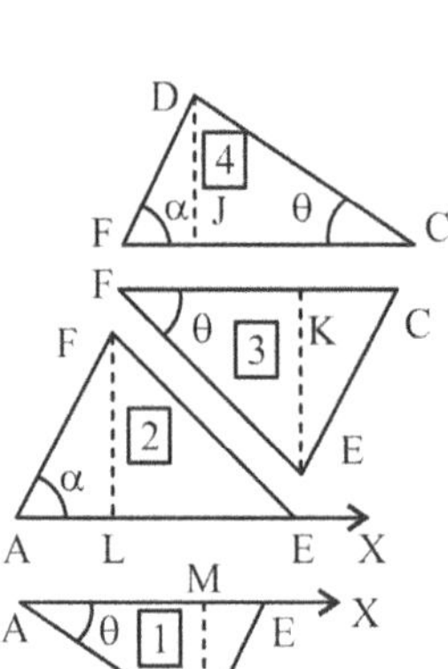

In ΔCDF,

DJ =CD sin θ = 6cm

CF = CD/cos θ = 12.5cm

FD = DJ/sin α or FC cos α = 7.5cm

In ΔAEF, FL= AF sin α = (AD – FD) sin α = 6cm

In ΔCEF, EK = FL = 6cm

$I_{XX} = (I_{XX})_{ABE} + (I_{XX})_{AEF} + (I_{XX})_{FCE} + (I_{XX})_{FCD}$

For ΔABE and ΔAEF,

M.I. about base (X-axis) = $b \times h^3/12$

For ΔFCE and ΔFCD, M.I. about X-axis

= M.I. about its centroidal X-axis + Area × d^2

$$\begin{aligned} \text{Thus, } I_{XX} &= (AE \times BM^3 / 12) + (AE \times FL^3 / 12) \\ &\quad + [(FC \times EK^3 / 36) \\ &\quad + (FC \times EK / 2) \times (2 \times EK / 3)^2] \\ &\quad + [(FC \times DJ^3 / 36) + (FC \times DJ / 2) \times \{EK + (DJ/3)\}^2] \\ &= (12.5 \times 6^3/12) + (12.5 \times 6^3/12) \\ &\quad + [(12.5 \times 6^3/36) + (12.5 \times 6/2) \times (2 \times 6/3)^2] \\ &\quad + [(12.5 \times 6^3/36) + (12.5 \times 6/2) \times \{6 + (6/3)\}^2] = 3600 \text{ cm}^4 \end{aligned}$$

Method-2: This problem can also be solved using ***rotation of axes***.

Considering AB as X'-axis and AD as Y'-axis,

$I_{X'X'} = b\,d^3 / 3 = 10 \times 15^3 / 3 = 11250$ cm^4

$I_{Y'Y'} = b^3\,d / 3 = 10^3 \times 15/3 = 5000$ cm^4

and $I_{X'Y'} = b^2\,d^2 / 4 = 10^2 \times 15^2 / 4 = 5625$ cm^4

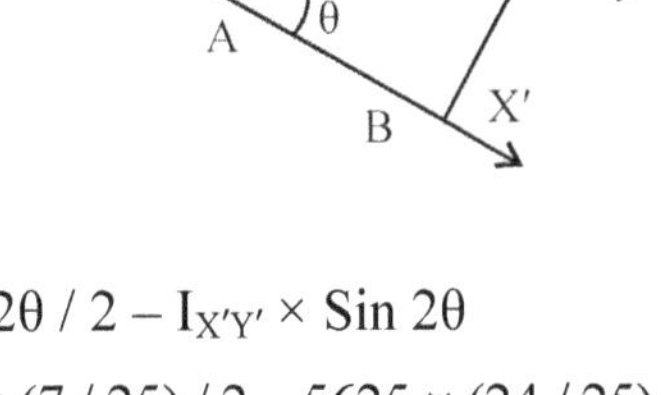

The angle between X-axis and X′-axis, $\theta = 90 - \alpha$

$\cos 2\theta = \cos^2\theta - \sin^2\theta = 7/25$

$\sin 2\theta = 2 \sin\theta \times \cos\theta = 24/25$

Using the relation for rotation of axes,

$$I_{XY} = [I_{X'X'} + I_{Y'Y'}] / 2 + [I_{X'X'} - I_{Y'Y'}] \times \text{Cos } 2\theta / 2 - I_{X'Y'} \times \text{Sin } 2\theta$$

$$= [11250 + 5000] / 2 + [11250 - 5000] \times (7/25) / 2 - 5625 \times (24/25)$$

$$= 8125 + 875 - 5400 = 3600 \text{ cm}^4$$

Method-3: This problem can also be solved using three triangles GCK, ABG and AKD, by subtracting ΔGCK from the combination of ΔABG and ΔAKD.

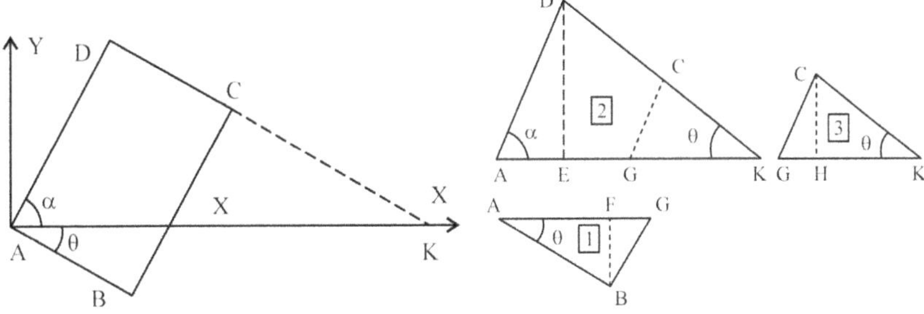

AG = AB / cos θ = 12.5 cm ; BF = AB sin θ = 6 cm

AK = AD / cos α = 25 cm ; GK = AK – AG = 12.5 cm

DE = AD sin α = 12 cm ; BG = AB tan θ = 7.5 cm

GC = BC – BG = 7.5 cm ; CH = GC sin α = 6 cm

$$I_{XX} = (I_{XX})_{ABG} + (I_{XX})_{AKD} - (I_{XX})_{GCK}$$

$$= AG \times BF^3 / 12 + AK \times DE^3 / 12 - GK \times CH^3 / 12$$

$$= [\,12.5 \times 6^3 + 25 \times 12^3 - 12.5 \times 6^3\,] / 12 = 3600 \text{ cm}^4$$

7.7 RADIUS OF GYRATION

It is a function of moment of inertia and cross sectional area and is an important property of rotating components. Corresponding to the three moments of inertia, there are three radii of gyration. It has the dimensions of length. It can be interpreted as the distance 'k' at which the complete area of cross section 'A' is squeezed and kept as a thin rectangle, parallel to the axis such that there is no change in its moment of inertia.

Then, $I = A \times k^2$

or $k_{XX} = \sqrt{I_{XX} / A}$; $k_{YY} = \sqrt{I_{YY} / A}$; $k_{ZZ} = \sqrt{I_{ZZ} / A}$

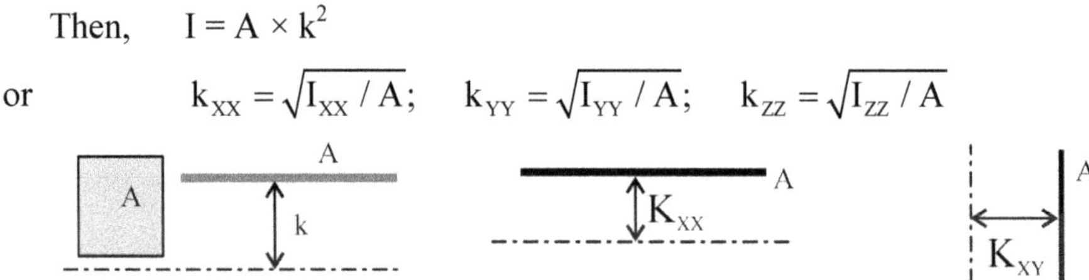

Radius of gyration is not defined for product of inertia

These values are given below for a few simple sections

(i) Rectangular section of breadth 'b' and depth 'd', through its centroid :

$$k_{XX} = \sqrt{\frac{b \times d^3 / 12}{b \times d}} = \sqrt{d^2 / 12} = d / \sqrt{12}$$

$$k_{YY} = \sqrt{\frac{d \times b^3 / 12}{b \times d}} = \sqrt{b^2 / 12} = b / \sqrt{12}$$

$$k_{ZZ} = \sqrt{\frac{b \times d \times (b^2 + d^2) / 12}{b \times d}} = \sqrt{\left(b^2 + d^2\right) / 12}$$

(ii) Circular cross section of diameter 'd', axis through diameter :

$$k_{XX} = k_{YY} = \sqrt{\frac{\pi d^4 / 64}{\pi d^2 / 4}} = \sqrt{d^2 / 16} = d / 4$$

$$k_{ZZ} = \sqrt{[\pi d^4 / 32] / [\pi d^2 / 4]} = \sqrt{d^2 / 8} = d / \sqrt{8}$$

(iii) Semi-circular cross section of diameter 'd', axis through diameter

$$k_{XX} = k_{YY} = \sqrt{\frac{\pi d^4 / 32}{\pi d^2 / 2}} = \sqrt{d^2 / 16} = d / 4$$

$$k_{ZZ} = \sqrt{\frac{\pi d^4 / 16}{\pi d^2 / 2}} = \sqrt{d^2 / 8} = d / \sqrt{8}$$

(iv) Hollow circle of outer diameter 'D' and inner diameter 'd', axis through its diameter :

$$k_{XX} = k_{YY} = \sqrt{\frac{\pi(D^4 - d^4)/64}{\pi(D^2 - d^2)/4}} = \sqrt{(D^2 + d^2)/16}$$

$$k_{ZZ} = \sqrt{\frac{\pi(D^4 - d^4)/32}{\pi(D^2 - d^2)/4}} = \sqrt{(D^2 + d^2)/8}$$

Example 10

Two semi-circular areas of radius 'a' are cut from a square area of side '4a', as shown. Find polar moment of inertia and polar radii of gyration through the centroid of the remaining area.

Solution:

Because of symmetry about X and Y axes, centroid lies at the center of the square. Using suffix 'SQ' for square and 'SC' for semi-circle,

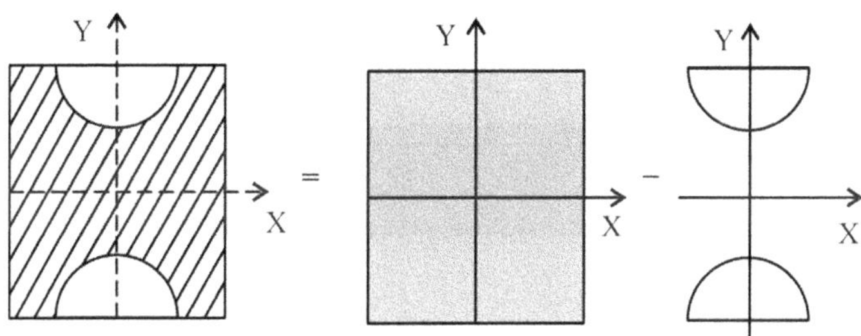

$$A = (4a) \times (4a) - 2\,[\,\pi \times a^2/2] = 12.86\ a^2$$

$$I_{XX} = [I_{XX}]_{SQ} - 2[I_{XX}]_{SC}$$

$$= (4a)^4/12 - 2\{\,0.11a^4 + (\pi a^2/2)\,[2a - (4a/3\pi)]^2\}$$

$$= 13.31\ a^4$$

$$I_{YY} = [I_{YY}]_{SQ} - 2[I_{YY}]_{SC} = (4a)^4/12 - 2[\pi a^4/8] = 20.54\ a^4$$

Polar M.I., $I_{ZZ} = I_{XX} + I_{YY} = 13.31\ a^4 + 20.54\ a^4 = 33.85\ a^4$

Radius of gyration about X-axis, $k_{XX} = \sqrt{I_{XX}/A} = \sqrt{13.31\ a^4/12.86\ a^2}$

$$= 1.017\ a$$

Radius of gyration about Y-axis, $k_{YY} = \sqrt{I_{YY}/A} = \sqrt{20.54\ a^4/12.86\ a^2}$

$$= 1.264\ a$$

Radius of gyration about Z-axis, $k_{ZZ} = \sqrt{I_{ZZ} / A} = \sqrt{33.85\ a^4 / 12.86\ a^2}$

$= 1.622\ a$

Example 11

A triangular area of width 3 cm is cut from the left end of the area bounded by the curve $y^3 = a\ x^2$; X-axis and x = 9cm line, as shown. Find polar moment of inertia and polar radius of gyration about Z-axis passing through 'O'.

Solution:

OB = 6 cm ; BC = b = 3 cm ; OC = OB + BC = 9cm ; CA = d = 6 cm

From the coordinates of point A, $a = y^3/x^2 = 6^3/9^2 = 8/3$

This can be considered as area (2) of triangle subtracted from the area (1) below the curve

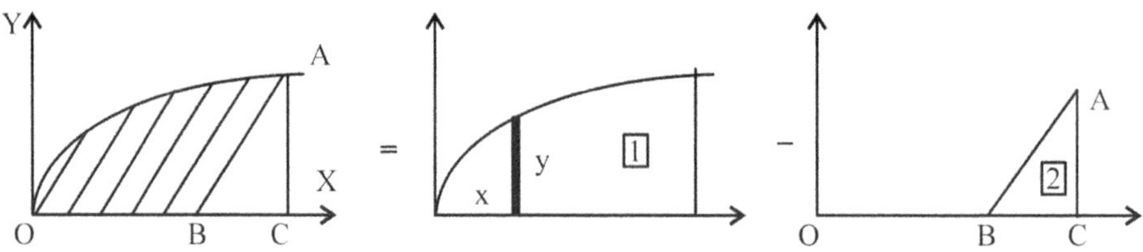

Considering a small area (y × dx) at a distance of 'x' from O,

For area-1 below the curve,

$I_{XX} = \int (y \times dx) \times (y/2)^2 = (1/4) \int a \times x^2\ dx = a \times x^3/12$ for $0 \le x \le 9$

$= 162\ cm^4$

$I_{YY} = \int (y \times dx) \times x^2 = \int a^{1/3} \times x^{8/3}\ dx = a^{1/3} \times x^{11/3} / (11/3)$ for $0 \le x \le 9$

$= 1192.9\ cm^4$

Polar Moment of inertia of the remaining area (area-1 – area-2),

$I_{ZZ} = [I_{XX} + I_{YY}]_1 - [I_{XX} + I_{YY}]_2$

$= [I_{XX} + I_{YY}]1 - [b \times d^3/12 + \{d \times b^3/12 + (b \times d/2) \times (OC - BC/3)^2\}\ ]$

$= (162 + 1192.9) - [3 \times 6^3/12 + \{6 \times 3^3/12 + (3 \times 6/2) \times (9 - 3/3)^2\}\ ]$

$= 1354.9 - (54 + 13.5 + 576) = 711.4\ cm^4$

Remaining area, $A = \int y \times dx - (b \times d)/2$

$$= \int (a \times x^2)^{1/3} dx - (b \times d)/2 = a^{1/3} \times x^{5/3}/(5/3) - (b \times d)/2$$

$$= (8/3)^{1/3} \times 9^{5/3}/(5/3) - (3 \times 6)/2 = 23.4 \text{ cm}^2$$

Radius of gyration about Z-axis, $k_{ZZ} = \sqrt{I_{ZZ} / A} = \sqrt{[711.4 / 23.4]}$

$$= 5.5138 \text{ cm}$$

SUMMARY

1. Moment of inertia (also called 2nd moment of area) about any axis is defined as the area of the cross section multiplied by square of the distance of the area from the axis.

   If the area is in X-Y plane,

   - For a small area A at a distance of x and y from the origin, $I_{XX} = A \times y^2$; $I_{YY} = A \times x^2$
   - For a large area of arbitrary shape, $I_{XX} = \int (dA \times y^2)$; $I_{YY} = \int (dA \times x^2)$
   - Moment of inertia about Z-axis (also called **Polar moment of inertia**), is defined as the area of section multiplied by square of its distance 'r' from the origin.

     $I_{ZZ} = A \times r^2$ or $\int (dA \times r^2) = \int (dA \times y^2) + \int (dA \times x^2)$ or $I_{ZZ} = I_{XX} + I_{YY}$

     This is also called **perpendicular axis theorem**.

2. Since area is always +ve and square of any distance is also +ve, M.I. is always +ve. It depends on the axis about which it is calculated and is minimum about an axis through centroid of the section

3. Moment of inertia I_{XX} calculated about an axis (X-X) through its centroid can be transferred to a parallel axis A-A using the relation, $I_{AA} = I_{XX} + A \times y^2$ where, A is the area of the plane and 'y' is the distance between the two parallel axes X-X and A-A. This is called Parallel axes theorem or Transfer of axis theorem.

   B Y X X A B Y A

   Similarly, $I_{BB} = I_{YY} + A \times x^2$

4. Moment of inertia of a composite section is the algebraic sum of moments of inertia of its simpler components, all transferred to a common axis using parallel axes theorem.

   $I_{XX} = \sum [I_{XX}]_i$ and $I_{YY} = \sum [I_{yy}]_i$

5. Product of inertia is defined as $I_{XY} = \int dA \times (x \times y)$. Depending on the nature of cross section and coordinate axes, I_{XY} can take +ve or –ve values. As the coordinate axes through centroid are rotated, the product of inertia value changes from +ve to –ve or vice versa. Therefore, with reference to some particular axes, the product of inertia value becomes zero. These axes are called ***Principal axes***. Maximum and minimum Moments of Inertia are always seen on the principal axes.

6. Product of inertia is zero, if any one axis or both the axes divide the section into symmetric parts i.e., ***line of symmetry is a principal axis.***

7. Parallel axes theorem is also applicable to product of inertia with appropriate change.

   i.e., $I_{X'Y'} = I_{XY} + A \times (a \times b)$,

   where a and b are the distances between the two sets of axes.

8. M.I. of any section about axes rotated by θ are given by

   $I_{X'X'} = (I_{XX} + I_{YY}) / 2 + (I_{XX} - I_{YY}) \cos 2\theta / 2 - I_{XY} \sin 2\theta$

   and $I_{Y'Y'} = (I_{XX} + I_{YY}) / 2 - (I_{XX} - I_{YY}) \cos 2\theta / 2 + I_{XY} \sin 2\theta$

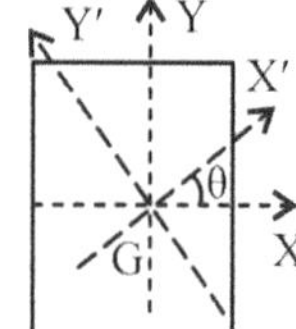

   Adding and subtracting these two equations, we get,

   $$I_{X'X'} + I_{Y'Y'} = I_{XX} + I_{YY} = I_{ZZ}$$

   and $I_{X'X'} - I_{Y'Y'} = (I_{XX} - I_{YY}) \times \text{Cos } 2\theta - 2 I_{XY} \times \text{Sin } 2\theta$

   Product of inertia w.r.t. the new axes,

   $$I_{X'Y'} = (I_{XX} - I_{YY}) \times \text{Sin } 2\theta / 2 + I_{XY} \times \text{Cos } 2\theta$$

9. Corresponding to $I_{X'Y'} = 0$, on the principal axes, Max & min moments of inertia are

   $$I_{X'X'}, I_{Y'Y'} = (I_{XX} + I_{YY}) / 2 \pm \sqrt{[(I_{XX} - I_{YY}) / 2]^2 + I_{XY^2}}$$

10. Mohr's circle is a graphical method of determining principal axes, max & min M.I., with I_{XX} and I_{YY} on X-axis and I_{XY} on Y-axis

    $OE = I_{XX}$; $OA = I_{YY}$; $DE = GH = I_{XY}$

    Distance of center, $OC = I_{Average}$

    Ends on X-axis diameter, $OB = I_{Max}$; $OA = I_{Min}$

    Radius of circle, CH = CD = Max Product of inertia $(I_{XY})_{Max}$

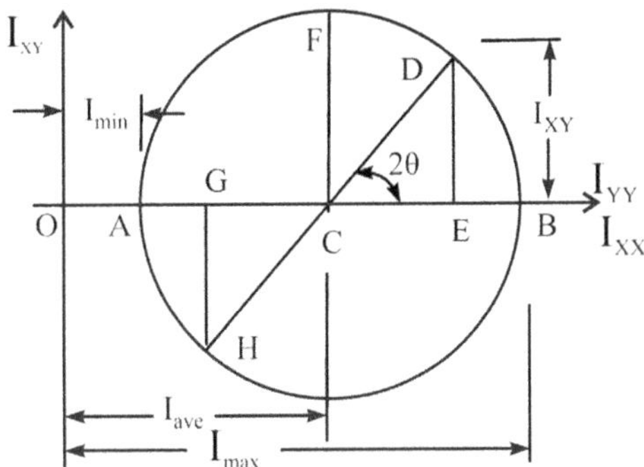

11. **Radius of gyration** is the distance from the axis, at which the entire area is assumed to be concentrated, giving the same moment of inertia

i.e., $I_{XX} = A \times k_{XX}^2$ or $k_{XX} = \sqrt{(I_{XX} / A)}$

Similarly, $k_{YY} = \sqrt{(I_{YY} / A)}$ and $k_{ZZ} = \sqrt{(I_{ZZ} / A)}$

SOME USEFUL VALUES

1. For a rectangle of width b and height h through centroid,

   $I_{XX} = b\,h^3/12$; $I_{YY} = b^3\,h/12$

2. For a rectangle of width b and height h about its sides,

   $I_{XX} = b\,h^3/3$; $I_{YY} = b^3\,h/^3$

3. For a triangle of base b and height h about its base, $I_{XX} = b\,h^3/12$
4. For a triangle about its centroidal axis parallel to base, $I_{XX} = b\,h^3/36$
5. For a quarter circle through diameter, $I_{XX} = I_{YY} = \pi\,r^4/16$
6. For an ellipse of major axis '2a' and minor axis '2b' through its axes,

   $I_{XX} = \boldsymbol{\pi\,a\,b^3/4}$; $\boldsymbol{I_{YY} = \pi\,a^3\,b/4}$

   For a quarter circular ring of radius R and thickness t,

   $\boldsymbol{I_{XX} = I_{YY} = I_{ZZ}/8 = \pi R^3 t/4}$

PROBLEMS FOR PRACTICE

1. Determine moment of inertia of a C - section of width 20cm, Height 30cm and thickness of web and flanges 4cm about its centroidal X and Y axes.

(***Ans:*** $I_{XX} = 14002.7\text{cm}^4$, $I_{XX} = 9084.4\text{cm}^4$)

2. A circular hole of radius 'r' is cut centrally from a square cross section of side '4r'. Calculate moment of inertia about a diagonal of the square.

(***Ans:*** $20.548\ r^4$)

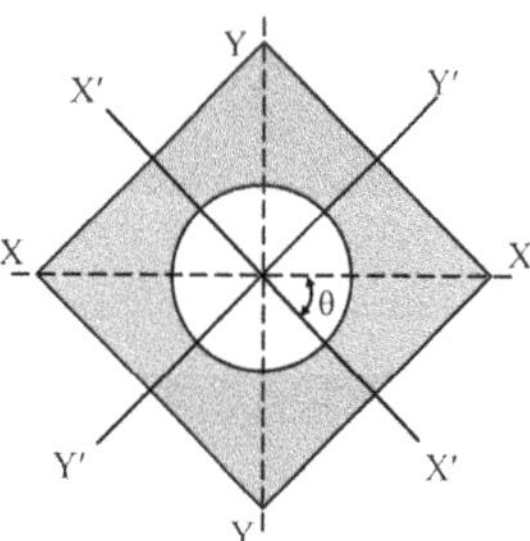

3. An isosceles triangle has a horizontal base of width 3m and height 4m. What is the area M.I. about a horizontal axis passing through its vertex.

(***Ans:*** 48m^4)

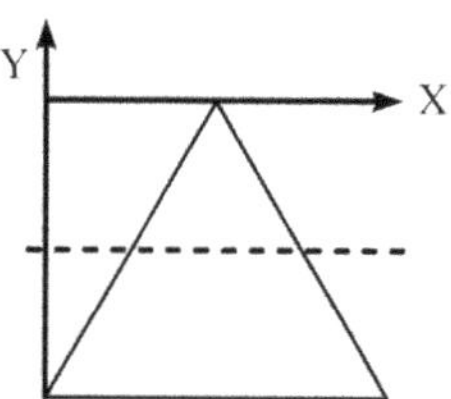

4. Find the moment of inertia I_{XX} of the shaded area, obtained by removing the triangular area from the semicircular area of radius 0.4 m

(***Ans:*** 0.0579m^4)

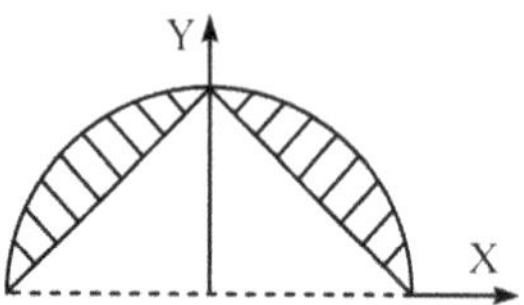

CHAPTER 8

MASS MOMENT OF INERTIA

Mass moment of inertia (or second moment of mass about an axis) is very much similar to area moment of inertia, with area replaced by mass of volume. As such, its units are 'kg mm^2 or kg cm^2'. It is a property, associated with dynamic response of bodies. Mass moment of inertia about any axis is defined as the mass of volume of the body multiplied by square of the distance of the center of mass from the axis. It can be represented by $(I_m)_{XX}$, $(I_m)_{YY}$,.. But, it is usually represented by I_{XX}, I_{YY}, etc.. and the engineer should distinguish it from area moment of inertia. Throughout this chapter, we use 'ρ' to represent mass density of the body in 'kg/mm^3 or kg/cm^3'

For a plane lamina of area A and thickness t having a mass 'M' at a distance of 'x' from Y-axis and at a distance of 'y' from X-axis,

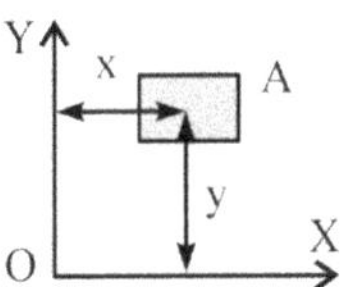

$$I_{XX} = M \times y^2 = (\rho\, t\, A) \times y^2$$

$$= (\rho\, t\,) \times (\text{Area moment of inertia})$$

where t is the thickness of a plane lamina

Similarly, $I_{YY} = M \times x^2 = (\rho\, t) \times (\text{Area moment of inertia})$

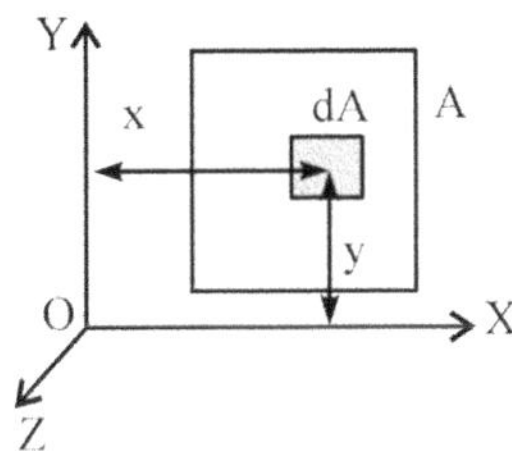

Moment of inertia of a large body can be obtained by integration as

$$I_{XX} = \int (dm \times y^2) = \rho\, t \int (dA \times y^2)$$

Similarly, $$I_{YY} = \int (dm \times x^2) = \rho\, t \int (dA \times x^2)$$

$$I_{ZZ} = \int (dm \times r^2) = \rho\, t \int (dA \times r^2)$$

$$= \rho\, t \int (dA \times x^2) + \rho\, t \int (dA \times y^2) = I_{XX} + I_{YY}$$

$\mathbf{I_{ZZ} = I_{XX} + I_{YY}}$ is also called **perpendicular axes theorem**.

Thus, mass moment of any lamina is equal to the area moment of inertia, multiplied by the constant '$\rho\, t$'

Since mass moment of inertia is defined for volumes of bodies, there exist ***three products of inertia*** defined by $I_{XY} = \int dm \times x \times y$; $I_{YZ} = \int dm \times y \times z$; $I_{ZX} = \int dm \times z \times x$

8.1 PARALLEL AXES THEOREM

Parallel axes theorem defined for area moment of inertia is also applicable to mass moment of inertia and is defined as

$$I_{X'X'} = I_{XX} + M \times y^2$$

where, M is the mass of the body

and y is the distance between the two parallel axes X-X (through centroid) and $X' - X'$

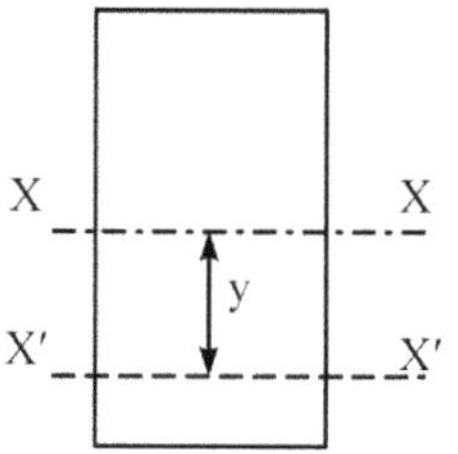

This is also called <u>*Transfer of axis theorem*</u>

(a) Mass moment of inertia of a ***slender bar*** of length 'L' and area of cross section 'A', ***about Y-axis*** at a distance 'L_1' from one end and 'L_2' from the other end,

$$I_{YY} = \int dm \times x^2 = \int (\rho\, A\, dx) \times x^2 = [\rho\, A\, x^3 / 3] \quad \text{for } -L_1 \le x \le L_2$$

$$= \rho\, A\, (L_2^3 - L_1^3) / 3$$

Case 1: $L_1 = L_2 = L/2$ Then, $I_{YY} = \rho\, A\, L^3 / 12 = M \times L^2 / 12$

where $M = \rho\, A\, L$, is the total mass of the bar

Case 2: $L_1 = 0$ and $L_2 = L$ Then, $I_{YY} = \rho\, A\, L^3 / 3 = M \times L^2 / 3$

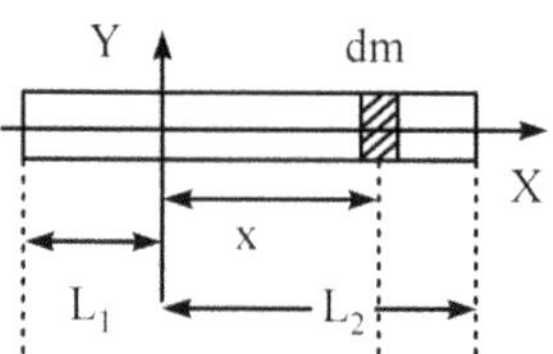

(b) Mass moment of inertia of a ***thin rectangular plate*** of dimensions $b \times d \times t$, through its centroid, from the area moment of inertia is

$$I_{XX} = \rho\, t \times (b \times d^3/12) = M \times d^2/12$$

where $M = \rho\, t \times (b \times d)$ is the total mass of the plate

Alternatively, from fundamental principles,

$$I_{XX} = \int dm \times y^2 = \int\int (\rho\, t \times dx \times dy) \times y^2$$

$$= \rho\, t \times x \times [y^3/3]$$

$$\text{for} \quad -b/2 \le x \le +b/2\,; \quad -d/2 \le y \le +d/2$$

$$= \rho\, t \times b \times d^3/12 = M \times d^2/12$$

Similarly, $I_{YY} = \rho\, t\,(b^3 \times d/12) = M \times b^2/12$

and $I_{ZZ} = \rho\, t\ b \times d\,(b^2 + d^2)/12 = M \times (b^2 + d^2)/12$

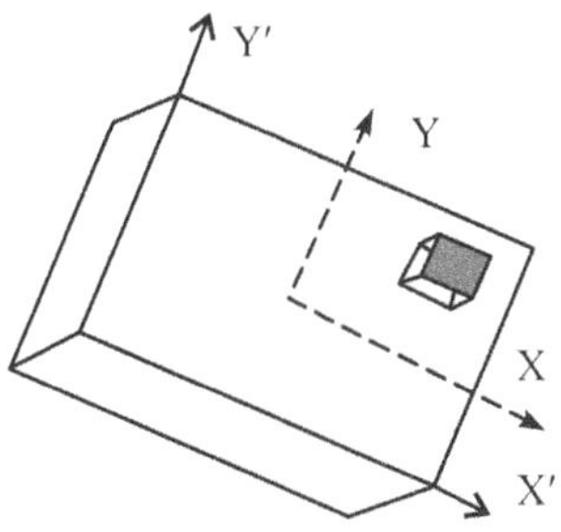

Mass moment of inertia about the edges,

$$I_{X'X'} = \int dm \times y^2 = \int (\rho \times t \times dx \times dy) \times y^2$$

$$= \rho \times t \times x \times [y^3/3] \qquad \text{for } 0 \le x \le +b \quad \text{and} \quad 0 \le y \le +d$$

$$= \rho \times t \times b \times d^3/3 = M \times d^2/3$$

Similarly, $I_{Y'Y'} = M \times b^2/3$ and $I_{Z'Z'} = M \times (b^2 + d^2)/3$

(c) Mass moment of inertia of a ***circular plate*** of radius 'r' and thickness 't', about its diameter,

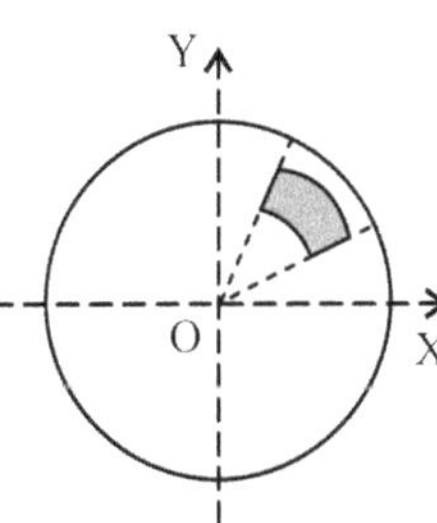

$$I_{XX} = \rho \times t \times (\pi \times R^4/4) = M \times R^2/4$$

Where, $M = \rho\, t\, \pi \times R^2$ is the total mass of the plate

Similarly, $I_{YY} = \rho \times t \times (\pi R^4/4) = M \times R^2/4$

and $I_{ZZ} = \rho \times t \times (\pi R^4/2) = M \times R^2/2$

Alternatively, mass moment of inertia about an axis perpendicular to the plate through O,

$$I_{ZZ} = \int dm \times r^2 = \int\int [\rho \times t \times (r\,d\theta) \times dr\,]\,r^2$$

for $0 \le r \le R$ and $0 \le \theta \le 2\pi$

$$= \rho \times t \times (\int r^3\,dr) \times (\int d\theta)$$

since r and θ are mutually independent

$$= \rho \times t \times (2\pi) \times (R^4/4) = M \times R^2 / 2$$

with $M = \rho \times t \times \pi R^2$ is the mass of the plate

and $I_{XX} = I_{YY} = I_{ZZ} / 2 = M \times R^2/4$ from perpendicular axis theorem

Case 1: Mass moment of inertia of a **semi-circular plate** of radius 'R'

$$\mathbf{I_{ZZ}} = \int dm \times r^2 = \int\int [\rho \times t \times (r\,d\theta) \times dr\,]\,r^2$$

for $0 \le r \le R$ and $0 \le \theta \le \pi$

$$= \rho \times t \times \pi \times (R^4/4) = M \times R^2 / 2 \qquad \text{with } M = \rho \times t \times \pi \times R^2/2$$

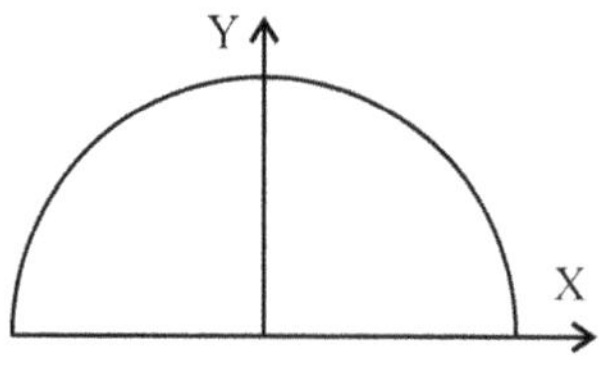

Case 2: Mass moment of inertia of a **hollow circular plate** of outer radius 'R_1' and inner radius 'R_2'

$$\mathbf{I_{ZZ}} = \int dm \times r^2 = \int\int [\rho \times t \times (r\,d\theta) \times dr\,]\,r^2$$

for $R_2 \le r \le R_1$ and $0 \le \theta \le 2\pi$

$$= \rho \times t \times \pi \times (R_1^4 - R_2^4)/2$$

$$= [\rho \times t \times \pi \times (R_1^2 - R_2^2)] \times (R_1^2 + R_2^2) / 2$$

$$= M \times (R_1^2 + R_2^2) / 2 \qquad \text{with } M = \rho \times t \times \pi \times (R_1^2 - R_2^2)$$

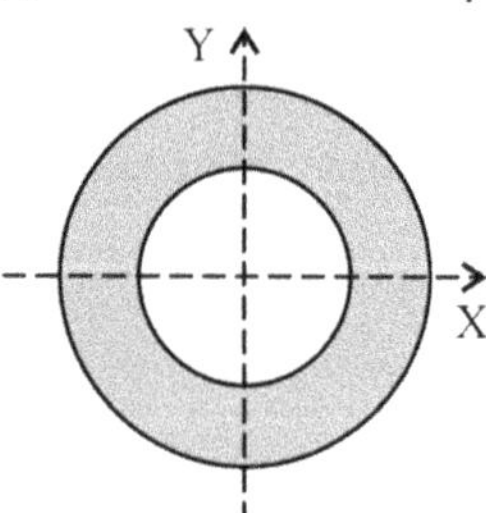

(d) Mass moment of inertia of a ***circular ring*** of radius 'R' and area of cross section 'A', $I_{XX} = \rho \times A \times (\pi R^3) = M \times R^2/2$

where $M = \rho \times A \times 2\pi R$, is the total mass of the plate

Similarly, $I_{YY} = \rho \times A \times (\pi R^3) = M \times R^2/2$

and $I_{ZZ} = \rho \times A \times (2\pi \times R^3) = M \times R^2$

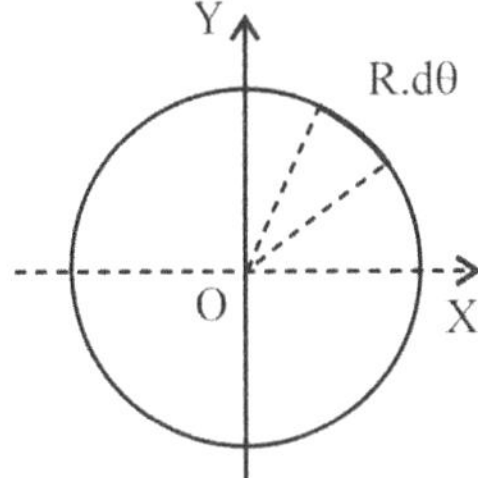

Alternatively, it can also be calculated from fundamental principles by integrating moment of inertia of a small mass 'dm' of arc length 'R dθ' for the full circle as

$$I_{ZZ} = \int dm \times r^2 = \int [\rho \times A \times (R\, d\theta)]\, R^2 \quad \text{for } 0 \le \theta \le 2\pi$$

$$= \rho \times A \times R^3 \times (2\pi) = M \times R^2$$

and $I_{XX} = I_{YY} = I_{ZZ} / 2 = M \times R^2/2$ from perpendicular axis theorem

(e) Mass moment of inertia of an ***elliptical plate*** of semi-major axis 'a' and semi-minor axis 'b', from area moment of inertia is

$$I_{XX} = \rho\, t\, (\pi \times a \times b^3 / 4) = M \times b^2/4$$

where $M = \rho\, t\, \pi \times a \times b$, is the total mass of the plate

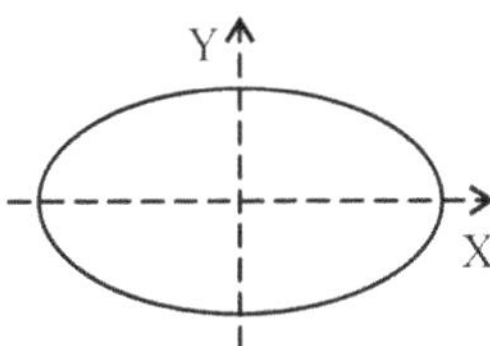

Similarly, $I_{YY} = \rho\, t\, (\pi \times a^3 \times b / 4) = M \times a^2/4$

and $I_{ZZ} = \rho\, t\, \pi \times a \times b \times (a^2 + b^2) / 4 = M \times (a^2 + b^2)/4$

8.2 MASS MOMENT OF INERTIA OF THREE-DIMENSIONAL BODIES

(a) *Parallelopiped* or Rectangular prism or cuboid of sides 'L × B × D'

Mass moment of inertia about centroidal X-axis is obtained by integration of mass moment of inertia of a small lamina of dimensions B, D and dx, whose normal is along X-axis.

$I_{XX} = \int dI = \int [\ \rho \times B \times D \times (B^2 + D^2) / 12\ ]\ dx$ with $-L/2 \le x \le L/2$

(from I_{ZZ} of rectangular plate, about its normal axis)

$= [\rho \times B \times D \times (B^2 + D^2) / 12] \times \int dx = \rho \times B \times D \times L \times (B^2 + D^2) / 12$

$= M\ (B^2 + D^2) / 12$

where, $M = \rho \times L \times B \times D$ is the total mass of rectangular prism

Similarly, $I_{YY} = M\ (L^2 + D^2) / 12$ and $I_{ZZ} = M\ (L^2 + B^2) / 12$

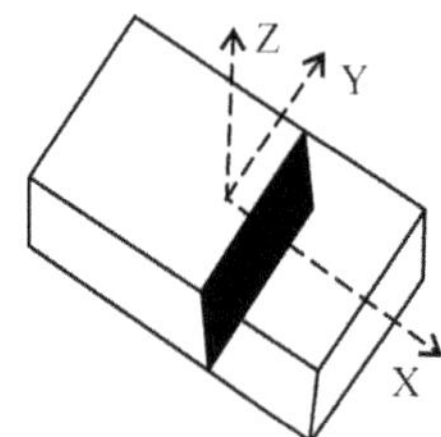

Alternatively, considering a small elemental volume dV,

$I_{XX} = \int dm \times r^2 = \int dm \times (y^2 + z^2)$

$= \int\int\int (\rho \times dx \times dy \times dz) \times (y^2 + z^2)$

with $-L/2 \le x \le L/2$; $-B/2 \le y \le B/2$; $-D/2 \le z \le D/2$

$= \rho \int dx\ [\ \int y^2 dy \times \int dz + \int dy \times \int z^2 dz\ ]$

since integrations about X, Y and Z are mutually independent

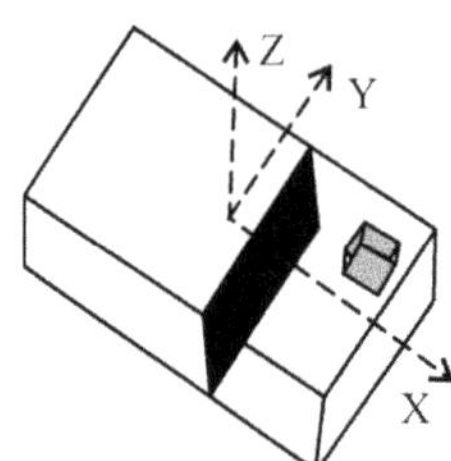

$= \rho \times [x] \times \{\ [y^3/3] \times [z] + [y] \times [z^3/3]\ \}$ within the limits

$= \rho \times L \times (B^3D + BD^3)/12 = \rho \times L \times B \times D \times (B^2 + D^2)/12$

$= M \times (B^2 + D^2)/12$ since $M = \rho \times L \times B \times D$

(b) ***Solid cylinder of radius'R' and height 'H'***

Considering a small lamina of thickness 'dy', whose normal is the Y-axis, Moment of inertia about Y-axis, passing through O, is

$I_{YY} = \int dI = \int \rho\ \pi\ R^4\ dy / 2$ with $0 \le y \le H$ (from I_{ZZ} of circular plate)

$= \rho\ \pi\ R^4 \times H / 2 = M\ R^2 / 2$ since $M = \rho\ \pi\ R^2 \times H$

$I_{XX} = I_{ZZ} = M \times (4H^2 + 3R^2) / 12$

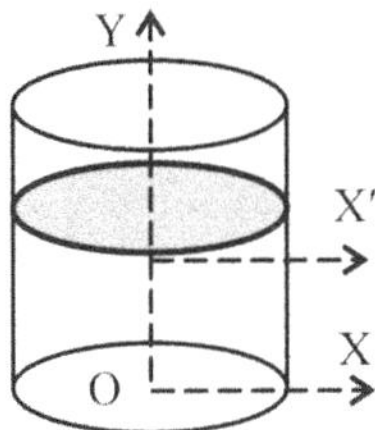

Alternatively, considering a small elemental volume dV, of arc length 'r dθ', radial length 'dr', and thickness 'dy',

$I_{YY} = \int dm \times r^2$

$= \int\int\int [\rho \times (r\,d\theta) \times dr \times dy] \times r^2$

with $0 \le \theta \le 2\pi$; $0 \le y \le H$; $0 \le r \le R$

$= \rho \times \int r^3\,dr \times \int d\theta \times \int dy$

since integrations about r, θ and y are mutually independent

$= \rho \times (R^4/4) \times (2\pi) \times H = \rho \times \pi \times H \times R^4/2$

$= M \times R^2/2$ since $M = \rho \times \pi \times R^2 \times H$

Mass moment of inertia about X-X through base,

$I_{XX} = \int dI = \int dm \times y^2 = \int (\rho\,\pi\,R^2\,dy) \times y^2$ with $0 \le y \le H$

$= \rho\,\pi\,R^2\,[y^3/3] = \rho\,\pi\,R^2 H^3 / 3 = M\,H^2 / 3$

since Mass of cylinder, $M = \rho\,\pi\,R^2 H$

Mass moment of inertia about X'-X' through centroid,

$I_{X'X'} = \int dI = \int dm \times y^2 = \int (\rho\,\pi\,R^2\,dy) \times y^2$ with $-H/2 \le y \le H/2$

$= \rho\,\pi\,R^2\,[y^3/3] = \rho\,\pi\,R^2 H^3/12 = M\,H^2/12$

since Mass of cylinder, $M = \rho\,\pi\,R^2 H$

Alternatively, from parallel axes theorem,

$I_{X'X'} = I_{XX} - M \times (H/2)^2 = MH^2/3 - MH^2/4 = MH^2/12$

(c) *Hollow cylinder* of radius 'R', thickness 't' and height 'H' :

Let us consider a small ring of thickness 'dy' at a height 'y' from X-axis

$I_{YY} = \int dI = \int dm \times R^2 = \int [\,\rho\,t \times (2\pi R) \times dy\,]\,R^2$

$= \rho \times t \times R^3 \times (2\pi)\,[\,y\,]$ with $0 \le y \le H$

$= \rho \times t \times R^3 \times (2\pi)\,H = M\,R^2$

where $M = \rho \times t \times (2\pi R) \times H$ is the total mass

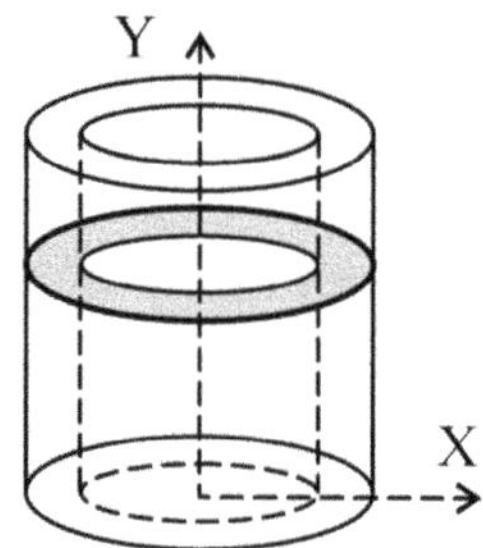

Similarly, if the cylinder has inner radius 'r' and outer radius 'R', instead of thickness 't', then

$I_{YY} = \rho\,\pi\,(R^4 - r^4)\,H/2 = [\,\rho\,\pi\,(R^2 - r^2) \times H\,] \times (R^2 + r^2)/2$

$= M \times (R^2 + r^2)/2$ where, $M = \rho\,\pi(R^2 - r^2)\,H$

(d) Solid sphere of radius 'R', about its diameter

Consider a small elemental volume ($\pi\ x^2\ dy$) at a distance of 'y' from X-axis through its diameter

$I_{YY} = \int dm \times x^2/2 = \int (\rho\,\pi\,x^2\,dy) \times x^2/2$ and $x^2 = R^2 - y^2$

(from I_{ZZ} of circular lamina)

$= (\rho\,\pi/2) \int (R^2 - y^2)^2\,dy$

$= (\rho\,\pi/2)\,[(R^4y) - (2R^2\,y^3/3) + (y^5/5)]$ with $-R \le y \le R$

$= \rho\,\pi\,R^5\,(15 - 10 + 3)/\,15 = (8/15)\,\rho\,\pi\,R^5$

$= (2/5)\,M\,R^2$ where, mass of sphere $M = (4/3)\,\rho\,\pi\,R^3$

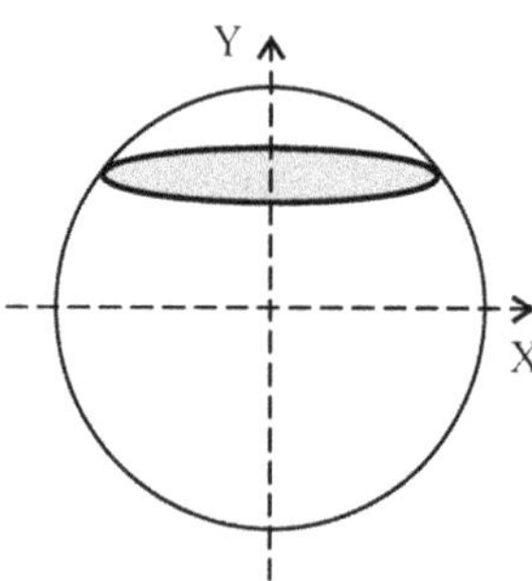

Similarly, for a **solid hemisphere**,

$I_{YY} = (4/15)\,\rho\,\pi\,R^5 = (2/5)\,MR^2$

where, mass of hemisphere $M = (2/3)\,\rho\,\pi\,R^3$

(e) Solid cone of base radius 'R' and height 'H'

Consider a small elemental lamina of thickness 'dy' and radius 'x' at a distance 'y' from the vertex of the cone. Then,

$I_{YY} = \int dI = \int [\, \rho\, (\pi x^2)\, dy\,]\, (x^2/2)$ from I_{ZZ} of circular lamina

Variable x, a function of y, can only be integrated by replacing it with y from the relation x/R = y/H

Then, $I_{YY} = (\rho \times \pi/2) \times \int [Ry/H\,]^4\, dy$

$= (\rho \times \pi/2) \times (R^4/H^4)\, [\, y^5/5\,]$

with $0 \le y \le H$

$= \rho\, \pi R^4 \times H/10$

$= (3/10)\, M\, R^2$

where, mass of cone $M = \rho \times \pi R^2 \times H / 3$

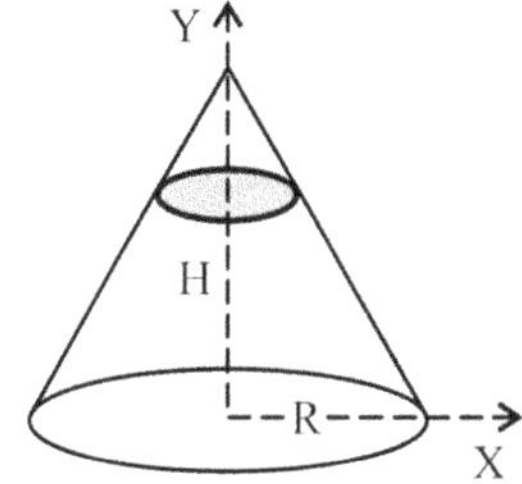

(f) Hollow sphere of radius 'R'

Let us consider a small ring of radius 'x' at a height 'y' from X-axis

$I_{YY} = \int dI = \int \rho \times t \times (2\pi\, R\sin\theta) \times (R \times d\theta) \times (R\sin\theta)^2$

(from I_{ZZ} of circular ring)

$= \int \rho\, t \times R^4 \times (2\pi)\, [\sin^3\theta\, d\theta\,]$

with $-\pi/2 \le \theta \le \pi/2$

$= (8/3)\, \rho\, t \times R^4 \times \pi = (2/3)\, M\, R^2$

where, mass of hollow sphere

$M = \rho \times t \times (4\pi R^2)$

Similarly, for a **hollow hemisphere**,

$I_{YY} = (2/3)\, MR^2$ where, $M = \rho \times t \times (2\pi\, R^3)$

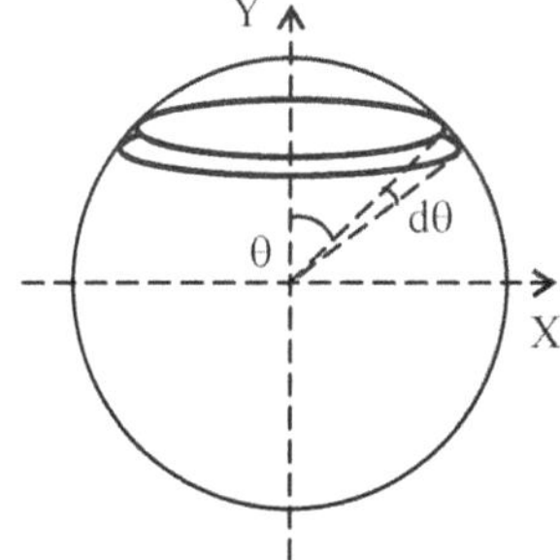

(g) *Hollow cone* of base radius 'R' (open base) and height 'H'

$I_{YY} = \int dI = \int dm \times x^2 = \int [\, \rho\, (2\pi x) \times t \times ds\,] \times x^2$

where, $ds = dy \times \sec\alpha$

$= \int [\, \rho\, 2\pi t \times \sec\alpha \times [R\, y / H\,]^3 \times dy$ since $x/R = y/H$

$= [\, \rho\, (2\pi t \times \sec\alpha \times R^3 / H^3\,] \times H^4/4$ with $0 \le y \le H$

$= M\, R^2/ 2$ where $M = \rho \times \pi R \times t \times L$ and $L = H \times \sec\alpha$

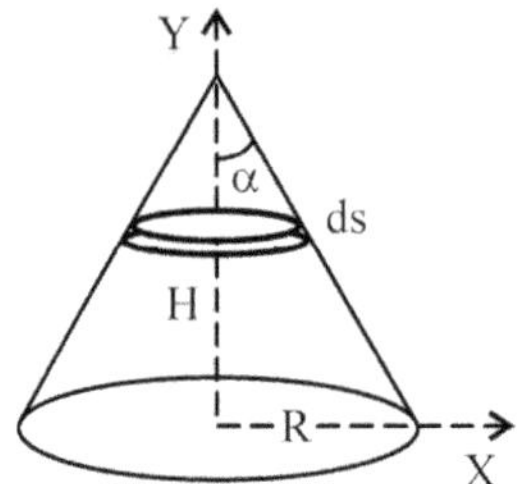

8.3 MOMENT OF INERTIA OF COMPOSITE SECTIONS

Moment of inertia of a composite section is the algebraic sum of moments of inertia of its simpler components, about the same axis.

$$I_{XX} = \sum [\, I_{xx} \,]_i \;\; ; \quad I_{YY} = \sum [\, I_{yy} \,]_i$$

Example 1

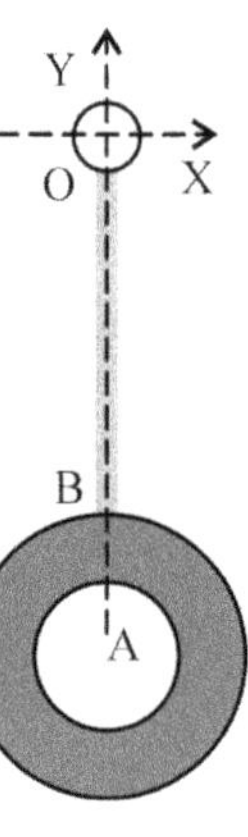

A clock pendulum consists of (i) a slender rod of length 20 cm, cross sectional area 50 mm^2 and mass density 7000 kg/m^3 and (ii) a circular disc of 10 cm radius with a central hole of 5 cm radius, thickness 5 mm and mass density 8000 kg/m^3. Calculate mass moment of inertia of the pendulum about its axis of rotation, perpendicular to the plane of the disc.

Solution:

Masses of the two parts (larger disc and cut smaller disc) are

$$M_1 = 8000 \times 10^{-9} \times (\pi \times 100^2) \times 5 = 1.257 \text{ kg}$$

$$M_2 = 8000 \times 10^{-9} \times (\pi \times 50^2) \times 5 = 0.314 \text{ kg}$$

Mass moment of inertia about Z-axis has to be transferred from A to O using parallel axes theorem, before adding to the inertia of the bar about O,

Moments of inertia of the two parts about O are,

For the rod,

$$[I_{ZZ}]_1 = M \times L^2/3 = (7000 \times 10^{-9} \times 50 \times 200) \times 200^2/3 = 933 \text{ kg mm}^2$$

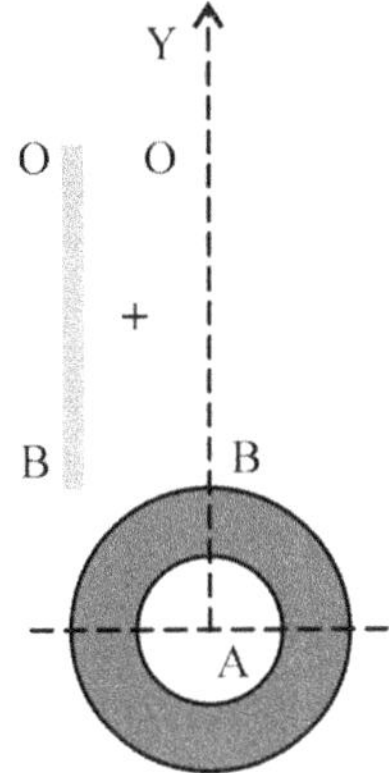

For the disc,

$$[I_{ZZ}]_2 = M_1 \times R_1^2/2 - M_2 \times R_2^2/2 + (M_1 - M_2) \times OA$$
$$= 1.257 \times 100^2/2 - 0.314 \times 50^2/2 + (1.257 - 0.314) \times (200 + 100)$$
$$= 6175.4 \text{ kg mm}^2$$

Mass moment of inertia of the pendulum,

$$I_{ZZ} = [I_{ZZ}]_1 + [I_{ZZ}]_2 = (933.0 + 6175.4) = 7108.4 \text{ kg mm}^2$$

Example 2

A spherical bob of radius R and mass M_B is attached to a slender rod of length L and mass M_R. Calculate the moment of inertia of the assembly about the axis of rotation (Z-axis).

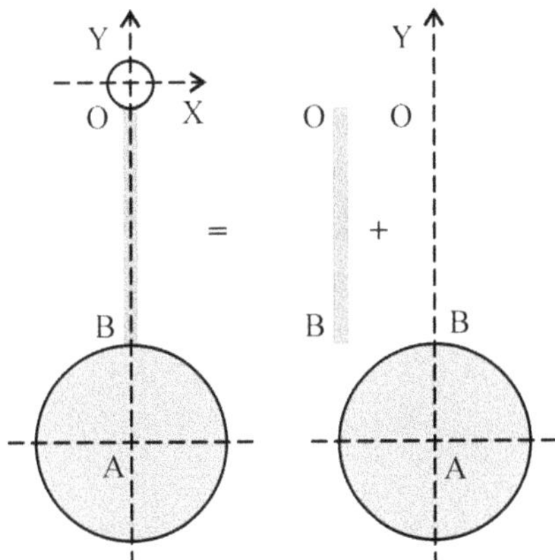

Solution :

$$I_{ZZ} = [I_{ZZ}]_{ROD} + [I_{ZZ}]_{BOB}$$

For the rod, $I_{ZZ} = M_R \times L^2/3$

For the Bob, $I_{ZZ} = (2/5)\ M_B \times R^2 + M_B \times AO^2$

$= (2/5)\ M_B \times R^2 + M_B \times (L + R)^2$

Thus, $I_{ZZ} = M_R \times L^2/3 + (2/5)\ M_B \times R^2 + M_B \times (L+R)^2$

Example 3

Calculate mass moment of inertia of the slender rod C-O-A-B, bent as shown, about X, Y and Z axes. Assume mass density of the rod per unit length ($m = \rho \times A$) as 3 kg/m, OC = 2m, OA = 3m and AB = 4m

Solution:

$I_{XX} = [I_{XX}]_{OA} + [I_{XX}]_{AB} + [I_{XX}]_{OC}$

$= (m \times L_{OA}) \times (R^2/2) + (m \times L_{AB}) \times (L_{AB}{}^2/3) + (m \times L_{OC}) \times (L_{OC}{}^2/3)$

where R is the radius of the rod.

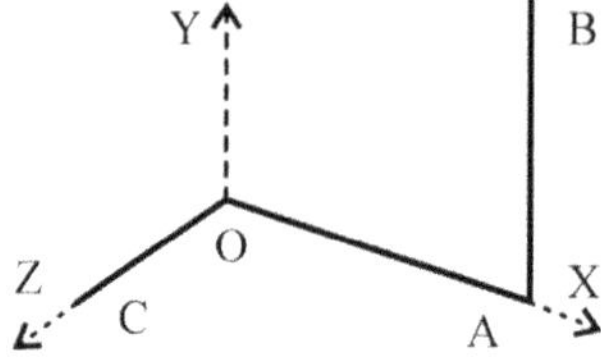

Assuming $R^2 \approx 0$,

$I_{XX} = (m \times L_{AB}) \times (L_{AB}{}^2/3) + (m \times L_{OC}) \times (L_{OC}{}^2/3) = 72\ \text{kg m}^2$

Similarly,

$I_{YY} = (m \times L_{OA}) \times (L_{OA}{}^2/3) + (m \times L_{OC}) \times (L_{OC}{}^2/3)$

$+ (m \times L_{AB}) \times L_{OA}{}^2 = 143\ \text{kg m}^2$

and $I_{ZZ} = (m \times L_{OA}) \times (L_{OA}{}^2/3) + (m \times L_{AB}) \times (L_{AB}{}^2/3) = 91\ \text{kg m}^2$

Example 4

Two semi-circular cuts of radius 40 cm are made throughout the depth of a block of sides 80 cm, 100 cm and 120 cm along X, Y and Z axes, as shown. Calculate mass moment of inertia of the remaining solid about X-axis, with mass density of ρ kg/m^3

Solution :

The body is treated as a rectangular prism with two semi-circular (or half) cylinders removed from it, as shown.

Mass of the rectangular prism , $M_1 = \rho \times L \times B \times H = 0.96\ \rho$ kg

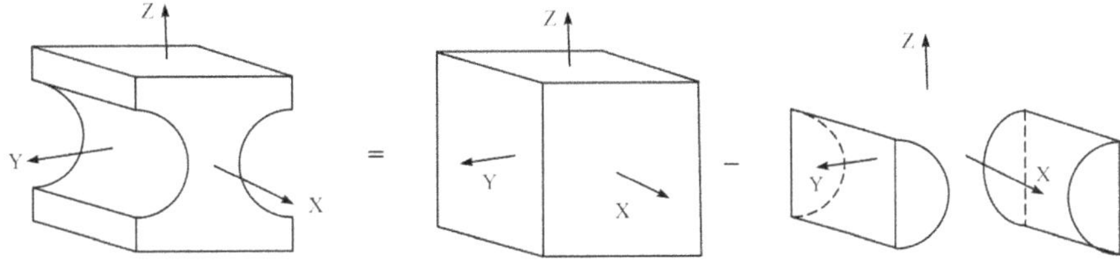

Mass of half cylinder, $M_2 = \rho \times (\pi \times R^2/2) \times L = 0.201\ \rho$ kg

$I_{XX} = M_1 \times (B^2 + D^{2)}/12 - 2 \times M_2 \times [\ (R^2/2) - (3R/8)^2 + (B/2 - 3R/8)^2\ ]$

$= \rho\ [0.96 \times (1.0^2 + 1.2^2) / 12]$

$- 2\ \rho \times 0.201 \times [\ (0.4^2 / 2) - (0.15)^2 + (0.5 - 0.15)^2]$

$= 0.31484 \times \rho$ kg m^2

Example 5

A missile consists of cylinder of length 400 mm and radius 200 mm and a cone of length 300 mm attached to it. Calculate mass moment of inertia of the missile about its axis of revolution, assuming mass density to be 8500 kg/m^3 for the cone material and 2560 kg/m^3 for the cylinder material.

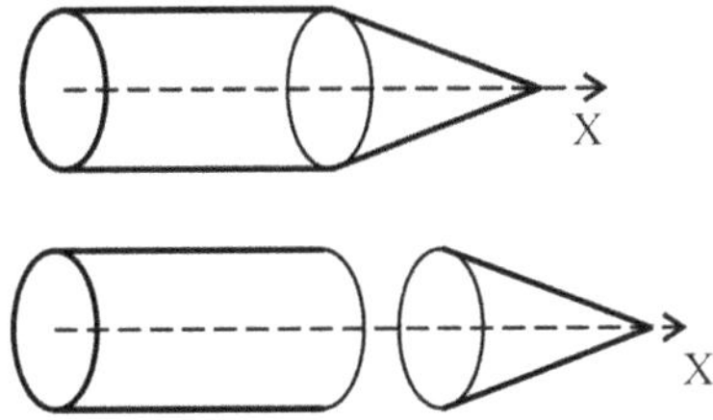

Solution :

Let mass of cylinder and cone be M_1 and M_2

Then, $M_1 = \pi\ \rho_1 \times R^2 \times H_1 = \pi \times 2560 \times 0.2^2 \times 0.4 = 128.68$ kg

$M_2 = \pi\ \rho_2 \times R^2 \times H_2/3 = \pi \times 8500 \times 0.2^2 \times 0.3/3 = 106.81$ kg

$$I_{XX} = [I_{XX}]_{CYL} + [I_{XX}]_{CONE}$$
$$= M_1 \times (R^2/2) + M_2 \times (3R^2/10)$$
$$= 128.68 \times 0.2^2 / 2 + 106.81 \times 3 \times 0.2^2 / 10$$
$$= 3.8553 \text{ kg m}^2$$

8.4 PRODUCT OF INERTIA

While Product of inertia for area is limited to one plane, product of inertia of a mass deals with volumes and hence, the following three products exist.

$$I_{XY} = \int dm \times x \times y \quad ; \quad I_{YZ} = \int dm \times y \times z \quad ; \quad I_{ZX} = \int dm \times z \times x$$

Depending on the nature of cross section and coordinate axes, I_{XY} can take +ve or -ve values. As the coordinate axes through centroid are rotated, the product of inertia value changes from +ve to –ve or vice versa. Therefore, with reference to some particular axes, the product of inertia value becomes zero. These axes are called ***Principal axes***. These axes have significance in terms of maximum stress in a cross section, as can be seen in detail in 'Mechanics of solids'.

Obviously, the product of inertia about any geometrically symmetric axis is zero.

Parallel axis theorem is also applicable to product of inertia with appropriate change

i.e., $$I_{X'X'} = I_{XX} + M \times (a \times b)$$

where a and b are the distances between the two sets of axes.

8.5 RADIUS OF GYRATION

It is a function of moment of inertia and cross sectional area and is an important property for rotating components. Corresponding to the three moments of inertia, there are three radii of gyration. It has the dimensions of length. It can be interpreted as the distance 'k' at which the complete mass of the body 'M' is squeezed and kept such that there is no change in its mass moment of inertia.

$$I = A \times k^2$$

or $$k_{XX} = \sqrt{(I_{XX} / M)}\,; \quad k_{YY} = \sqrt{(I_{YY} / M)}\,; \quad k_{ZZ} = \sqrt{(I_{ZZ} / M)}$$

These values are given below for a few simple sections

Rectangular section, through its centroid :

$$k_{XX} = d / \sqrt{12} \;;\; k_{YY} = b / \sqrt{12} \;;\; k_{ZZ} = \sqrt{(b^2 + d^2)} / \sqrt{3}$$

Circular cross section, axis through diameter :

$k_{XX} = k_{YY} = d / 4;$

$k_{ZZ} = d / \sqrt{8}$

Semi-circular cross section, axis through diameter : $k_{AB} = d/4$

Hollow circle, axis through its diameter : $k_{XX} = k_{YY} = \sqrt{(D^2 + d^2)/4}$

Triangle, axis along base : $k_{XX} = h/\sqrt{6}$; $(k_{XX})_G = h/\sqrt{18}$

SUMMARY

1. Mass moment of inertia is defined, for masses or volumes of bodies ,as the product of mass and square of its distance from the axis. It has units of kg × m^2 and is also indicated by I or I_m.

   Thus, $I_{XX} = \int dm \times y^2$; $I_{YY} = \int dm \times x^2$; $I_{ZZ} = \int dm \times r^2$

2. $I_{ZZ} = I_{XX} + I_{YY}$ for plates of small thickness – called **perpendicular axes theorem**

3. Mass moment of any lamina (of uniform thickness 't' and density 'ρ') is equal to the area moment of inertia, multiplied by the constant 'ρ × t'

4. Since mass M.I. is defined for volumes, there exist ***three products of inertia*** defined by $I_{XY} = \int dm \times x \times y$; $I_{YZ} = \int dm \times y \times z$; $I_{ZX} = \int dm \times z \times x$

5. **Parallel axes theorem** is also applicable to mass moment of inertia and is defined as

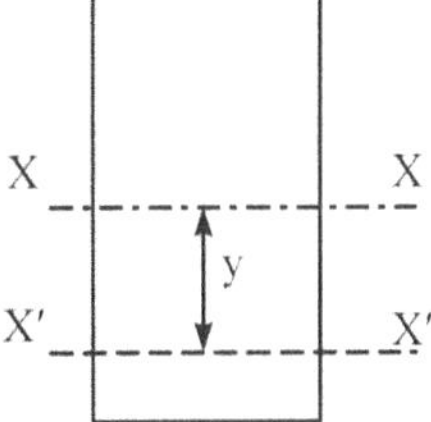

$$I_{X'X'} = I_{XX} + M \times y^2$$

where, M is the mass of the body

and y is the distance between the two parallel axes X-X (through centroid) and X'- X'

This is also called *Transfer of axis theorem*

SOME USEFUL RESULTS

1. Mass M.I. of a bar of length 'L' about an axis perpendicular to length and through centroid, $I_{YY} = M \times L^2/12$. Mass M.I. about a parallel axis through its end,

$$I_{AA} = M \times L^2/3$$

2. Mass M.I. of a thin rectangular plate of width 'b', depth 'd' and thickness 't', about an axis parallel to width through centroid,

$$I_{XX} = (\rho \times t) \times (b \times d^3/12) = M \times d^2/12$$

Similarly, about an axis parallel to depth through centroid, $I_{YY} = M \times b^2/12$ and about an axis perpendicular to plate through centroid,

$$I_{ZZ} = M \times (b^2 + d^2)/12$$

3. Mass M.I. of a thin circular plate of radius 'R' about its axis,

$$I_{ZZ} = \rho \times t \times \pi \times R^4/2 = M\,R^2/2$$

4. Mass M.I. of a thin hollow circular plate of outer radius 'R' and inner radius 'r' about its axis, $I_{ZZ} = \rho \times t \times \pi \times (R^4 - r^4)/2 = M \times (R^2 + r^2)\ /\ 2$ since $M = \rho \times t \times \pi \times (R^2 - r^2)$

5. Mass M.I. of a thin elliptical plate of major and minor axes '2a' and '2b' about its major axis and minor axis respectively are,

$$I_{XX} = M \times b^2/4\ ;\ I_{YY} = M \times a^2/4$$

6. Mass M.I. of a rectangular prism or cuboid of sides L, B, H along X, Y and Z axes respectively, about its centoidal axes,

$$I_{XX} = M \times (B^2 + H^2)/12$$

$$I_{YY} = M \times (L^2 + H^2)/12$$

$$I_{ZZ} = M \times (L^2 + B^2)/12$$

7. Mass M.I. of a solid cylinder of radius 'R' and height 'H' about its axis,

$$I_{XX} = \rho \times H \times \pi R^4/2 = M\,R^2\ /\ 2$$

8. Mass M.I. of a hollow cylinder of outer radius 'R' and inner radius 'r' about its axis,

$$I_{ZZ} = \rho \times t \times \pi \times (R^4 - r^4)/2 = M \times (R^2 + r^2)\ /\ 2$$

since $M = \rho \times t \times \pi \times (R^2 - r^2)$

9. Mass M.I. of a sphere or hemisphere through center, $I_{XX} = (2/5)\ M \times R^2$

10. Mass M.I. of a solid cone of base radius 'R' about its axis,

$$I_{XX} = (3/10)\ M \times R^2$$

CHAPTER 9

TRUSSES

9.1 FRAME

Frame is a discrete Structure, made up of distinctly identifiable members. Frames are commonly used in many structures. Structures with all members lying in one 2-D plane (usually vertical) are called plane frames. Structures with members lying in different orientations in 3-D space are called space frames. Each member is considered as a **one-dimensional** (1-D) element, identified by its length. Frames are designed to withstand different types of loads (concentrated, distributed) applied on it at the joints or on the individual members. Accordingly, three equations of equilibrium applicable to a planar frame (force equilibrium along two mutually perpendicular directions and moment equilibrium about one axis) or six equations of equilibrium applicable to a space frame (force equilibrium along three mutually perpendicular directions and moment equilibrium about three axes) are used for calculating member forces in these structures.

Frames are broadly classified as trusses and beams based on the type of loads and their deformations.

9.2 TRUSS

A truss is a **discrete structure,** analysed with the following assumptions –

- The members are treated as **pin-jointed** or **connected with hinges** at their ends and are free to have **relative rotation** between the members at any joint. In many trusses, *members may be welded or joined with multiple rivets*, preventing relative rotation of members
- The members **carry loads only at their ends or joints** with no distributed loads along the members. Distributed loads such as Wind load, self weight etc are negligible

- Load distribution is considered along the centroidal axes of the members, *neglecting overlap of some members*. Axial loaded members are called **spars**
- In order to ensure that members are not subjected to bending, only *one joint is rigidly supported with a hinge* and *all other joints are supported on rollers*, supporting only vertical loads with vertical reaction.
- Multiple plane trusses (2-D) are commonly used in a single structure in industries, function halls etc. Space trusses (3-D) are commonly used in power transmission towers, small houses without concrete slabs etc.

  For ease of analysis, physical trusses (Ref. Fig 9.1) are simplified ignoring joint details, cross sections of individual members, overlap of members etc., (Ref. Fig 9.2). The results may therefore differ at specific locations.

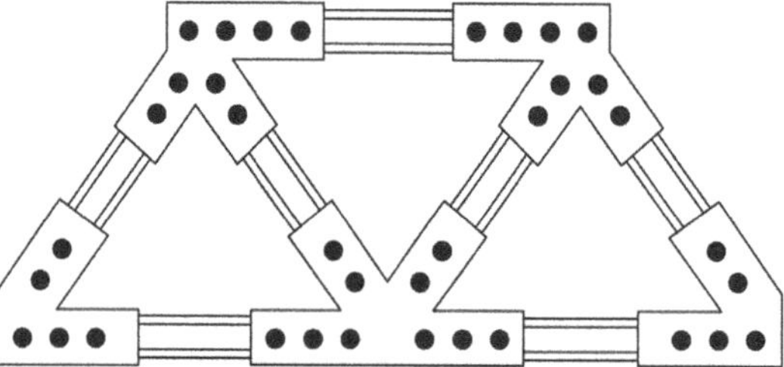

FIGURE 9.1 Physical structure of a plane truss

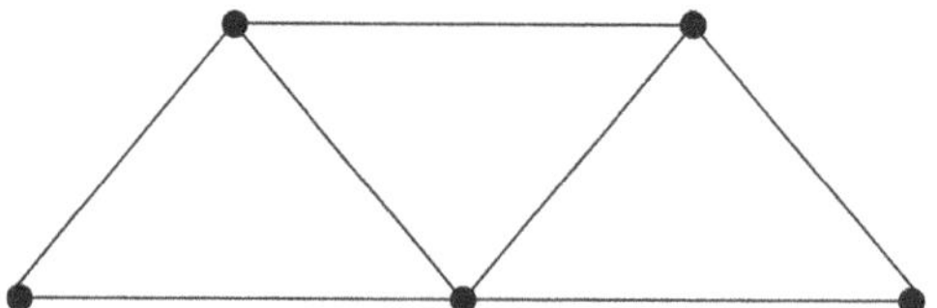

FIGURE 9.2 Geometric model of a plane truss

9.2.1 TYPES OF TRUSSES

(a) **Determinate trusses or Perfect frames:** Trusses are classified as statically determinate structures or perfect frames, if all member forces can be evaluated using the equations of static equilibrium ($\Sigma F_X = 0$; $\Sigma F_Y = 0$ in plane trusses and $\Sigma F_X = 0$; $\Sigma F_Y = 0$; $\Sigma F_Z = 0$ in space trusses). Thus, two unknown axial forces in members of a plane truss and three unknown axial forces in members of a space truss can be evaluated using the equations of static equilibrium at a joint and all member forces can be evaluated analyzing all joints, one joint at a time, in proper sequence.

Number of members 'm' and number of joints 'j' in a ***statically determinate plane truss*** are governed by the rule m = 2 j – 3. From the basic triangle (m=3 & j=3), every new joint adds two new members. Some examples of determinate plane trusses, with identification of members and joints are shown in Fig 9.3.

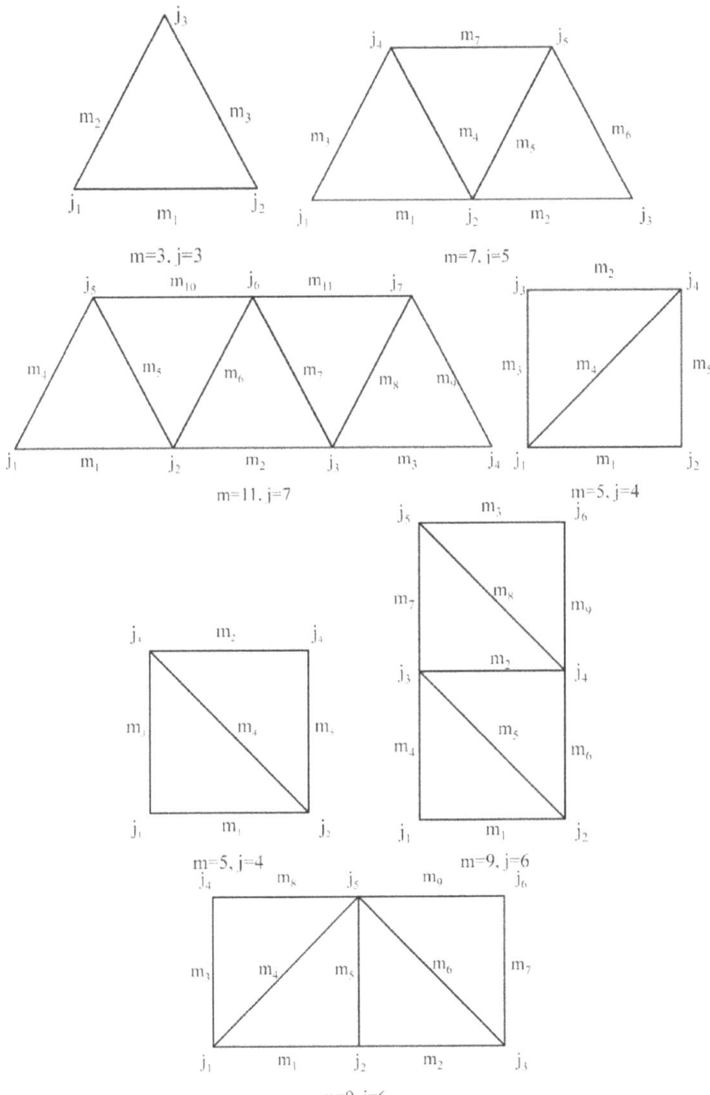

FIGURE 9.3 Some examples of determinate planar trusses

Space frames (3-D structures) may also be categorized on the same lines as statically determinate structures (satisfying the three equations of static equilibrium $\Sigma F_X = 0$; $\Sigma F_Y = 0$ and $\Sigma F_X = 0$ at each joint). If the number of members is ‘m’ and number of joints is ‘j’, statically determinate space frames satisfy the relation m = 3j – 6

(b) Indeterminate trusses or Imperfect frames: Member forces in structures with more members can not be determined by the equations of static equilibrium alone. Such structures are called ***statically indeterminate structures*** or ***imperfect frames*** or ***internally redundant frames.*** The difference r = m – (2j – 3) is called the degree of indeterminacy or degree of redundancy. Some examples of indeterminate planar trusses, with identification of members and joints are shown in Fig 9.4.

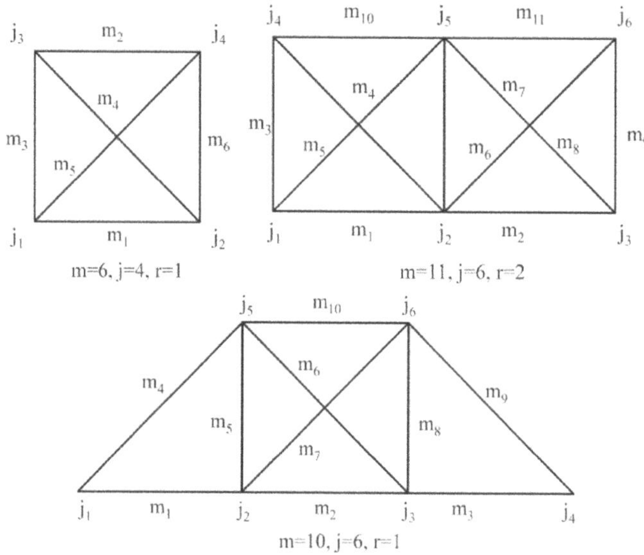

FIGURE 9.4 Some examples of indeterminate planar trusses

9.3 ANALYSIS OF PLANE TRUSSES

Member forces of a determinate truss are calculated by the rigid body mechanics approach, with no displacements or deformations, assuming that the orientations of the members remain practically unchanged. Stress in each member is obtained by dividing force in the member with the corresponding area of cross section ($\sigma = P/A$). Stress is assumed to be uniform across the cross section in each member. Strain and change in length (elongation/compression) of each member are calculated from stress, modulus of elasticity and length of each member ($\varepsilon = \sigma / E$ and $\delta = \varepsilon \times L$)

Planar trusses are broadly classified as cantilever trusses and simply supported trusses. The method of solution differs in these two cases, in the way reactions are calculated.

Determinate **simply supported truss** (Fig a) and **cantilever truss** (Fig b) are shown with loads and boundary conditions in Fig 9.5.

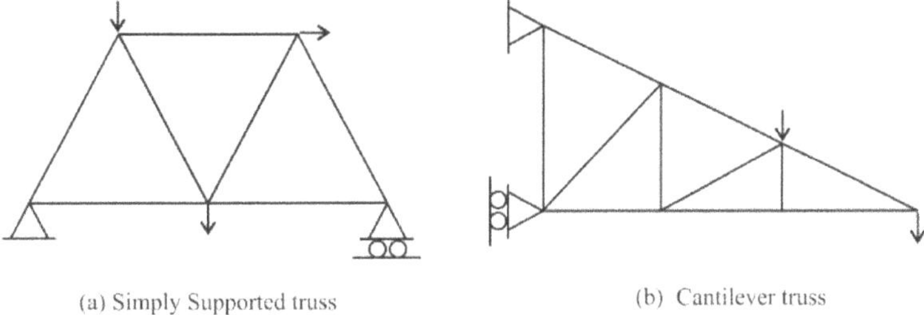

FIGURE 9.5 Determinate planar trusses with boundary conditions

Indeterminate plane truss is shown with loads and boundary conditions in Fig 9.6

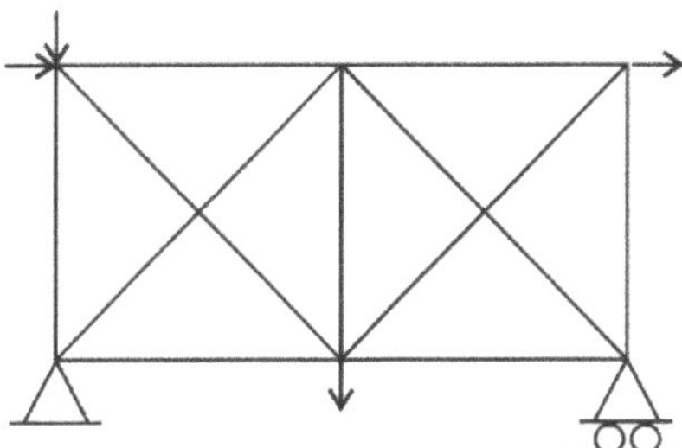

FIGURE 9.6 Indeterminate planar truss with boundary conditions

9.3.1 ANALYSIS BY METHOD OF JOINTS

In this method, the analysis for calculation of member forces is carried out by applying equations of static equilibrium for concurrent forces at each joint of the truss, choosing the joints in a sequence such that the system of equations matches with the number of unknowns in the set of equations.

Thus, in a plane truss in X-Y plane, the equations of equilibrium ($\Sigma F_X = 0$ and $\Sigma F_Y = 0$) can be used at each joint to evaluate two unknown member forces. Similarly, in a space truss, the equations of equilibrium ($\Sigma F_X = 0$, $\Sigma F_Y = 0$ and $\Sigma F_Z = 0$) can be used at each joint to evaluate three unknown member forces. Each spar member will be in tension or compression. Accordingly, direction of member force will be represented by opposite directions at the two ends.

A member in tension is represented by inward arrow at each joint while a member in compression is represented by outward arrow at each joint. The force directions in each member are internal reactions and, hence, the tensile force directions appear to produce compression in the member and compressive force directions appear to produce elongation of the member. Thus, if a member I-J has its internal force (shown in figure by solid arrow) oriented towards J, while analysing joint I, then internal force is taken oriented towards I while analysing joint J. Similarly, if the member I-J has its internal force oriented towards I while analysing joint I, then the internal force is taken oriented towards J while analysing joint J.

The following example illustrates practical application of this method.

Unlike in the case of cantilever plane truss, a single joint with only two member forces to start with the method of joints is unlikely in the case of any simply supported plane truss. It is therefore necessary to calculate the three unknown reactions first by considering the three equations of equilibrium of planar non-concurrent forces. The procedure is explained through the following two examples.

Example 1

Find forces in all the members of the cantilever plane truss, shown below, for the vertical loads of 20 kN applied at joints C and D. DE = EF = FG = 3 m ; AG = 4 m

Solution:

Let $\angle CDE = \theta$. Considering equilibrium of each joint, two unknown forces can be evaluated using the two equations of equilibrium

$$\sum F_X = 0 \text{ and } \sum F_Y = 0$$

$$\tan\theta = AG/DG = AG/(DE + EF + FG) = 4/(3+3+3) = 4/9$$

$$\Rightarrow \quad \sin\theta = 4/\sqrt{97}\,; \quad \cos\theta = 9/\sqrt{97}$$

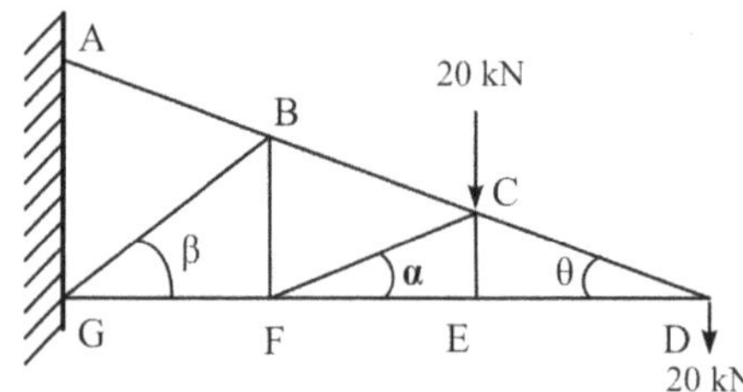

At D, $\sum F_Y = F_{CD}\sin\theta - 20 = 0$

$$\Rightarrow \quad F_{CD} = 20/\sin\theta = 20/\left(4/\sqrt{97}\right) = 49 \text{ kN}$$

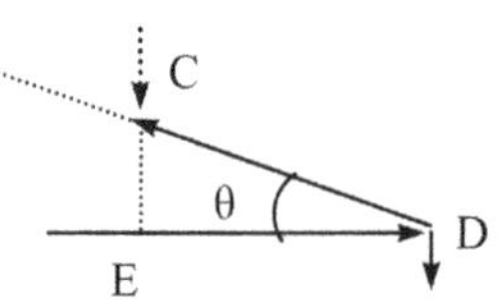

$$\sum F_X = F_{ED} - F_{CD} \cos \theta = 0$$

$$\Rightarrow \quad F_{ED} = F_{CD} \cos \theta = 49 \times \left(9/\sqrt{97}\right) = 45 \text{ kN}$$

At E, $\sum F_X = F_{ED} - F_{EF} = 0$

$\Rightarrow \quad F_{EF} = F_{ED} = 45\text{kN}$

and $\sum F_Y = F_{CE} = 0$

At C, $\sum F_Y = -F_{CD} \sin \theta + F_{FC} \sin \alpha + F_{BC} \sin \theta + F_{CE} - 20 = 0$

where, $CE = ED \tan \theta = 3 \times (4/9) = 4/3$

$\tan \alpha = CE/FE = (4/3)/3 = 4/9$

$\sin \alpha = 4/\sqrt{97}$; $\cos \alpha = 9/\sqrt{97}$

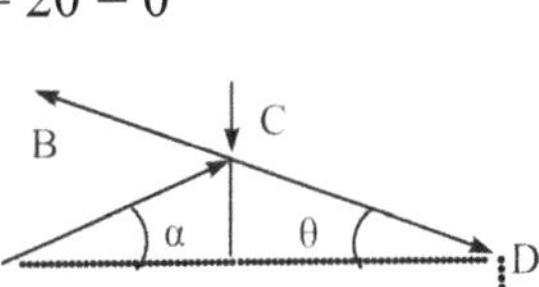

and $\sum F_X = -F_{BC} \cos \theta + F_{CD} \cos \theta + F_{FC} \cos \alpha = 0$

Solving these two equations F_{BC} and F_{FC} can be obtained as

$F_{BC} = 73.87$ kN

and $F_{FC} = 24.62$ kN

At F, $\sum F_X = -F_{FC} \cos \alpha - F_{FE} + F_{GF} = 0$

$\sum F_Y = -F_{FC} \sin \alpha + F_{FB} = 0$

from which F_{FB} and F_{GF} can be obtained as

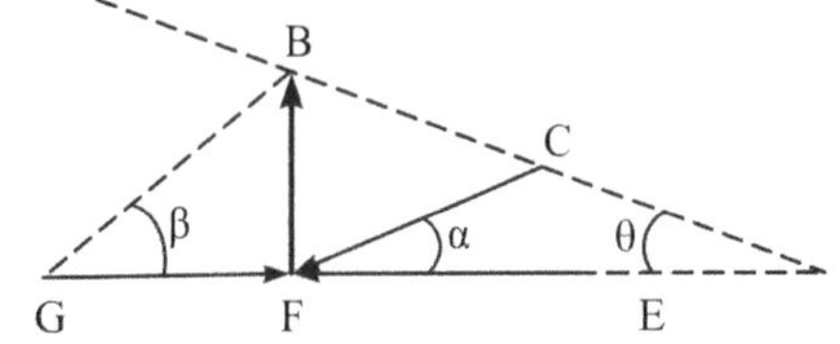

$$F_{FB} = F_{FC} \sin \alpha = 24.62 \times \left(4/\sqrt{97}\right) = 10 \text{ kN}$$

and $$F_{GF} = F_{FC} \cos \alpha + F_{FE} = 24.62 \times \left(9/\sqrt{97}\right) + 45 = 67.5 \text{ kN}$$

At B, $\sum F_Y = -F_{BC} \sin \theta + F_{AB} \sin \theta + F_{GB} \sin \beta - F_{FB} = 0$

and $\sum F_X = F_{BC} \cos \theta - F_{AB} \cos \theta + F_{GB} \cos \beta = 0$

where, $BF = FD \tan \theta = 6 \times (4/9) = 8/3$

$\tan \beta = BF/GF = (8/3)/3 = 8/9$

$\Rightarrow \quad \sin \beta = 8/\sqrt{145}$; $\cos \beta = 9/\sqrt{145}$

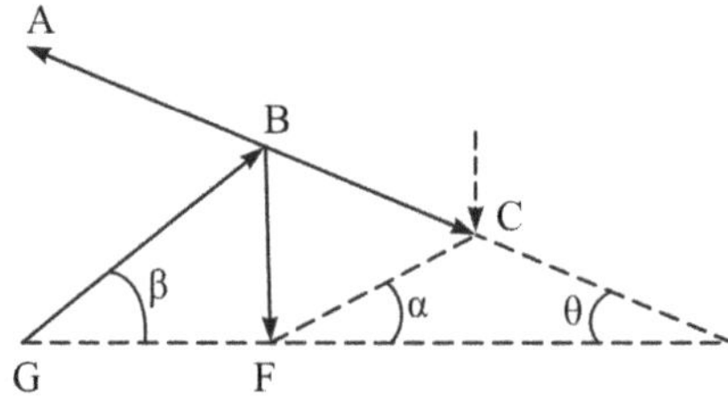

Solving these two equations, F_{GB} and F_{AB} can be obtained as

$$F_{GB} = 82 \text{ kN} \qquad \text{and} \qquad F_{AB} = 10 \text{ kN}$$

Please note that

- Equations of equilibrium are written for the assumed directions of member forces, (indicating tensile or compressive forces) as shown in the figures. If the assumed direction is wrong, the calculated value of member force will be negative. It is assumed in the above equations that forces directed towards +ve X-axis (to the right, in the horizontal direction) are +ve and those directed towards –ve X-axis are –ve. Similarly, forces directed towards +ve Y-axis (upwards, in the vertical direction) are +ve and those directed along –ve Y-axis are –ve.
- A member with zero force, in this example CE, is called a ***null member***
- At the end A, the member force and reaction form a system of collinear forces along member axis and, hence, application of equations of equilibrium will give reaction which is equal and opposite to the member force. At the end G, resultant of member forces and reaction form a system of collinear forces and are equal and opposite, for ensuring equilibrium
- The sequence of joints is so chosen as to have only two unknowns in the equations of equilibrium at each joint. For example, in the above problem, if joint C is considered after joint D, with unknown member forces in CD, CE and BC then the two equations of equilibrium are not adequate to obtain a solution. Similarly, if joint B is considered after joint E, with unknown member forces in BF, BG and BA, then also the two equations are not adequate to obtain a solution

Force distribution in the complete truss is shown in Fig 9.7 with members in tension represented by positive forces, members in compression represented by negative forces and a null member (CE) with member force of zero magnitude.

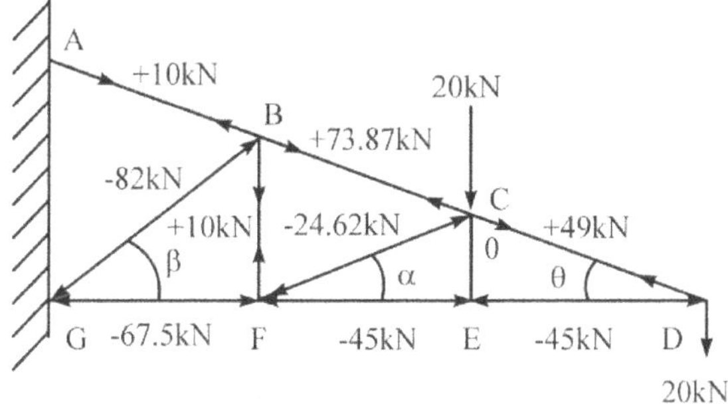

FIGURE 9.7 Member forces in a plane truss

Example 2

Calculate forces in all the members of the simply supported plane truss, shown below, for the loads of P_{EX} = 3kN (applied at joint E along +X direction) and P_{BY} = – 5kN (applied at joint B along – Y direction). Joint A is on hinge support and joint C is on roller supports. Each member is of length 3m.

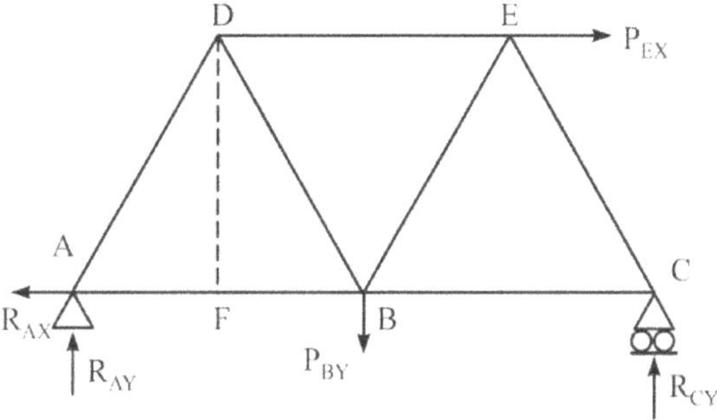

Solution:

Unlike in the case of cantilever plane truss, a single joint with only two unknown forces, to start with the method of joints does not exist in this case.

It is therefore necessary to calculate unknown reactions first by considering equations of equilibrium of planar non-concurrent forces on the truss. Since all members are of same length, $\angle DAB = \angle ECB = 60^0$.

Height of the truss, DF = AD sin 60 = $3 \times \left(\sqrt{3}/2\right)$

$= 2.598$m

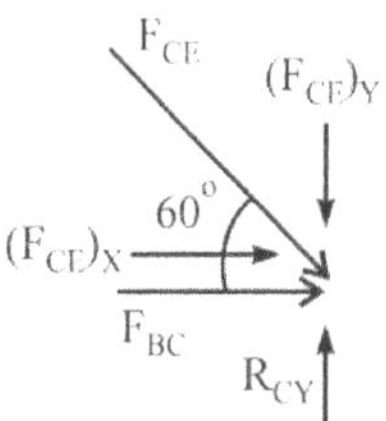

Reaction along + Y direction ($R_c = R_{CY}$) only is possible at roller support C, while component of reaction R_{AX} and R_{AY} are possible at hinged support A.

Taking moments about joint A,

$$\Sigma M_A = R_{CY} \times AC - P_{EX} \times DF - P_{BY} \times AB$$
$$= 0 \text{ for equilibrium}$$

$$\Rightarrow \quad R_{CY} = (P_{EX} \times DF + P_{BY} \times AB) / AC$$
$$= (3 \times 2.598 + 5 \times 3) / 6 = 3.8\text{kN}$$

Using equations of force equilibrium at joint C,

$$\Sigma F_Y = R_{CY} - (F_{CE})_Y = R_{CY} - F_{CE} \times \sin 60 = 0 \text{ for equilibrium}$$

$$\Rightarrow \quad F_{CE} = 3.8 \;/\; \left(\sqrt{3}/2\right) = 4.387\text{kN}$$

Alternatively, force F_{CE} can be resolved into components along X (horizontal) and Y (vertical) directions. Then,

$$\Sigma F_Y = R_{CY} - (F_{CE})_Y = 0$$

and $$\Sigma F_X = (F_{CE})_X + F_{BC} = F_{CE} \times \cos 60 + F_{BC} = 0 \text{ for equilibrium}$$

$$\Rightarrow \quad F_{BC} = -F_{CE} \times \cos 60 = -4.387/2 = -2.193\text{kN}$$

Member forces at the two ends of a member are represented in opposite directions.

Thus, F_{CE} is represented as directed towards joint E.

Using equations of force equilibrium at joint E,

$$\Sigma F_Y = F_{CE} \times \sin 60 + F_{BE} \times \sin 60 = 0$$

$$\Rightarrow \quad F_{BE} = -F_{CE} = -4.387\text{kN}$$

and $$\Sigma F_X = P_{EX} + F_{DE} + F_{BE} \times \cos 60 - F_{CE} \times \cos 60 = 0$$

$$\Rightarrow \quad F_{DE} = F_{CE} \times \cos 60 - P_{EX} - F_{BE} \times \cos 60$$
$$= 4.387/2 - 3 + 4.387/2 = +1.387\text{kN}$$

Using equations of force equilibrium at joint B,

$$\Sigma F_Y = F_{BD} \times \sin 60 + F_{BE} \times \sin 60 - P_{BY} = 0$$

$$\Rightarrow \quad F_{BD} = (P_{BY} - F_{BE} \times \sin 60) / \sin 60$$
$$= [5 - 4.387 \times (\sqrt{3}/2)] / (\sqrt{3}/2)$$
$$= 1.387 \text{ kN}$$

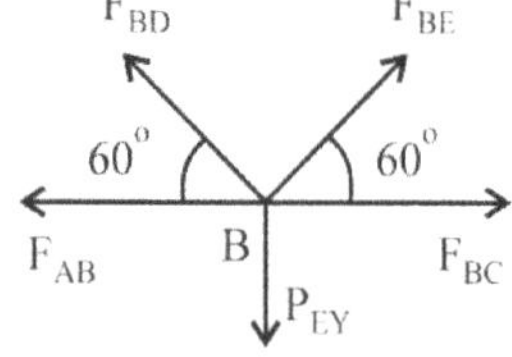

and $$\Sigma F_X = F_{BC} + F_{BE} \times \cos 60 - F_{BD} \times \cos 60 - F_{AB} = 0$$

$$\Rightarrow \quad F_{AB} = F_{BC} + F_{BE} \times \cos 60 - F_{BD} \times \cos 60$$
$$= 2.193 + (4.387/2) - (1.387/2) = +3.693\text{kN}$$

Using equations of force equilibrium at joint D,

$$\Sigma F_Y = F_{AD} \times \sin 60 - F_{BD} \times \sin 60 = 0$$

$\Rightarrow$ $F_{AD} = F_{BD} = 1.387\text{kN}$

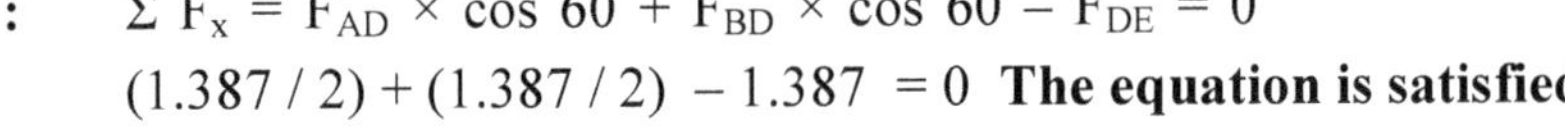

Check : $\Sigma F_x = F_{AD} \times \cos 60 + F_{BD} \times \cos 60 - F_{DE} = 0$

$\Rightarrow$ $(1.387 / 2) + (1.387 / 2) - 1.387 = 0$ **The equation is satisfied**

Equations of force equilibrium at joint A give the reactions

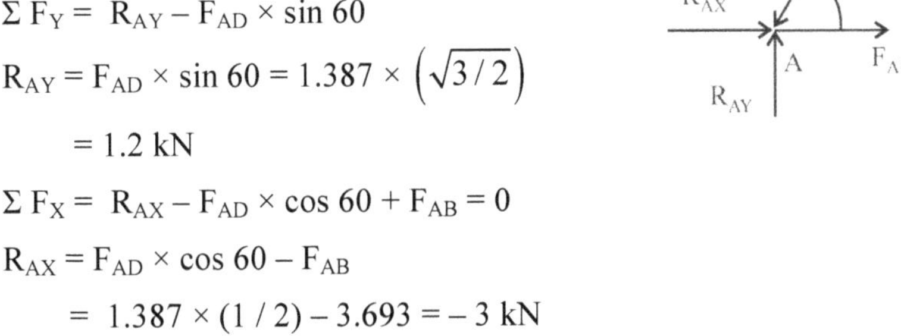

$$\Sigma F_Y = R_{AY} - F_{AD} \times \sin 60$$

$$R_{AY} = F_{AD} \times \sin 60 = 1.387 \times \left(\sqrt{3/2}\right)$$

$$= 1.2 \text{ kN}$$

$$\Sigma F_X = R_{AX} - F_{AD} \times \cos 60 + F_{AB} = 0$$

$$R_{AX} = F_{AD} \times \cos 60 - F_{AB}$$

$$= 1.387 \times (1/2) - 3.693 = -3 \text{ kN}$$

Check: Considering the total truss, all applied loads should be balanced by the reactions at the supports.

$$\Sigma F_Y = R_{AY} + R_{CY} - P_{BY} = 1.2 + 3.8 - 5 = 0$$

The equation is satisfied

$\Sigma F_X = R_{AX} + P_{EX} = -3 + 3 = 0$ **The equation is satisfied**

Since numerical errors are very common by considering wrong direction of member forces at the joints, such checks are essential to validate the calculated member forces & reactions. These values, in kN, are represented in the figure. +ve values indicate tensile member forces, represented by →—← and –ve values indicate compressive member forces, represented by ←—→. The four members connected to joint B, where vertical downward load is applied, are in tension while other three members are in compression.

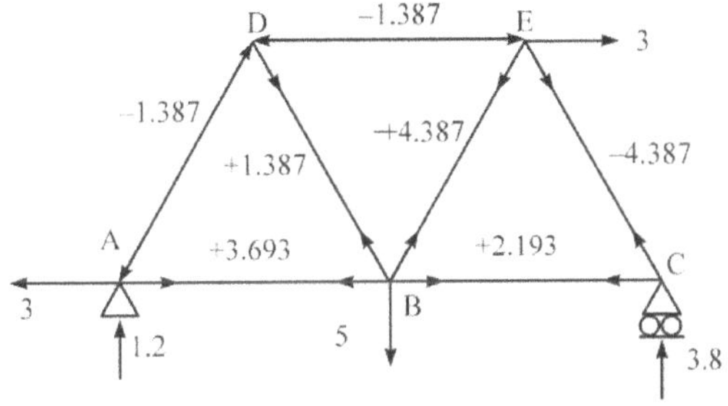

9.3.2 ANALYSIS BY METHOD OF SECTIONS

In this method, the truss is cut by imaginary section planes (straight or curved) and equations of static equilibrium are used to calculate member forces acting on the free body diagram of one part of the cut truss. The forces on the free body diagram are usually non-concurrent and, hence, applicable equations of equilibrium are considered. This method is explained through the following example, which was already solved by the method of joints.

Example 3

Find forces in all the members of the cantilever plane truss, shown below, for the vertical loads of 20kN applied at joints C and D. DE = EF = FG = 3m ; AG = 4m

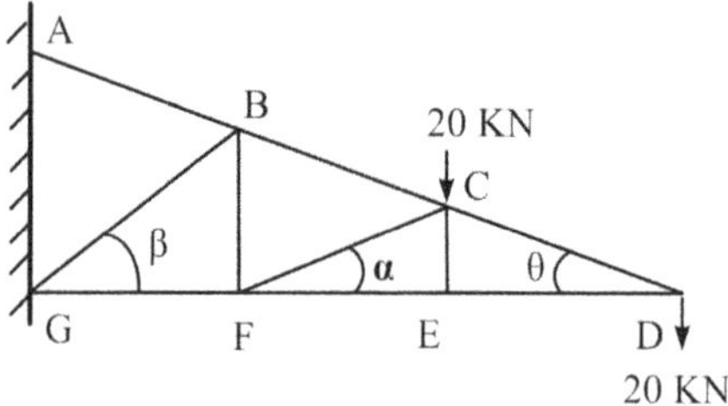

Solution:

Let $\angle CDE = \theta$. Considering equilibrium of any part of structure, two unknown forces can be evaluated using the two equations of equilibrium.

$$\sum F_X = 0 \text{ and } \sum F_Y = 0$$

$$\tan\theta = AG/GD = AG/(GF + FE + ED) = 4/(3 + 3 + 3) = 4/9$$

$$\Rightarrow \quad \sin\theta = 4/\sqrt{97}\,; \quad \cos\theta = 9/\sqrt{97}$$

If the truss is cut by an imaginary section plane-1, cutting members CD and DE as shown, for the equilibrium of the free body, we get

$$\sum F_Y = F_{CD}\sin\theta - 20 = 0$$

$$\text{or} \quad F_{CD} = 20/\sin\theta = 20/\left(4/\sqrt{97}\right) = 49 \text{ kN}$$

$$\sum F_X = F_{ED} - F_{CD}\cos\theta = 0$$

$$\Rightarrow \quad F_{ED} = F_{CD}\cos\theta = 49 \times \left(9/\sqrt{97}\right) = 45 \text{ kN}$$

If the truss is cut by an imaginary section plane-2, cutting members CD, CE and FE as shown, for the equilibrium of the free body, we get

$$\sum F_Y = F_{CD} \sin\theta + F_{EC} - 20 = 0$$

$$\Rightarrow \qquad F_{EC} = 20 - F_{CD} \sin\theta = 20 - 49 \times \left(4/\sqrt{97}\right) = 0$$

and $$\sum F_X = F_{EF} - F_{CD} \cos\theta = 0$$

$$\Rightarrow \qquad F_{EF} = F_{CD} \cos\theta \times \left(9/\sqrt{97}\right) = 45 \text{ kN}$$

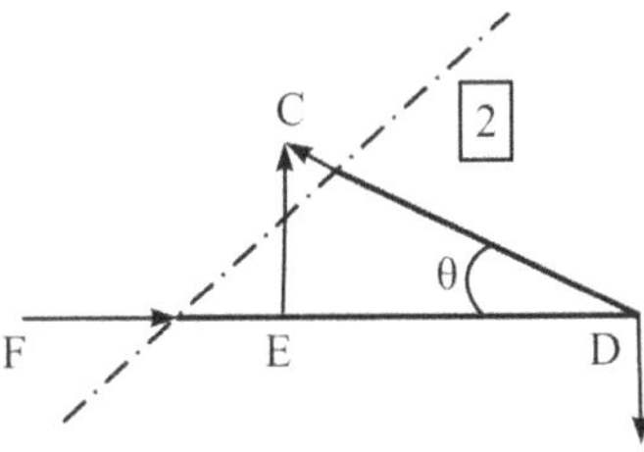

If the truss is cut by an imaginary section plane-3, cutting members FE, FC and BC as shown, for the equilibrium of the free body, we get

$$\sum F_Y = F_{FC} \sin\alpha + F_{BC} \sin\theta - 20 - 20 = 0$$

$$CE = DE \tan\theta = 3 \times (4/9) = 4/3$$

$$\tan\alpha = CE/FE = (4/3)/3 = 4/9$$

$$\Rightarrow \qquad \sin\alpha = 4/\sqrt{97} \ ; \cos\alpha = 9/\sqrt{97}$$

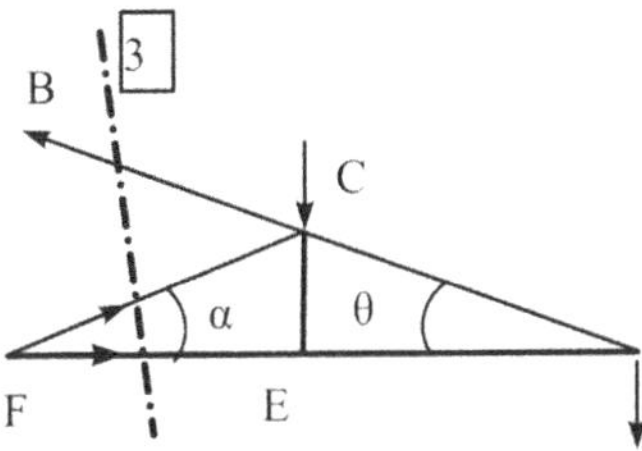

and $$\sum F_X = -F_{BC} \cos\theta + F_{FC} \cos\alpha + F_{FE} = 0$$

Solving the two conditions, F_{BC} and F_{FC} can be obtained as

$$F_{BC} = 73.87 \text{ kN}$$

and $$F_{FC} = 24.62 \text{ kN}$$

If the truss is cut by an imaginary section plane-4, cutting members GF, BF and BC as shown, for the equilibrium of the free body, we get

$$\sum F_X = F_{GF} - F_{BC} \cos \theta = 0$$

and $$\sum F_Y = F_{FB} + F_{BC} \sin \theta - 20 - 20 = 0$$

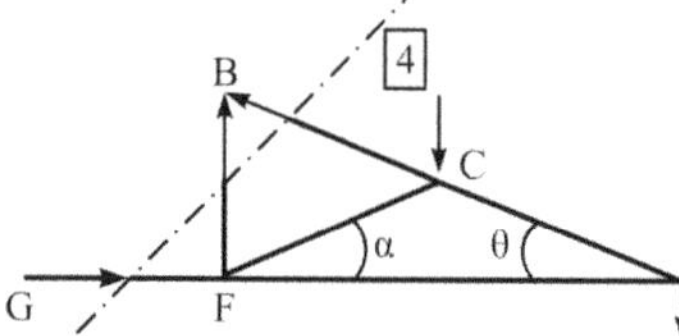

from which F_{FB} and F_{GF} can be obtained as

$$F_{GF} = F_{BC} \cos \theta = 73.87 \times \left(9/\sqrt{97}\right) = 67.5 \text{ kN}$$

and $$F_B = 20 + 20 - F_{BC} \sin \theta = 40 - 73.87 \times \left(4/\sqrt{97}\right) = 10 \text{ kN}$$

If the truss is cut by an imaginary section plane-5, cutting members GF, GB and AB as shown, for the equilibrium of the free body, we get

$$\sum F_Y = F_{AB} \sin \theta + F_{GB} \sin \beta - 20 - 20 = 0$$

and $$\sum F_X = F_{GF} - F_{AB} \cos\theta + F_{GB} \cos\beta = 0$$

where, $$BF = FD = \tan \theta = 6 \times (4/9) = 8/3$$

$$\tan \beta = BF/GF = (8/3)/3 = 8/9$$

$$\sin \beta = 8/\sqrt{145}\ ;\ \cos \beta = 9/\sqrt{145}$$

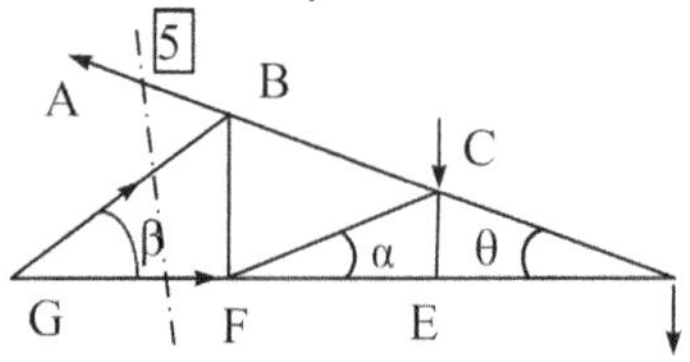

Solving these two equations, F_{GB} and F_{AB} can be obtained as

$$F_{GB} = 82 \text{ kN} \text{ and } F_{AB} = 10 \text{ kN}$$

Please note that

- Equations of equilibrium are written for the assumed directions of member forces, as shown in the figures. If the assumed direction is wrong, the calculated value of member force will be negative. It is assumed in the above equations that forces directed towards +ve X-axis (to the right, in the horizontal direction) are +ve and those directed towards –ve X-axis are –ve. Similarly, forces directed towards +ve Y-axis (upwards, in the vertical direction) are +ve and those directed along –ve Y-axis are –ve.

- The free body diagram of the truss obtained by cutting with the section plane-1 is acted upon by a system of concurrent forces; whereas the parts of truss obtained by cutting with the section planes 2 to 5 are acted upon by non-concurrent forces. It is therefore possible to choose different section planes so that three unknown member forces can be obtained from the three equations of equilibrium (for example, free body diagram with section plane-2 can be used to evaluate forces in three members BC, FC and EC; similarly free body diagram with section plane-4 can be used to evaluate forces in AB, GB and FB). There is ***no unique set of section planes*** for this analysis.

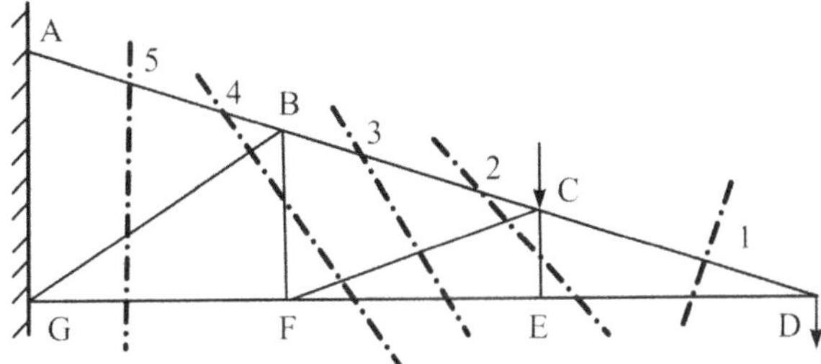

- It is possible to consider any one of the two parts of the truss, obtained by cutting with the section plane, for evaluating unknown member forces so long as the number of equilibrium equations match with the number of unknown member forces.

It can be seen that the solutions by both the methods are identical.

SUMMARY

- Frame is a discrete Structure, made up of distinctly identifiable axial-loaded members
- Structures with all members lying in one 2-D plane (usually vertical) are called plane frames. Structures with members lying in different orientations in 3-D space are called space frames.
- Truss is a frame consisting of axial load carrying members, (also called spars), with loads acting at the joints only and spars are connected with hinged joints, which permit relative rotation of members
- They are broadly classified as 'statically determinate trusses' or 'perfect frames', if the all member forces can be calculated using equations of static equilibrium. The number of members 'm' and number of joints 'j' are related by the rule $m = 2j - 3$ in plane trusses and $m = 3j - 6$ in space trusses; otherwise, they are called as 'statically indeterminate trusses' or 'imperfect frames'
- They are also classified as cantilever trusses or simply supported trusses

- In order to ensure that members are not subjected to bending, only *one joint is rigidly supported with a hinge* and *all other joints are supported on rollers*, supporting only vertical loads with vertical reaction
- A truss is analysed by calculating reactions with the application of equations of equilibrium on the entire structure and then calculating member forces by repeated application of equations of equilibrium

  $$\Sigma F_X = 0,\ \Sigma F_Y = 0 \text{ and } \Sigma F_Z = 0$$

  at the joints or on imaginary cut sections, in sequence
- Normal or longitudinal stress is assumed to be uniform across the cross section and calculated by dividing member force with area of cross section ($\sigma = P / A$); Strain and change in length (elongation/compression) of each member are calculated from stress, modulus of elasticity and length of each member ($\varepsilon = \sigma / E$ and $\delta = \varepsilon \times L$)

CHAPTER 10

BEAMS AND MECHANICS

10.1 BEAM

Most common examples of beams in regular use are cantilever wall brackets (with one end fixed to the wall or column and the other end free) for suspending name boards, display boards etc in Railway stations, bus stations or airports and simply supported bars used by athletes for pull-up exercises or used for weight measuring balances or see-saw in children's play grounds etc. In a big frame, beam is a long, slender and homogeneous member, similar to a spar in a truss. It differs from a spar by

- the nature of loads (concentrated loads or moments at any point along the length and/or loads distributed along the length), producing deflection of the beam (Ref Fig 10.1 a & b)

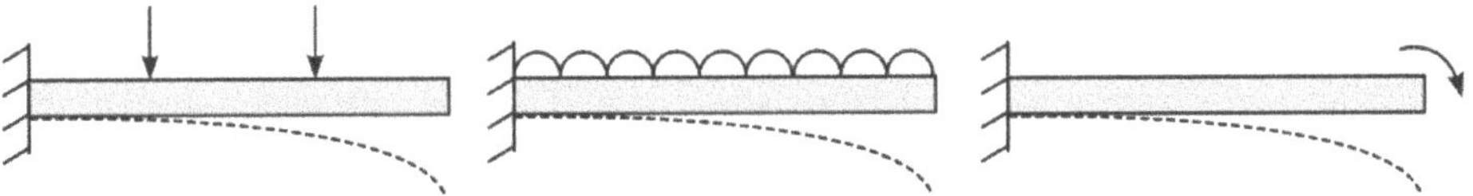

FIGURE 10.1(a) Cantilever beam with different types of loads

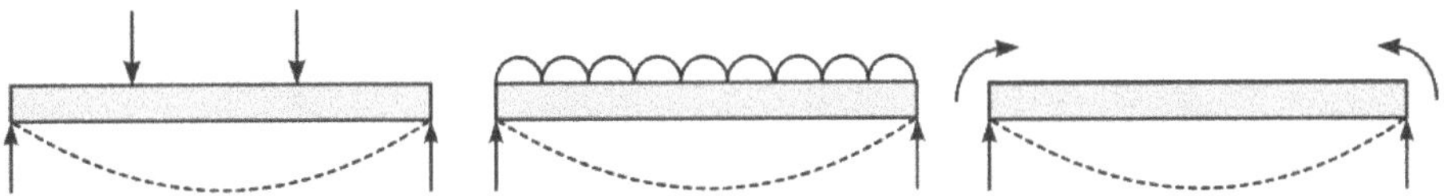

FIGURE 10.1(b) Simply supported beam with different types of loads

- type of joints (rigid, preventing relative rotation of members at a joint so that all members at a joint will have the same slope - Ref Fig 10.1 c)

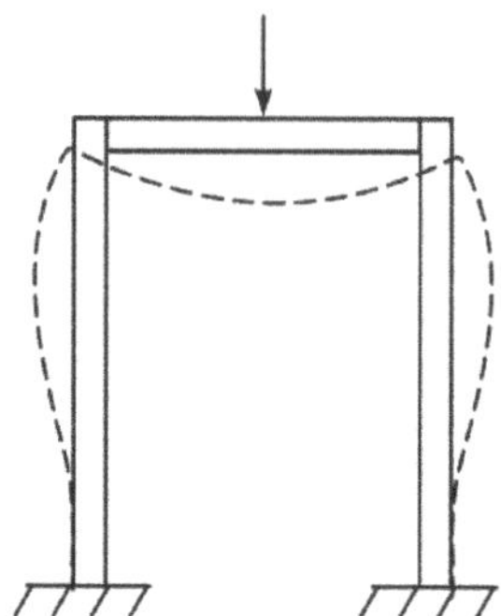

FIGURE 10.1(c) A beam with rigid joints at its ends

- nature of deformations (axial and normal deflection/slope)
- stresses (axial and shear, which vary across the cross section) and
- end condition which may be ***free*** with no restraint on displacement and slope or may have a ***simple*** (or hinged) support with restraint only on deflection or a ***fixed*** support with restraint on both deflection and slope

Cantilever beams and simply supported beams in a plane with three unknown reactions are classified as ***determinate beams*** (reactions can be calculated from the three equations of static equilibrium) while propped cantilever beams, continuous beams, fixed beams etc with more than three unknown reactions are classified as ***indeterminate or redundant beams*** (reactions can not be calculated from the three equations of static equilibrium)

10.2 SHEAR FORCE AND BENDING MOMENT

Concentrated or distributed loads and moments acting on a beam produce shear force and bending moment all along the beam. They can be calculated at each section by considering equilibrium of any one part of the beam as a free body, imagining a cut in the beam at that section. The behaviour of a beam in terms of deflection and stress depends upon the distribution of internal shear force and bending moment along the length of a beam member.

Throughout this chapter, beams are considered in X-Y plane, with X-axis along the neutral axis of the beam. The following sign convention (Ref Fig 10.2) is followed for shear force and bending moment.

Plot of variation of shear force and bending moment along the length of the beam are called ***shear force diagram*** **(SFD)** and ***bending moment diagram*** **(BMD)** respectively.

A few examples are given below for understanding these diagrams

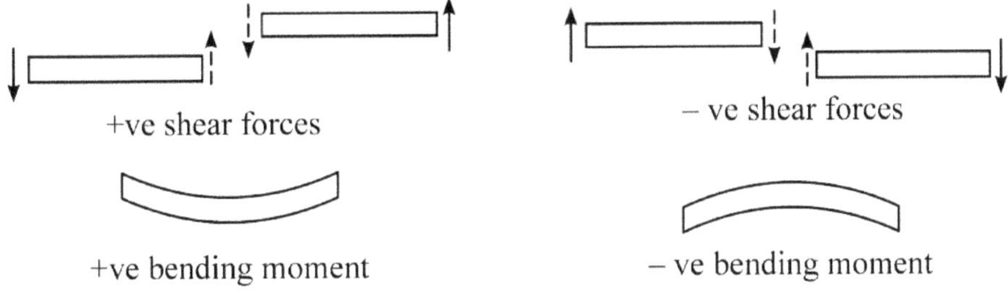

FIGURE 10.2 Sign conventions for shear force & bending moment

Example 1

SFD and BMD for a cantilever with a concentrated load P at the free end

For equilibrium of the free body of the cantilever beam

$\sum F_Y = S - P = 0$ or $S = P$ and $\sum M_Z = M - P \times x = 0$ or $M = P \times x$

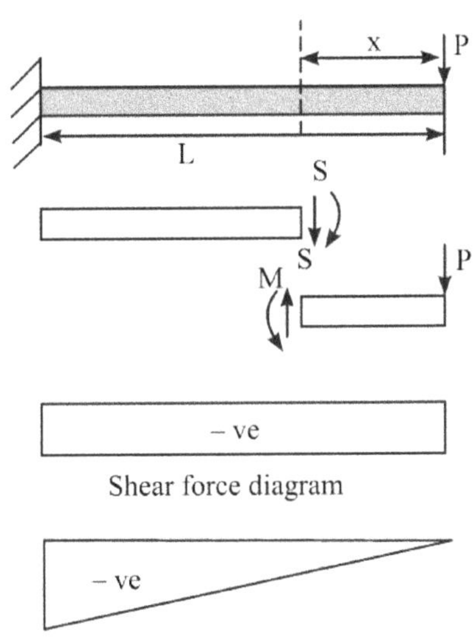

We notice, for this case, shear force S is constant all along the beam while bending moment M varies linearly from

$M = 0$ at the free end to $M = M_{MAX} = P \times L$ at the fixed end

Example 2

SFD and BMD for a simply supported beam with uniformly distributed load p

For equilibrium of the free body of the simply supported beam,

$\sum F_X = S - R_1 + p \times x = 0$ or $S = R_1 - p \times x$ where $R_1 = p \times L / 2$

$$\sum M_Z = M - R_1 \times x + (p \times x) \times x/2 = 0$$

or $$M = R_1 \times x - p \times x^2/2 = p \times x \times (L - x) / 2$$

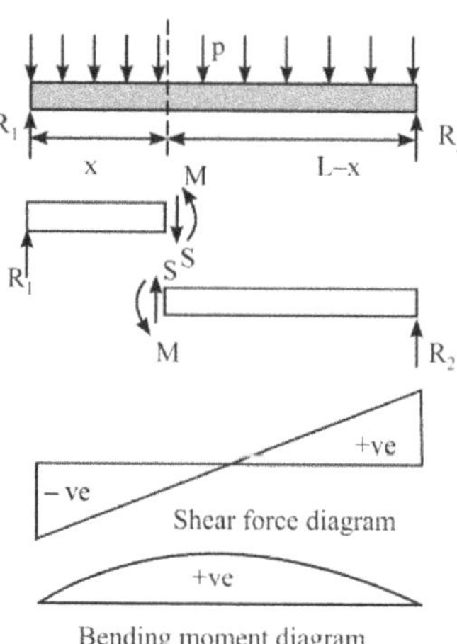

For this particular case, therefore, shear force S varies linearly along the beam while bending moment M varies parabolically from M = 0 at the support ends to $M = M_{Max} = p \times L^2/8$ at x = L/2 or at mid-point along the axis of the beam.

10.2.1 RELATIONSHIP BETWEEN SHEAR FORCE AND BENDING MOMENT

Consider an infinitesimal length 'dx' of a beam, with shear force 'S' and moment 'M' on the right side, which increase to 'S+dS' and 'M+dM' on the left side due to applied normal load 'p(x)' along the length, as shown in Fig 10.3.

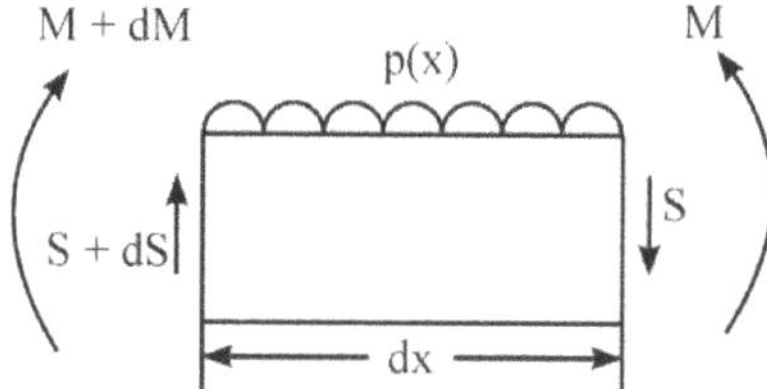

FIGURE 10.3 Shear force and bending moment over a small length of a beam

Equilibrium equations for this segment of beam are:

$$\sum F_Y = (S + dS) - p(x) \times dx - S = 0 \quad \Rightarrow dS/dx = p(x)$$

and $$\sum M_Z = M - p(x) \times dx \times (dx/2) + S \times dx - (M + dM) = 0$$

$$\Rightarrow dM/dx = S, \text{ neglecting higher order terms}$$

Therefore, $p(x) = dS/dx = d^2M/dx^2$

It can therefore be concluded that

(a) when $p(x)=0$, S is constant and M varies linearly

(b) when $S = 0$, M is maximum or minimum

SOME EXAMPLES

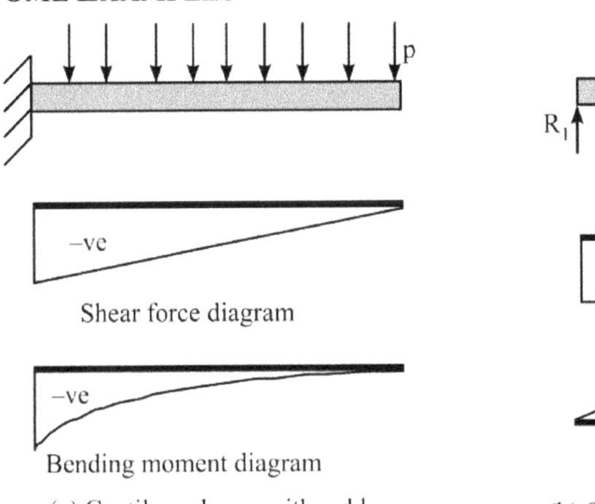

(a) Cantilever beam with u.d.l.

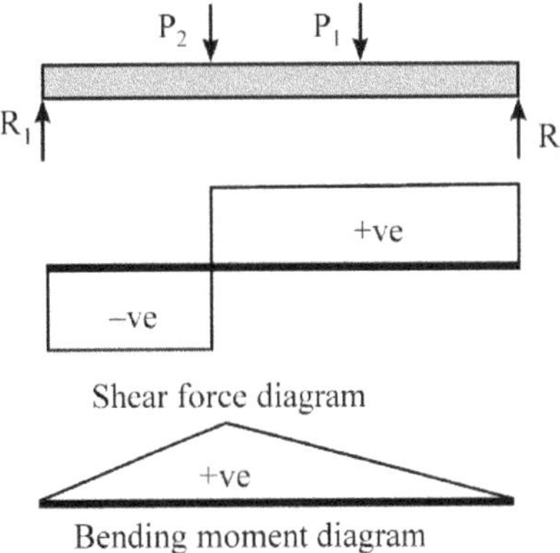

(b) Simply supported beam with point loads

10.2.2 POINTS OF CONTRA-FLEXURE

The points in a beam (other than the extreme ends) at which the bending moment is zero, are called ***points of contra-flexure*** or ***points of inflexion*** (See Fig 10.4). They usually occur in overhang beams.

Some examples of contra-flexure or reversal of bending moment

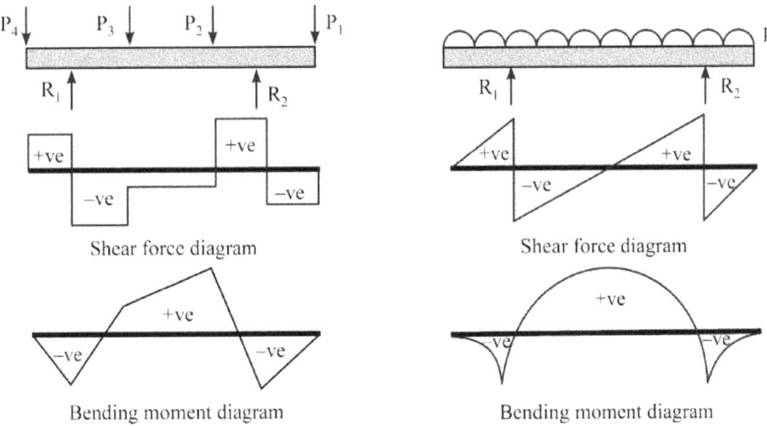

(a) Overhanging beam with concentrated loads (b) Overhanging beam with distributed load

FIGURE 10.4 Points of contra-flexure in a overhanging beam

Example 3

A 6m long cantilever carries a uniformly distributed load of 0.5kN/m for a length of 4m from its fixed end and a point load of 2kN at free end. Draw SFD and BMD

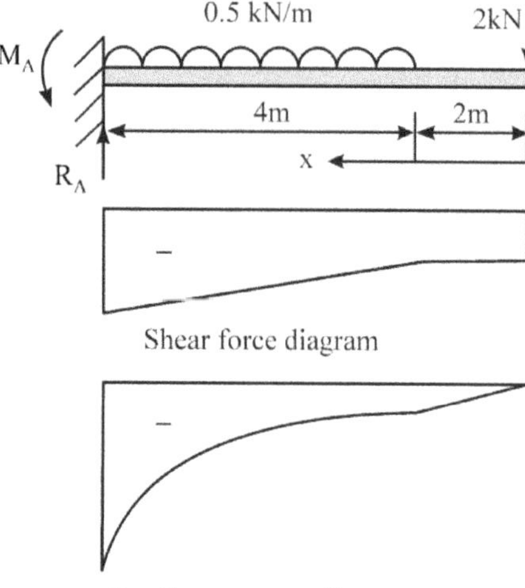

Shear force diagram

Bending moment diagram

Solution:

Reaction, $R_A = 2 + 0.5 \times 4 = 4$ kN

Fixed end moment, $M_A = 2 \times 6 + 0.5 \times 4 \times (4/2) = 18$ kNm

On a section at distance x from free end,

$$S = 2 + 0.5 \times (x - 2)$$

$$M = 2x + 0.5 \times (x - 2) \times (x - 2) / 2$$

Example 4

Draw SFD and BMD of a overhang beam, with udl of 9kN/m over 3m length from A and 3kN/m over the remaining length as well as a point load of 5kN, as shown

Solution:

Reactions are

$$R_B = (9 \times 3 \times 4.5 + 3 \times 3 \times 1.5 + 5 \times 1.8 - 3 \times 1.5 \times 0.75) / 4.5$$

$$= 31.25 \text{ kN}$$

$$R_C = (9 \times 3 + 3 \times 4.5 + 5) - R_B = 14.25 \text{ kN}$$

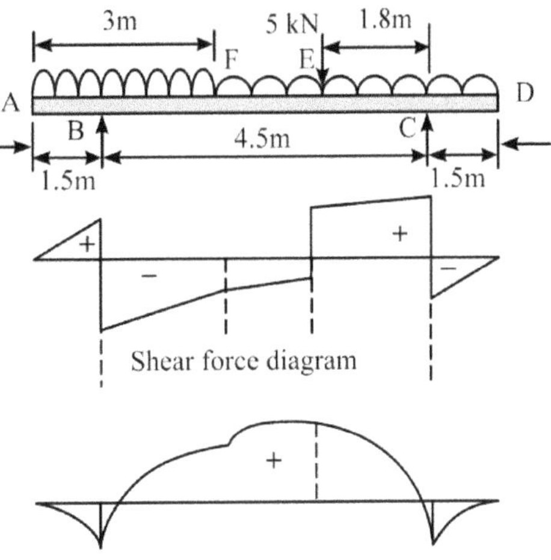

Shear forces at different points are :

$S_D = 0$

$S_C = -3 \times 1.5 + R_C = -4.5\text{kN} + R_C$

$S_E = -3 \times (1.5 + 1.8) + R_C$

$= 9.9 + 14.25 = 4.35$ kN

$S_F = -3 \times 4.5 - 5 + R_C$

$= -18.5 + 14.25 = 4.25$ kN

$S_B = 9 \times 1.5 = 13.5$ kN ; $S_A = 0$

Bending moments at different points are :

$M_D = 0$;

$M_C = -3 \times 1.5 \times 0.75$

$= -3.375$ kNm

$M_E = -3 \times 3.3 \times 1.65 + R_C \times 1.8$

$= 9.315$ kNm

$M_F = -3 \times 4.5 \times 2.25 - 5 \times 1.2 + R_C \times 3 = 6.375\text{kNm}$

$M_B = -9 \times 1.5 \times 0.75 = -10.125\text{kNm}$; $M_A = 0$

Example 5

Draw SFD and BMD of a 5m long simply supported beam carrying a load of 200N through a bracket welded to the beam, as shown

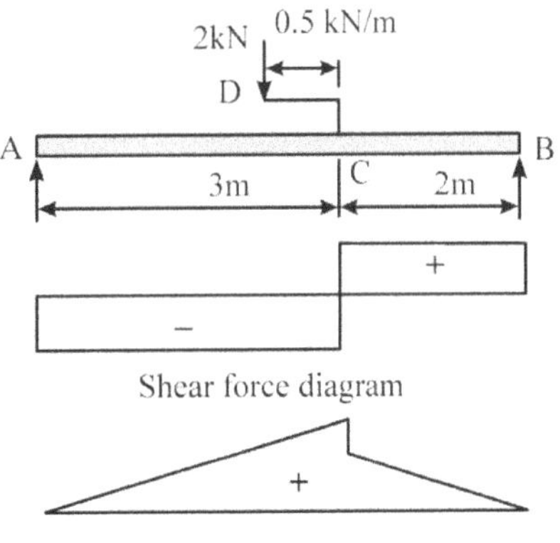

Solution:

Load of 2kN at D is equivalent to applying load of 2kN and moment of 1 kNm (2kN × 0.5m) at C. This moment produces reaction of 1/5 kN at A and B in opposite directions to balance moment at C.

Reaction, $R_A = (2 \times 2/5) + 1/5 = 1$ kN

Reaction, $R_B = (2 \times 3/5) - 1/5 = 1$ kN

At a section at x_1 from A (in CB portion),

$$S = R_B = 1 \text{ kN} \quad ; \quad M = R_B \times (5 - x_1) = 5 - x_1$$

At a section at x_2 from A (in AC portion),

$$S = R_B - 2 = 1 \text{ kN}\ ; \quad M = R_B \times (5 - x_2) + 1 - 2 \times (3 - x_2) = x_2$$

10.3 DEFLECTION OF A SIMPLE BEAM

Due to load distributed along the beam or moment applied at the joint or free end, a beam deforms into a circular arc of radius 'R' (also called elastic curve). The magnitude of deflection and slope from its undeformed position varies with location, depending not only on the applied loads but also on the section and material properties of the beam. Compared to change in length (elongation or compression) of a spar in a truss, deflection of a beam is much larger. Magnitude of deflection can affect performance or utility of a structure. Hence, calculation of deflection of a beam, in addition to the stresses, for the applied loads is a necessary step in the design of any beam.

In a simple beam, let ABC be a beam of length 'L' which deflects into a circular arc with max deflection 'δ' (Ref Fig 10.5).

From ΔOAB, $R^2 = (L/2)^2 + (R - \delta)^2$ or $L^2/4 - 2R\delta + \delta^2 = 0$

⇒ Maximum deflection, $\delta \approx L^2 / 8R$, neglecting δ^2 term, which is very small

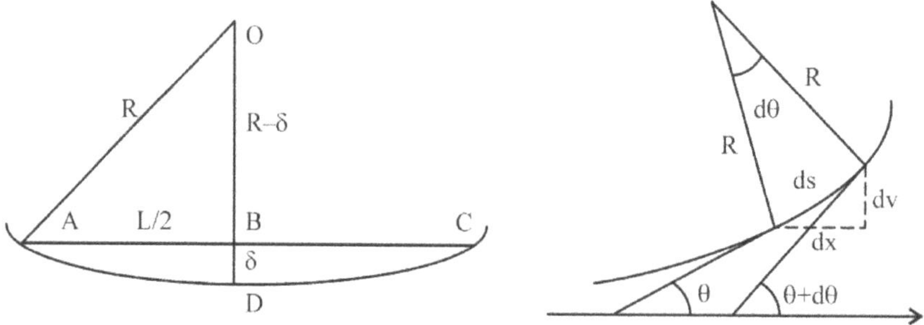

FIGURE 10.5 Deflection and curvature of a beam

Consider two points P(x) and Q(x+dx), on the elastic curve of a beam

$$\tan\theta \approx \theta = dv/dx$$

⇒ $\tan d\theta = ds/R \approx d\theta = dx/R$

or Curvature, $1/R = d\theta/dx = d^2v/dx^2$

The following assumptions are made while calculating deflections and stresses in a beam.

- Each layer in the beam deflects to form a circular arc
- Length of the beam is much larger than the cross sectional dimensions
- Shear deformation is neglected
- Plane sections before applying loads, remain plane after bending
- Maximum stress in any layer is within the elastic limit
- Applied loads do not produce torsion or twisting of the section

10.3.1 EULER-BERNOULLI BEAM THEORY

An initially straight beam is deflected to a circular arc when loaded in pure bending. Vertical planes before bending, ABCD and FGHJ, become radial planes A′B′C′D′ and F′G′H′J′ after bending, as shown in Fig 10.6.

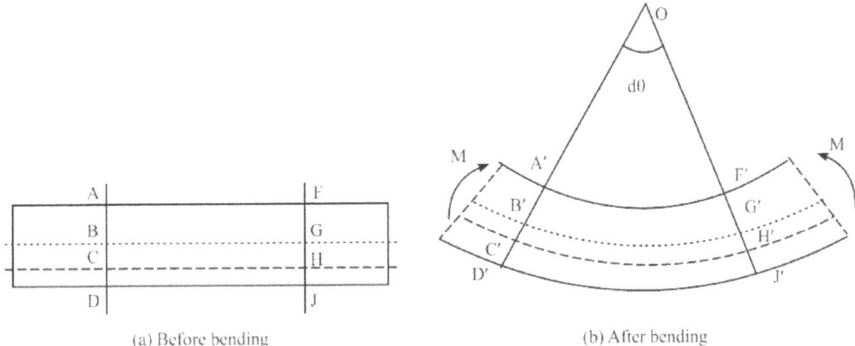

FIGURE 10.6 Deflection and curvature of a beam

Some planes shorten and some planes elongate. There is a layer or plane which has no strain and, hence, no stress. This is called a ***neutral plane.*** Intersection of neutral plane with the plane of bending is called ***neutral axis***

For small angle dθ, strain in the layer CH at distance y from neutral axis BG or B′G′,

$$\varepsilon_X = du/dx \approx (C'H' - CH) / CH = [(R + y) \times d\theta - R \times d\theta] / (R \times d\theta) = y/R$$

Thus, **strain in a beam is a linear function of distance y from neutral axis,** and changes its sign from tensile or +ve on one side of neutral axis B-G to compressive or –ve on the other side of the neutral axis, with maximum values on the outermost layers (for example, D-J and A-F in Fig).

10.3.2 NEUTRAL PLANE & NEUTRAL AXIS

Within the elastic limit, stress $\sigma_X = E \times \varepsilon_X = E \times y/R$ .....(10.1)

Thus, **bending stress is a linear function of distance y from neutral axis** and changes its sign from tensile or +ve when the layer CH is farther than neutral axis from center of curvature to compressive or –ve when the layer is closer to neutral axis from the center of curvature.

The resultant horizontal component of stress on any cross section due to bending loads is zero, so that the equilibrium equation $\sum F_X = 0$ is always satisfied

Total axial internal force in a beam across any cross section,

$$F_X = \int dF_X = \int \sigma_X \times dA = \int (E \times y/R) \times dA = 0$$

$\Rightarrow$ $\int y \times dA = 0$ since, E and R are constant

or BG, the **neutral axis, coincides with centroidal axis**

10.3.3 DEFLECTION BY DOUBLE INTEGRATION METHOD

Displacement v, normal to the beam axis, is calculated at any section by calculating bending moment at that section and using the relation

$$d^2v/dx^2 = M / EI$$

By integrating the above equation, we get slope of the elastic curve (deflected shape of beam) at that section

$$\theta = dv/dx = (1/EI) \times [\int M \times dx + C_1]$$

Integrating again, we get displacement at the section as

$$v = (1/EI) \times [\int (\int M \times dx) \times dx + \int C_1 \times dx + C_2]$$

From the above equations, it can be seen that the values of slope and displacement, naturally, depend on the values of the constants C_1 and C_2. These constants are evaluated from the end conditions (displacement and/or slope)

10.3.4 MACAULAY'S NOTATION

Integrating the equation EI (d^2v/dx^2) = M for different parts of the beam separately is tedious. It is also difficult to evaluate integration constants in each part. Macaulay has suggested a notation <......> for bending moment M which can be used over the entire beam. Expression in <.......> need to be considered only for positive values. Expression in <........> should not be expanded, during integration.

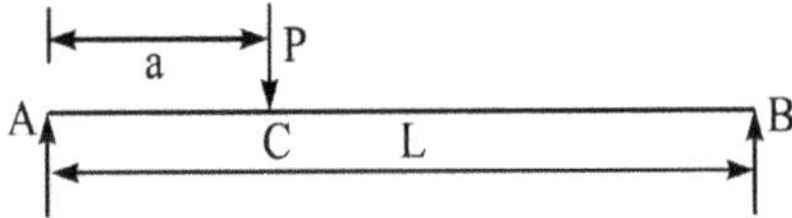

In the example shown here, $M_{AC} = R_A \times x$ for $0 \le x \le a$

$M_{CB} = R_A \times x - P \times (x - a)$ for $a < x \le L$

According to Macaulay's notation, $M_{AB} = R_A \times x - < P \times (x - a) >$

Expression in <.......> need to be considered only for $x > a$

Example 6

Calculate slope and deflection of a simply supported beam AB of length 'L', with a concentrated load P acting at C, at a distance of 'a' ($a < L/2$) from the left support, by double integration method. Deduce the values for the particular case of $a = L/2$

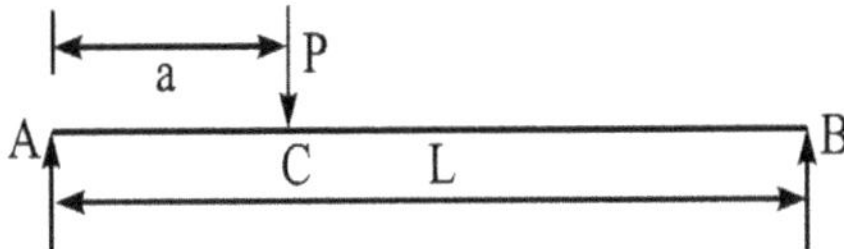

Solution:

Taking moments about B,

$$\sum M = R_A \times L - P \times (L - a) = 0 \quad \Rightarrow \quad R_A = P \times (L - a) / L$$

Then, $\sum F_Y = R_A + R_B - P = 0 \quad \Rightarrow \quad R_B = P \times a / L$

According to Macaulay's notation, $M_{AB} = R_A \times x - < P \times (x - a) >$

Expression in < > need to be considered only for $x > a$

Thus, $M_{AB} = R_A \times x - < P \times (x - a) > \; = R_A \times x = M_{AC}$ for $x < a$

$M_{AB} = R_A \times x - < P \times (x - a) > \; = R_A \times x - P \times (x - a) = M_{CB}$

for $x > a$

Integrating beam deflection equation, $E \times I \times (d^2v/dx^2) = M_{AB}$ once, we get

$$E \times I \times \theta = E \times I \times (dv/dx) = \int M_{AB}\, dx + C_1$$

$$= \int [\, R_A \times x - \; < P \times (x - a) >]\, dx + C_1$$

$$= [R_A \times x^2/2 - < P \times (x - a)^2/2 >] + C_1 \qquad \ldots\ldots (1)$$

Integrating the above expression once again, we get

$$E \times I \times v = \int [\, R_A \times x^2/2 - \; < P \times (x - a)^2/2 >]\, dx + C_1 \times x + C_2$$

$$= [\, R_A \times x^3/6 - \; < P \times (x - a)^3/6 >] + C_1 \times x + C_2 \qquad \ldots\ldots (2)$$

The end condition, $v = 0$ at $x = 0$, gives $0 = [\, R_A \times 0 \,] + C_1 \times 0 + C_2$

$\Rightarrow \quad C_2 = 0$

Here, $x - a$ is –ve for $x = 0$ and, hence, the 2nd term is ignored

The end condition, $v = 0$ at $x = L$, gives

$$0 = [\, R_A \times L^3/6 - P \times (L - a)^3/6 \,] + C_1 \times L + 0$$

from which C_1 can be evaluated if P, L and a are given

Here, $x - a$ is +ve for $x = L$ and, hence, the 2nd term is included

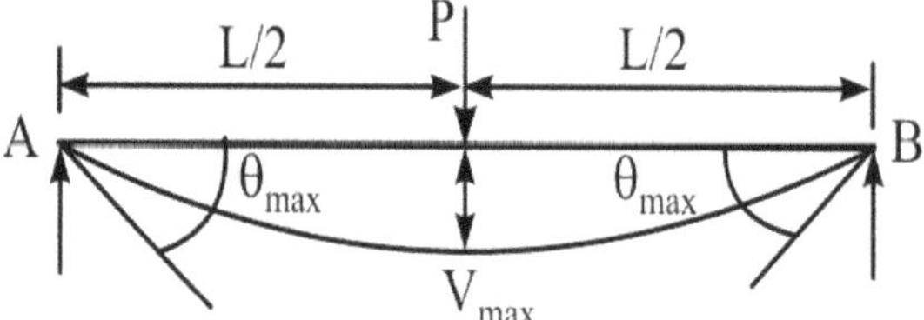

Special case :

When $a = b = L/2$, $R_A = R_B = P/2$

Corresponding to maximum deflection, $x = L / 2$, because of symmetry

Then, C_1 can be calculated from eq (1) for $\theta = 0$ at $x = L/2$, as

$$0 = R_A \times (L/2)^2 / 2 + C_1 \quad \Rightarrow \quad C_1 = -(P/2) \times (L/2)^2 / 2 = -P \times L^2 / 16$$

And maximum displacement can be calculated from eq (2) at $x = L/2$, as

$$v_{Max} = [\, R_A \times x^3/6 + C_1 \times x \,] / (E \times I)$$

$$= [\, (P/2) \times (L/2)^3 / 6 - (P \times L^2 / 16) \times (L/2) \,] / (E \times I)$$

$$= -P \times L^3 / (48E \times I)$$

Maximum slope occurs at A and B in opposite directions

$$\theta_A = C_1 / (E \times I) = -P \times L^2 / (16\,E \times I) \qquad \text{at} \quad x = 0$$

$$\theta_B = P \times (L/2) \times [L^2 - (L/2)^2] / (6E \times I \times L) = P \times L^2 / (16\,E \times I)$$

at $x = L$

Example 7

Calculate max deflection and its location in a simply supported beam AB of length 'L' with a uniformly distributed load of 'w' per unit length over a length 'a' from end B. Deduce the values for the particular case of $a = L$

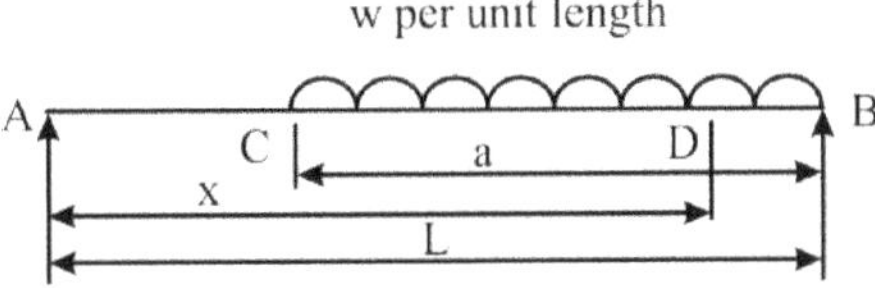

Solution :

$\sum M_A = 0$ gives $R_B = (w \times a) \times (L - a/2) / L$

$\sum M_B = 0$ gives $R_A = (w \times a) \times (a/2) / L$

Taking A as the origin and measuring x towards B, equation of elastic curve gives,

$$E \times I \times d^2v/dx^2 = M = R_A \times x - w \times \{x - (L - a)\}^2/2$$

Integrating, $E \times I \times (dv/dx) = R_A \times (x^2/2) - < w \times \{x - (L - a)\}^3 / 6 > + C_1$ .....(1)

Integrating again, $E \times I \times v = R_A \times (x^3/6) - < w \times \{x - (L - a)\}^4 /24 > + C_1 \times x + C_2$ .....(2)

$v = 0$ at $x = 0$ gives $C_2 = 0$ neglecting 2nd term which is -ve

$v = 0$ at $x = L$ gives

$$C_1 = - [R_A \times (x^3/6) - w \times \{x - (L - a)\}^4/24] / L$$
$$= - [\{w \times a^2 / (2L)\} \times (L^3/6) - w \times \{L - (L - a)\}^4 / 24] / L$$
$$= - w \times a^2 \times L/12 + w \times a^4 / (24L)$$

At the point of maximum deflection,

$$dv/dx = [R_A \times (x^2/2) - w \times \{x - (L - a)\}^3/6 - w \times a^2 \times L/12 + w \times a^4 / (24L)] / (E \times I) = 0$$

$$\Rightarrow \quad \{w \times a^2 / (2L)\} \times (x^2/2) - w \times \{x - (L - a)\}^3/6 - w \times a^2 \times L/12 + w \times a^4 / (24L) = 0$$

which gives the value of x for specific values of a and L

Using this value of x in the equation for deflection, maximum deflection can be evaluated

Special case :

When $a = L$, $R_A = R_B = w \times L/2$

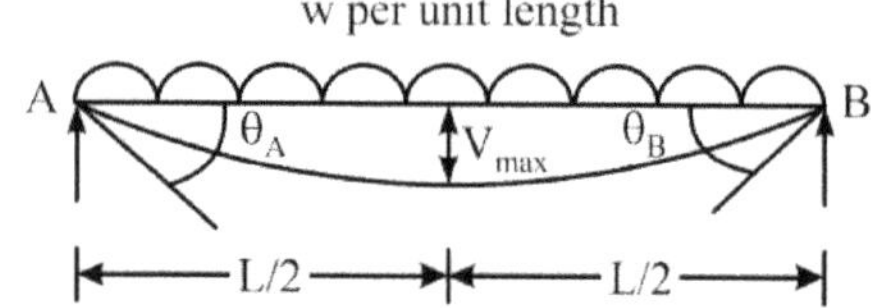

Corresponding to maximum deflection, x = L / 2, because of symmetry

Then, C_1 can be calculated from eq (1) for $\theta = 0$ at $x = L/2$, as

$$0 = R_A \times (L/2)^2 / 2 - w \times (L/2)^3 / 6 + C_1$$

$$\Rightarrow \quad C_1 = -(w \times L/2) \times (L/2)^2 / 2 + w \times (L/2)^3 / 6 = -w \times L^3 / 24$$

And maximum displacement can be calculated from eq (2) at $x = L/2$, as

$$v_{Max} = [\, R_A \times x^3/6 - w \times (L/2)^4 / 24 + C_1 \times x \,] / (E \times I)$$

$$= [\, (w \times L/2) \times (L/2)^3 / 6 - w \times (L/2)^4 / 24 - (w \times L^3 / 24) \times (L/2) \,] / (E \times I)$$

$$= -5w \times L^4 / (384\, E \times I)$$

Maximum slope occurs at A and B in opposite directions

$$\theta_A = C_1 / (E \times I) = -w \times L^3 / (24\, E \times I) \qquad \text{at } x = 0$$

$$\theta_B = [\, (w \times L/2) \times L^2 / 2 - w \times L^3 / 6 - w \times L^3 / 24 \,] / (E \times I)$$

$$= w \times L^3 / (24\, E \times I) \qquad \text{at } x = L$$

10.4 BENDING STRESS

In order that a beam does not fail when loads are applied, maximum stress in the beam is to be limited to the allowable limits.

10.4.1 THEORY OF BENDING

The resultant of compressive stresses and tensile stresses on a cross section form two equal and parallel forces (Ref Fig 10.7), thus producing a couple called ***moment of resistance***

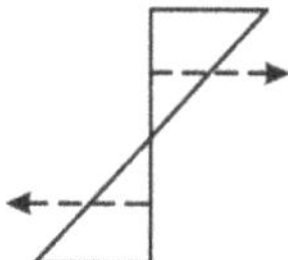

FIGURE 10.7 Moment of resistance on a cross section

Bending moment due to external applied loads is independent of beam section details while moment of resistance, for moment equilibrium, is equal to the bending moment and depends on shape of cross section and material of beam. If bending moment is more than moment of resistance, the beam fails

Moment of resistance at any section is given by,

$$M_Z = \int y \times dF_X = \int y \times (\sigma_X \times dA) = \int y \times (E \times y/R) \times dA$$
$$= (E/R) \times \int y^2 \times dA$$
$$= (E/R) \times I_Z \qquad .....(2)$$

Combining the two relations (1) and (2), we get $\mathbf{M_Z/I_Z = \sigma_X/y = E / R}$

This equation is called Engineer's ***theory of bending***

10.4.2 BENDING STRESS DISTRIBUTION

Bending stress distribution across a beam section can be obtained from the two conditions

$$\boldsymbol{\sigma_X = M_Z \times y / I_Z} \quad \textbf{and} \quad \boldsymbol{\sum F_X = \int \sigma_X \times dA = 0}$$

In beams with ***symmetric sections*** w.r.t. the neutral plane (such as rectangular, circular, I sections), maximum compressive stress is equal to the maximum tensile stress (Ref Fig 10.8 a).

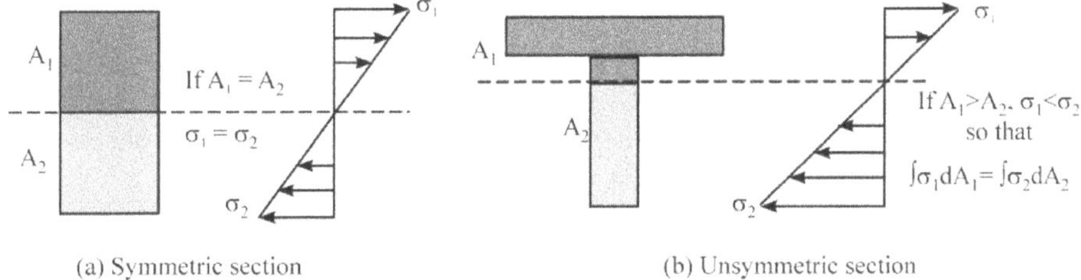

FIGURE 10.8 Bending Stress distribution on a section of a beam

In beams with ***unsymmetric sections*** (such as T, L sections), maximum compressive stress may not be equal to the maximum tensile stress Ref Fig 10.8 b). In these sections, smaller stress acting on the larger width will produce the same total force as larger stress on smaller width, satisfying the equilibrium equation $\Sigma F_X = 0$.

10.4.3 COMBINED BENDING AND AXIAL LOADS

A beam may be subjected to a combination of axial (tensile or compressive) load and bending load. Since bending load produces tensile stress on one side of neutral axis and compressive stress on the other side, bending stress $\boldsymbol{\sigma_b} = \mathbf{M \times y/I}$ can be superimposed on the axial (tensile or compressive) stress $\boldsymbol{\sigma_a} = \mathbf{P/A}$, as shown in Fig 10.9 a & b.

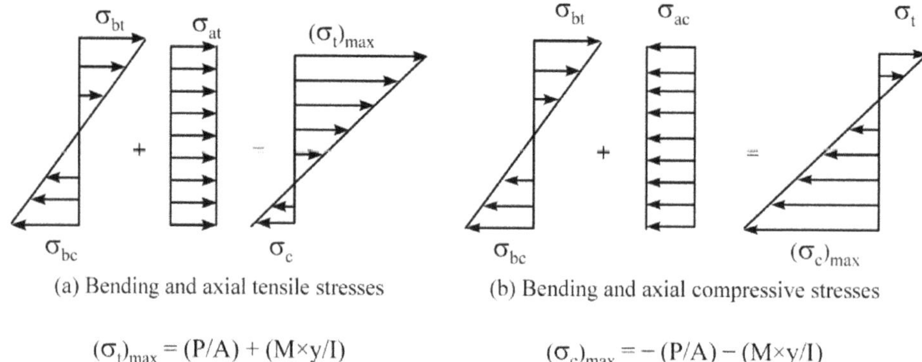

FIGURE 10.9 Combined Bending and Axial Stress distribution

Linear superposition of these stresses is valid only when the total stress is within the elastic limit

10.4.4 SHEAR STRESS IN A BEAM

Unlike in a spar element of a truss, varying normal (axial) stress along two parallel layers in a beam produces shear stress (τ_{YX}) between them, which can be seen to be zero on the outermost layers and maximum along the plane where bending stress changes from tensile to compressive

Let us consider a small section of beam of length 'dx' and an imaginary cut along a layer parallel to the neutral plane (Ref Fig 10.10). Let σ be the bending stress on a layer of thickness 'dy', at a distance 'y' from the neutral axis.

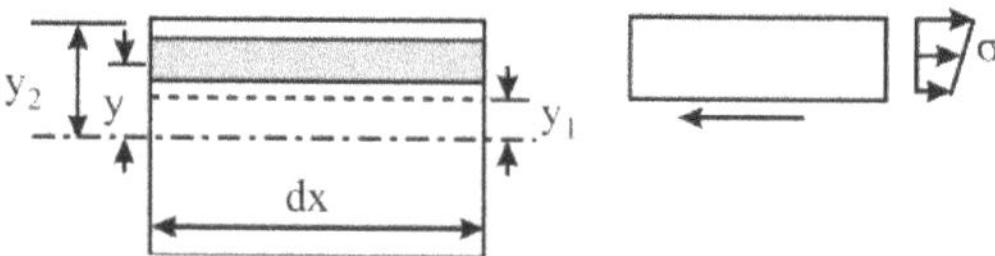

FIGURE 10.10 Shear stress in a beam

For the equilibrium of free body of the cut section of the beam, axial (or bending) stress on the free body should be equal to the shear force along the cutting plane (Ref Fig 10.11). Shear stress on that layer is then equal to the shear force divided by the area of the cutting plane. Let 'b' be the width of the beam in the Z-direction

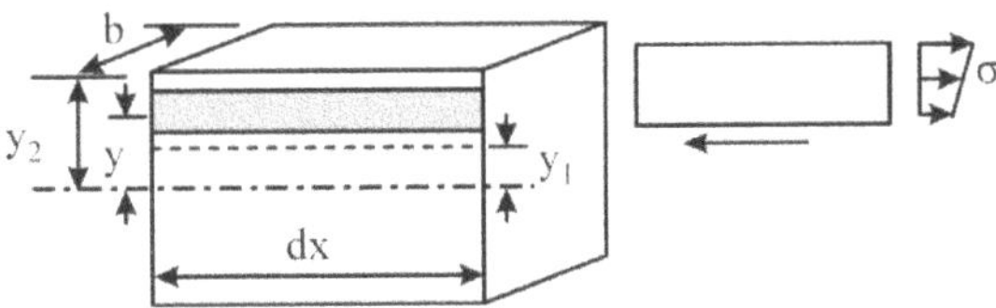

FIGURE 10.11 Relation between Bending stress and shear stress

$$\tau \times b \times dx = \int_{y_1}^{y_2} \sigma . dA = \int_{y_1}^{y_2} (dM \times y / I) \times dA = (dM / I) \int_{y_1}^{y_2} y \times dA = (dM / I) \times A \times \bar{y}$$

$$\tau = \frac{dM / I}{b \times dx} \times A \times \bar{y} = \frac{dM}{dx} \times \frac{1}{I.b} \times A \times \bar{y} = S_Y \times \frac{1}{I \times b} \times A \times \bar{y}$$

where $\bar{y}$ is the distance of centroid of part of beam section above or below neutral axis.

Example 8

Shear stress in a solid rectangular section

For a uniform rectangular section (width 'b' × depth 'd'), $y_2 = d/2$

$$\tau = S_Y \times \frac{1}{I \times b} \times A \times \bar{y}$$

$$= S_Y \times \frac{1}{I.b} \times [(d/2) - y_1] \times b \times \frac{[(d/2) + y_1]}{2} = (S_Y / 8I) \times (d^2 - 4y_1^2)$$

Maximum shear stress occurs when $y_1 = 0$

$$\tau_{Max} = \left[S_Y / (2I)\right] \times \left[(d/2)^2 - 0\right] = S_Y \times d^2 / (8I)$$

Minimum shear stress occurs when $y_1 = \pm\, d/2 \Rightarrow \tau_{min} = 0$

Average shear stress, $\tau_{Ave} = S_Y / A = S_Y / (b \times d)$

and $$\frac{\tau_{Max}}{\tau_{Ave}} = \frac{S_Y \times d^2 / 8I}{S_Y / (b \times d)} = \frac{d^3}{8I \times b} = \frac{d^3}{8 \times (b \times d^3 / 12) \times b} = 3/2$$

Example 9

Shear stress in a solid circular section of radius 'R'

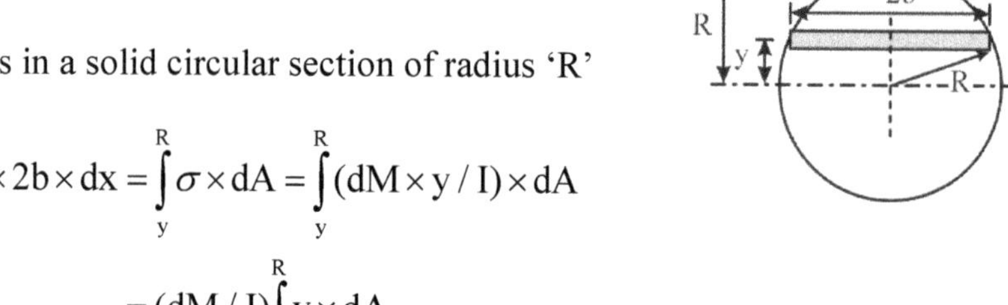

$$\tau \times 2b \times dx = \int_{y}^{R} \sigma \times dA = \int_{y}^{R} (dM \times y / I) \times dA$$

$$= (dM / I) \int_{y}^{R} y \times dA$$

$$= (dM / I) \int_{y}^{R} y \times \left[2 \times \sqrt{(R^2 - y^2)} \right] \times dy$$

By substituting $dA = 2b \times dy = \left[2 \times \sqrt{(R^2 - y^2)} \right] \times dy$

$$\tau = (dM / dx) \times (1 / I \times 2b) \times \left[-\left(R^2 - y^2\right)^{3/2} / (3/2) \right]$$

$$= S_Y \times \frac{1}{I.\left[2\sqrt{R^2 - y^2} \right]} \times \frac{2}{3} \times (R^2 - y^2)^{3/2}$$

$$= \frac{S_Y}{3I} \times (R^2 - y^2) = \frac{S_Y}{3 \times (\pi R^4 / 4)} \times (R^2 - y^2)$$

Maximum shear stress occurs when $y = 0$

$$\tau_{Max} = \left[\frac{4S_Y}{3\pi R^4} \right] \times \left[R^2 - 0 \right] = (4/3) \times (S_Y / \pi R^2)$$

Minimum shear stress occurs when $y = \pm R \Rightarrow \tau_{Min} = 0$

Average shear stress,

$$\tau_{Ave} = S_Y / A = S_Y / (\pi R^2)$$

and $$\frac{\tau_{Max}}{\tau_{Ave}} = \frac{(4/3) \times S_Y / (\pi R^2)}{S_Y / (\pi R^2)} = 4/3$$

10.4.5 SHEAR STRESS DISTRIBUTION

The stresses τ_{XY} and σ_X are related to the shear force and bending moment at a section of the beam subjected to bending in X-Y plane (X being the axial direction), by

$$S_Y = \int \tau_{XY} \times dA \quad \text{and} \quad M_Z = \int \sigma_X \times y \times dA$$

Distribution of shear stress across sections of uniform and non-uniform widths are shown in Fig 10.12 for a rectangular section and an I-section

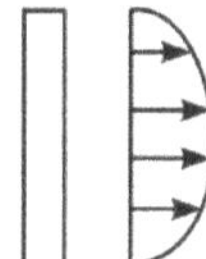

Shear stress distribution in beam section of uniform width

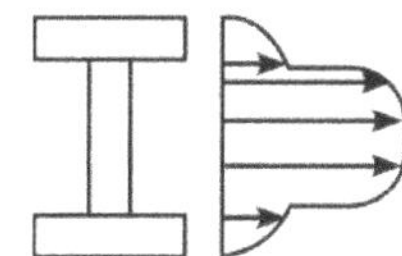

Shear stress distribution in beam section of non-uniform width

FIGURE 10.12 Shear stress distribution in beams of uniform & non-uniform width

Example 10

A wooden beam of rectangular section, 200×300mm is simply supported over a span of 4m. If allowable shear stress for the material is 50 N/cm^2, find maximum concentrated load that can be applied at its mid span

Solution :

Let the point load at mid-span of a simply supported beam be 'P'. Reaction at each end will be P/2 and the shear force S_Y is constant all along the span and equal to P/2

Moment of inertia, $I_Z = b \times d^3/12 = 200 \times 300^3/12 = 450 \times 10^6 \text{ mm}^4$

Max shear stress, $\tau_{Max} = S_Y \times d^2 / (8 \times I_Z)$

or $50\times10^{-2} = (P/2) \times 300^2 / (8\times450\times10^6)$

$\Rightarrow$ $P = 40000$ N or 40 kN

Example 11

A circular rod of 100mm diameter and 6m long is used as a cantilever. It carries a concentrated load of 200kN at the free end and a uniformly distributed load of 50kN/m over its entire length. Determine max shear stress and shear stress at a distance of 20mm from neutral axis

Solution:

Maximum shear force $S_Y = 200 + 50 \times 6 = 500$ kN

Average shear stress, $\tau_{Ave} = S_Y/A = S_Y / (\pi R^2) = 63.64$ N/mm^2

Maximum shear stress, $\tau_{Max} = (4/3) \times \tau_{Ave} = 84.84$ N/mm^2

Shear stress at a distance y of 20 mm from neutral axis,

$$\tau = [S_Y/(3 \times I_Z)] \times (R^2 - y^2) = [S_Y / (3\pi R^4/4)] \times (R^2 - y^2) = 71.3 \text{ N/mm}^2$$

10.5 MECHANISMS

Spars and beams when loaded may have small displacements in them but do not have gross displacements at any joint. They are parts of fixed structures, designed to withstand applied loads and transfer them to the supports as reactions. Some ***structures, designed for transfer of motion or power, are called mechanisms***. Members of such structures have significant gross displacements. Axial or bending deformations of members of these mechanisms are much less and hence they are assumed to be rigid. Number of independent coordinates (or variables) required to completely specify the relative movement of members of a mechanism (also called kinematic links) is called degrees of freedom (dof) of the mechanism. A typical example of a four bar mechanism or quadratic cycle chain is explained below.

Detailed study of mechanisms is outside the scope of this book

10.5.1 FOUR-BAR OR QUADRIC CYCLE CHAIN (R-R-R-R TYPE MECHANISM)

In a typical four bar or quadratic cycle chain mechanism (Ref Fig 10.13) with link-1 fixed, link-4 (the shortest link) is called ***driver***. Link-2 is called ***follower***. Link-3, which connects driver and follower, is called the ***connecting rod*** or ***coupler***. Fixed link (Link-1) is called the frame of the mechanism.

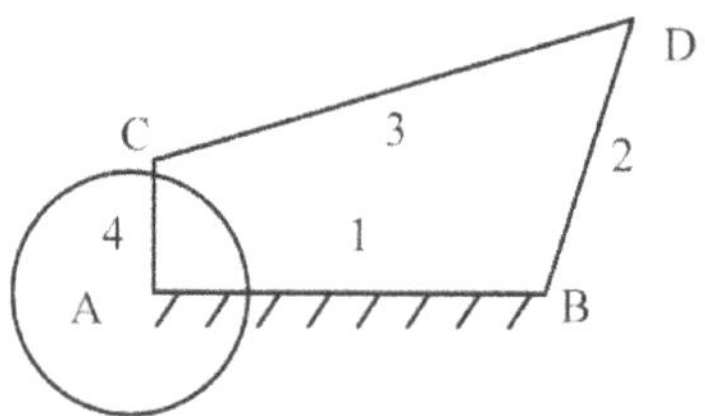

FIGURE 10.13 Four bar or quadratic cycle chain mechanism

Joints A and B are called ***fixed*** or ***ground hinges*** while joints C and D are called ***moving hinges***. Link-4 and Link-2 are called ***cranks***, if they make full rotation; while they are called ***rockers***, if they oscillate. According to ***Grashof's law***, sum of shortest link length (S) and longest link length (L) is equal to or less than the sum of the other two link lengths (a & b), if continuous relative motion is to exist between crank and follower. i.e., $L+S < a+b$

If Link-4 rotates and Link-2 oscillates, the mechanism is transforming rotary motion of Link-4 into oscillatory motion of Link-2. Depending on the relative lengths of the four links, it is also possible that both links rotate or both links oscillate, as shown in Fig 10.14

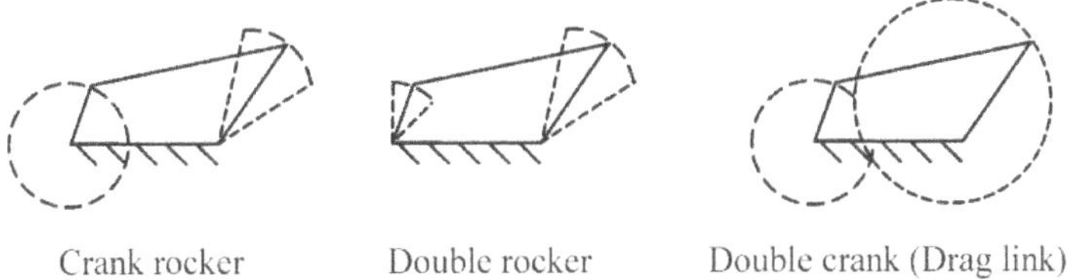

FIGURE 10.14 Four bar or quadratic cycle chain mechanism

A few other mechanisms, each for a specific purpose, are presented in the following sections.

10.5.2 SINGLE SLIDER CRANK CHAIN

This mechanism (Ref Fig 10.15) converts rotary motion of crank (2) into a reciprocating motion of piston (4) as in a water pump. Member 3 (also called connecting rod) is the connecting link between the two motions while 1 is the rigid supporting structure. Alternatively, it converts reciprocating motion of piston (4) into rotary motion of crank (2) as in an internal combustion engine.

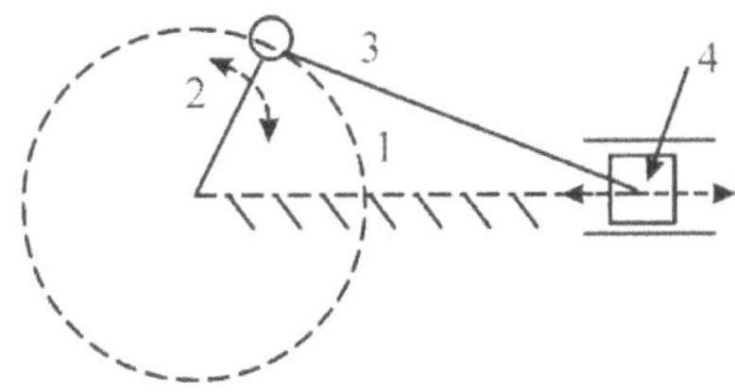

FIGURE 10.15 Slider crank mechanism

10.5.3 BEAM ENGINE OR CRANK AND LEVER MECHANISM

This mechanism shown in Fig 10.16 also converts rotary motion of crank into reciprocating motion of piston but uses an additional member (lever 4).

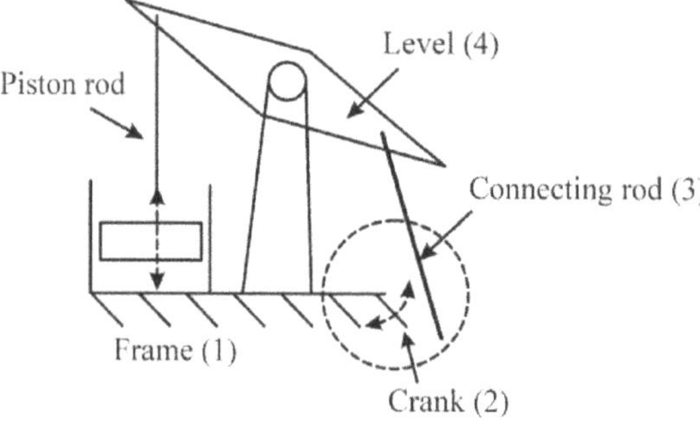

FIGURE 10.16 Crank and Lever mechanism

10.5.4 CRANK AND SLOTTED LEVER (QUICK RETURN MOTION MECHANISM)

This mechanism, shown in Fig 10.17, is mostly used in shaping machine, slotting machine etc.. Rotation of crank BC (2) oscillates bar ABP (4) about A with slider B (1) sliding in ABP. P is connected to tool at R through connecting rod P-R. As the crank rotates with B from its extreme position B_1 to B_2 in a clockwise direction, tool moves from R_1 to R_2 in cutting mode while the crank motion from B_2 to B_1 in clockwise direction moves tool in the return direction from R_2 to R_1, thus saving time of tool return and increasing productivity.

Time of cutting stroke / Time of return stroke $= (360 - \alpha) / \alpha$

and Tool travel, $R_1 - R_2 = P_1 - P_2 = 2\ AP \times CB / AC$

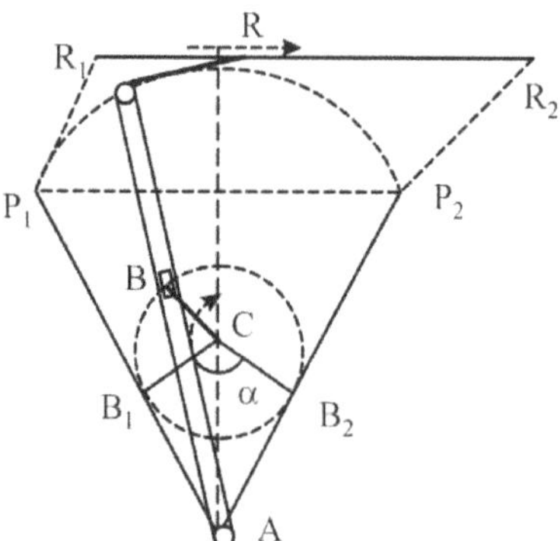

FIGURE 10.17 Crank and slotted Lever mechanism

10.5.5 SCOTCH YOKE MECHANISM

This mechanism, shown in Fig 10.18, is used for converting rotary motion of crank AB (1) about A to reciprocating motion of slider (3) and piston, as in a reciprocating air compressor. Thus, link AB (1) and frame (4) as well as link AB and slide at B (2) form revolute pairs while slide at B (2) and slider (3) form sliding pair.

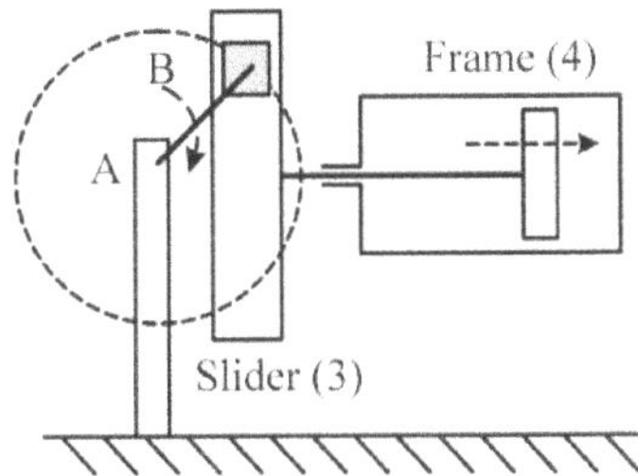

FIGURE 10.18 Crank and slotted Lever mechanism

10.5.6 OLDHAM COUPLING

Two parallel shafts are connected by a Oldham coupling (Ref Fig 10.19) in such a way that they both rotate at the same speed. This coupling is used when the two rotating shafts are parallel and offset by a small distance (non-collinear). Flange (1) and frame (4) as well as flange (3) and frame (4) form ***revolute pairs*** while flange (1) and intermediate piece (2) as well as flange (3) and intermediate piece (2) form ***sliding pairs***.

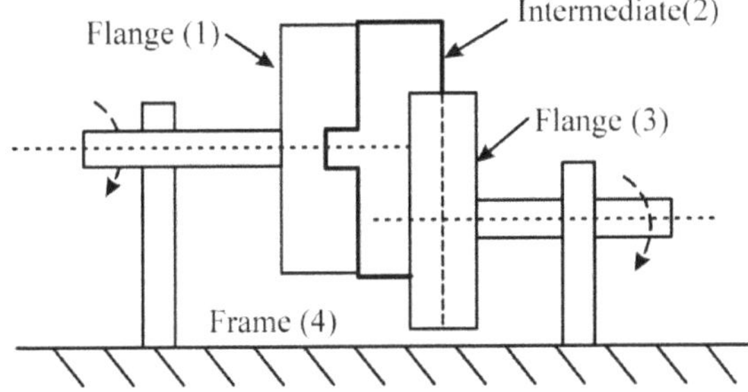

FIGURE 10.19 Crank and slotted Lever mechanism

CHAPTER **11**

KINEMATICS

11.1 MOTION OF A RIGID BODY

Dynamics is the study of motion of a body. In this subject, we limit ourselves to the study of rigid bodies. A ***rigid body*** is an idealization of the behaviour of a body, when its dimensions and relative positions of points within it do not change. Dynamics can further be subdivided into – ***kinematics***, which deals with motion of a body as a function of time (space-time relationship) and ***kinetics***, which deals with forces that need to be applied to produce a prescribed motion.

Motion of a body can be divided into ***rectilinear*** (along a straight line), ***curvilinear*** (along a curved path) and ***rotational*** (about an axis). For convenience, we limit our discussion to **plane motion**, i.e., motion limited to one plane.

11.2 RECTILINEAR MOTION

Study of dynamic motion involves rate of distance covered, specified in terms of some function. Nature of motion is defined by ***velocity*** and ***acceleration*** of the rigid body

Displacement 's' is the movement of a particle from a reference point.

Velocity, $v = ds/dt$ is the distance 'ds' covered by the body in time 'dt'

Acceleration, $a = dv/dt = d^2s/dt^2$ is the rate of change of velocity

The distance covered 's', velocity 'v' and acceleration 'a' are all vectors. Positive and negative velocities indicate motion of a rigid body along positive or negative axis. Negative acceleration is sometimes called '***deceleration***' or '***retardation***'. For a body moving with uniform velocity, acceleration is zero.

For example,

(i) $s = b + c \times t$ indicates initial displacement 'b' at time $t = 0$ and uniform rectilinear motion

Then, $v = ds / dt = c$ (constant) and $a = dv / dt = 0$

(ii) $s = g\, t^2/2$ indicates freely falling body

Then, $v = g \times t$, a linear function with time and $a = g$ (constant)

(iii) $s = c \times e^{-k \times t}$ Then, $v = -\, c \times k\, e^{-k \times t}$; $a = c \times k^2 \times e^{-k \times T}$

If the motion takes place along a line inclined to the Cartesian coordinate system, then velocity and acceleration can be defined X and Y axes, by taking components of distance covered 's' and velocity 'v' along the respective axes.

Rectilinear motion i.e., motion along a straight line can be classified based on variation of velocity or acceleration during the motion.

11.2.1 RECTILINEAR MOTION WITH CONSTANT ACCELERATION

If a particle starts from a point with initial velocity 'u' and moves a distance 's' in time 't' with uniform acceleration 'a' reaching final velocity 'v', then

$$a = (v - u) / t \quad \text{or} \quad \mathbf{v = u + a \times t} \qquad \text{.....(11.1)}$$

$$s = \text{average velocity} \times \text{time} = [(v + u)/2] \times t$$

$$= [\,(u + a \times t + u) / 2\,] \times t$$

or $$\mathbf{s = u \times t + a \times t^2 / 2} \qquad \text{.....(11.2)}$$

Also, $$s = \text{average velocity} \times \text{time} = [\,(v + u) / 2\,] \times t$$

$$= [\,(v + u) / 2\,] \times [\,(v - u) / a\,]$$

or $$\mathbf{v^2 - u^2 = 2 \times a \times s} \qquad \text{.....(11.3)}$$

Sometimes, velocity 'v' is defined as a function of distance 's'

Then, $a = dv / dt = (dv / ds) \times (ds / dt)$

or $$\mathbf{a = v \times (dv / ds)} \qquad \text{.....(11.4)}$$

Distance moved in nth second can be obtained as

s_n = distance traveled in 'n' sec – distance traveled in 'n – 1' sec

$$= [\,u \times n + a \times n^2/2\,] - [\,u \times (n - 1) + a \times (n - 1)^2/2\,]$$

$$= u + a \times n - a/2$$

or $$s_n = u + (2n - 1) \times a / 2 \qquad \text{.....(11.5)}$$

Acceleration of anybody moving up or down is the acceleration due to gravity, g

Thus, In the case of a body moving up, $a = -9.81$ m/sec^2

In the case of a body moving down, $a = 9.81$ m/sec^2

Example 1

A train, moving with constant acceleration, travels 7.2m during the 10th second of its motion and 5.4m during its 12th second of motion. Find its initial velocity

Solution :

$$s_{10} = u + a \times (2 \times 10 - 1)/2 \quad \text{or} \quad 7.2 = u + 9.5\,a$$

$$s_{12} = u + a \times (2 \times 12 - 1)/2 \quad \text{or} \quad 5.4 = u + 11.5\,a$$

Solving these two equations, we get $a = -0.9$ m/sec^2 and $u = 15.75$ m/sec

Example 2

A stone is dropped into a well and the sound of the splash is heard after 6 sec. Find the depth of water surface from the top, if velocity of sound is 340 m/sec

Solution :

Let t_1 be the time taken by the stone to reach water surface and t_2 the time taken by the sound to reach from water surface to the top of the well

Then, $s = u \times t_1 + a \times t_1^2/2 = 0 + 9.81 \times t_1^2 / 2$

Also $s =$ Velocity of sound $\times$ time $= 340 \times t_2$ and $t_1 + t_2 = 6$ sec

Therefore, $9.81 \times t_1^2 / 2 = 340 \times (6 - t_1) \Rightarrow t_1 = 5.555$ sec

Depth of water surface, $s = 340 \times (6 - 5.555) = 151.36$ m

Example 3

A stone is thrown vertically up from the top of a tower with certain initial velocity. It reaches ground in 5.64 sec. A second stone thrown down from the same tower with the same initial velocity reaches ground in 3.6 sec. Determine (i) height of the tower and (ii) initial velocity of the stones.

Solution:

Let us first consider motion of the first stone. It first moves upwards with negative acceleration 'g' till its velocity becomes zero and then moves down with positive acceleration 'g'. It, therefore, reaches the same velocity as the

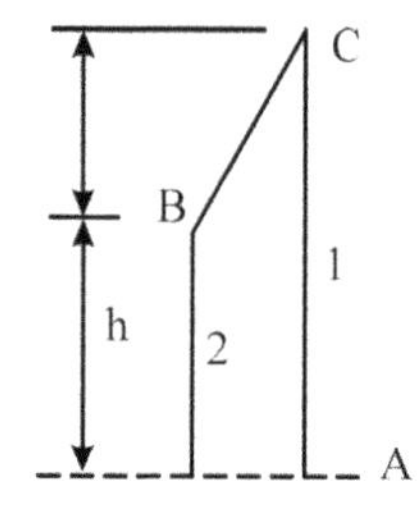

initial velocity, when it reaches top of tower in its downward motion. Let 'h' be height of the tower.

Let 's' be the distance covered upwards from the tower.

For travel of stone from B to C (upward),

$$s = (v_C^2 - u_B^2) / [2(-g)] = u_B^2/2g$$

For travel of same stone from C to B (downward),

$$s = (v_B^2 - u_C^2) / 2g = v_B^2/2g$$

Equating the above two conditions for 's',

downward velocity at B, $v_B = u_B$

Hence, time taken by the stone from top of tower 'B' to reach maximum height at 'C' and come down to the same level 'B' is same.

Therefore, $t = (t_1 - t_2) / 2 = (5.64 - 3.6) / 2 = 1.02$ sec

Then, initial velocity 'u_B' of the first stone thrown upwards from top of the tower is given by, $v_C = u_B - g \times t$ or $0 = u_B - 9.81 \times 1.02$

$\Rightarrow$ $u_B = 10$ m/sec

Considering travel of second stone from B to A,

Height of the tower, $h = u_B \times t_2 + g \times t_2^2/2 = 10 \times 3.6 + 9.81 \times 3.6^2/2$

$= 99.5$ m

Example 4

A stone is thrown up with a velocity of 20 m/sec. While coming down, it strikes a glass pane, held at half the height from which it is thrown up and loses half its velocity in breaking the glass. Find the velocity with which it will strike the ground.

Solution :

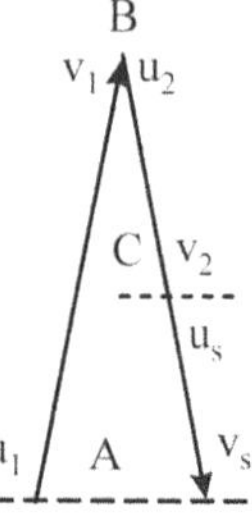

Let 's_1' be the maximum distance traveled by the stone from its starting position 'A' to its max height at 'B' (corresponding to its final velocity, $v_1 = 0$) is given by

$$v_1^2 - u_1^2 = 2 \times g \times s_1$$

or $s_1 = (0 - 20^2) / [2 \times (-9.81)] = 20.4$ m

Distance traveled by the stone from 'B' ($u_2 = 0$) before it strikes the glass pane at 'C' during its downward motion,

$$s_2 = s_1/2 = 20.4 / 2 = 10.2 \text{ m}$$

We know, $s_2 = (v_2^2 - u_2^2) / 2g = (v_2^2 - 0) / (2 \times 9.81)$

$\Rightarrow$ $v_2 = 14.14$ m/sec

During its remaining downward motion from 'C' to 'A',

$$s_3 = s_2 = 10.2\text{m} \quad \text{and} \quad u_3 = v_2 / 2 = 7.07 \text{ m/sec}$$

Therefore, final velocity with which the stone reaches the ground is obtained from,

$$s_3 = (v_3^2 - u_3^2) / 2g = (v_3^2 - 7.07^2) / (2 \times 9.81) \Rightarrow v_3 = 15.8 \text{ m/sec}$$

Example 5

A stone is dropped from the top of a tower 40 m high. At the same time, another stone is thrown upwards from the foot of the tower with a velocity of 20 m/sec. When and where the two stones cross each other and with what relative velocity.

Solution :

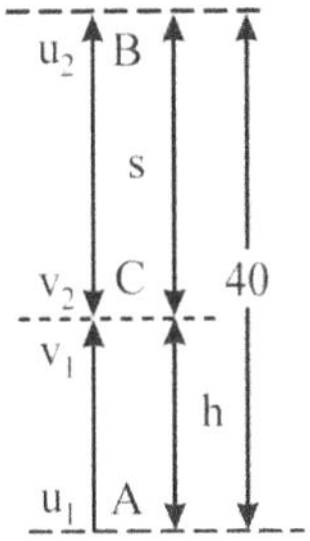

Since both the stones start at the same time, both the stones take the same time 't' before they cross each other. Let 'h' be the height from the ground where the two stones meet.

Distance travelled by the stone dropped from the top of tower,

$$s = u_2 \times t + g \times t^2 / 2$$

or $40 - h = 0 + 9.81 \times t^2 / 2 \Rightarrow h = 40 - 4.905 \times t^2$

and for stone thrown from the ground, $h = u_1 \times t - g \times t^2 / 2$

or $h = 20 \times t - 9.81 \times t^2 / 2 \Rightarrow h = 20 \times t - 4.905 \times t^2$

Equating these two relations, $h = 20 \times t - 4.905 \times t^2 = 40 - 4.905 \times t^2$

$\Rightarrow$ $t = 2$ sec and $h = 20 \times 2 - 4.905 \times 2^2 = 20.38$ m

At the point of crossing of the stones,

Velocity of stone falling down, $v_2 = u_2 + g \times t = 0 + 9.81 \times 2 = 19.62$ m/sec

Velocity of stone going up, $v_1 = u_1 - g \times t = 20 - 9.81 \times 2 = 0.38$ m/sec

Relative velocity, $v_R = v_1 + v_2 = 19.62 + 0.38 = 20$ m/sec

Example 6

Two trains A and B leave the same station on parallel tracks. Train A starts with a uniform acceleration of 0.15 m/sec^2 until it attains maximum speed of 27 kmph. Train B leaves 40 sec later with uniform acceleration of 0.3 m/sec^2 until it attains maximum speed of 54 kmph. When and where will B overtake A ?

Solution:

Maximum speed of train A, $v_A = 27$ kmph $= 27 \times 1000 / 3600 = 7.5$ m/sec

Maximum speed of train B, $v_B = 54$ kmph $= 15$ m/sec

Let 't_1' and 't_2' be the times taken by the two trains to reach their maximum speeds.

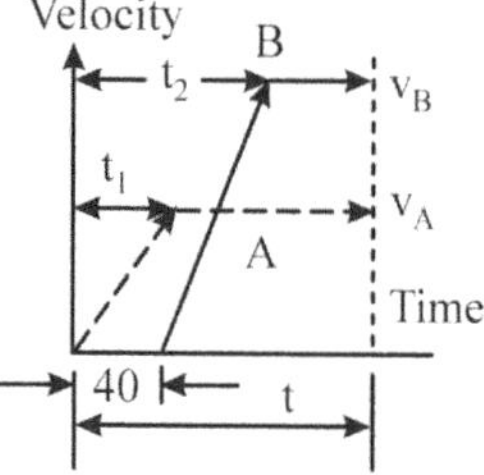

Then, $v_A = u_A + a_A \times t_1$

or $7.5 = 0 + 0.15 \times t_1 \quad \Rightarrow t_1 = 50$ sec

and $v_B = u_B + a_B \times t_2$

or $15 = 0 + 0.3 \times t_2 \quad \Rightarrow t_2 = 50$ sec

Let train A travel for 't' sec before train B overtakes it in (t–40) sec. Both the trains cover the same distance. Distance 'S' covered by both the trains includes two parts – one, with uniform acceleration till it reaches its maximum speed and the other, with constant velocity.

Thus, $S = S_1 + S_2$ for each train.

If, $S_A = (u_1 + v_1) \times t_1 / 2 + v_1 \times (t - t_1)$

and $S_B = (u_2 + v_2) \times t_2 / 2 + v_2 \times (t - t_2 - 40)$

Equating the two,

$$(0 + 7.5) \times 50 / 2 + 7.5 \times (t - 50)$$

$$= (0 + 15) \times 50 / 2 + 15 \times (t - 50 - 40)$$

or $7.5 \times t - 187.5 = 15 \times t - 975 \quad \Rightarrow t = 105$ sec

Therefore, $S = 7.5 \times 105 - 187.5 = 600$ m

Example 7

Three points A, B and C are marked at a distance of 120 m each along a straight road. A car starting from rest and with an uniform acceleration passes mark A and takes 12 sec to reach B and further 8 sec to reach C. Calculate

(i) velocity of car at A and B

and (ii) distance of C from starting point.

S_{O-A} u S_{A-B} S_{B-C}

O A t_1 B t_2 C

Solution :

Let velocity of car at A be 'u'

Between A and B, $S_{A\text{-}B} = u \times t_1 + a \times t_1^2 / 2$

i.e., $120 = u \times 12 + a \times 12^2 / 2 \Rightarrow u + 6a = 10$ .....(1)

Between A and C, $S_{A\text{-}C} = u \times (t_1 + t_2) + a \times (t_1 + t_2)^2 / 2$

i.e., $240 = u \times (12 + 8) + a \times (12 + 8)^2 / 2 \Rightarrow u + 10a = 12$ .....(2)

Solving equations (1) and (2), $a = 0.5$ m/sec^2 and $u = 7$ m/sec

Velocity of car at B, $v_B = u + a \times t_1 = 7 + 0.5 \times 12 = 13$ m/sec

Distance of A from starting point, $S_{O\text{-}A} = (u^2 - u_O^2) / (2a)$

$= (7^2 - 0) / (2 \times 0.5) = 49$ m

Distance of C from starting point,

$$S_{O\text{-}C} = S_{O\text{-}A} + S_{A\text{-}B} + S_{B\text{-}C} = 49 + 120 + 120 = 289 \text{ m}$$

Example 8

A particle is moving with uniform acceleration. In the 5th and 9th seconds from the beginning, it travels 16 m and 20 m respectively. Find (i) initial velocity and acceleration and (ii) distance covered in 12 seconds and in12th sec.

Solution :

Using the relation $S_n = u + (2n - 1) \times a / 2$ for the distance traveled in nth sec,

$S_5 = u + (2 \times 5 - 1) \times a / 2$ or $16 = u - 4.5a$ .....(1)

and $S_9 = u + (2 \times 9 - 1) \times a / 2$ or $20 = u - 8.5a$ .....(2)

Solving equations (1) and (2), $a = 1$ m/sec^2 and $u = 11.5$ m/sec

Distance traveled in 12th sec,

$$S_{12} = u + (2 \times 12 - 1) \times a / 2 = 11.5 + 23 \times 1 / 2 = 23 \text{ m}$$

Distance traveled in 12 sec,

$$S = u \times 12 + a \times 12^2 / 2 = 11.5 \times 12 + 1 \times 12^2 / 2 = 210 \text{ m}$$

11.2.2 GRAPHICAL REPRESENTATION OF VELOCITY AND DISTANCE TRAVELLED

If velocity is plotted as a function of time, acceleration ($a = dv/dt$) at any time is represented by the slope of the velocity curve at that time while the distance traveled ($s = \int v \times dt$) during the time t_1 to t_2 is represented by the area under the velocity curve during that time interval. Thus, a linear velocity line represents constant acceleration and the triangular area below the velocity line represents distance traveled.

Example 9

A cage goes down a main shaft 750 m deep in 45 sec. For the first quarter of the distance, the speed is uniformly accelerated and during the last quarter, the speed is uniformly retarded. Assuming acceleration and retardation to be the same, find the uniform speed of the cage during the central portion of the travel.

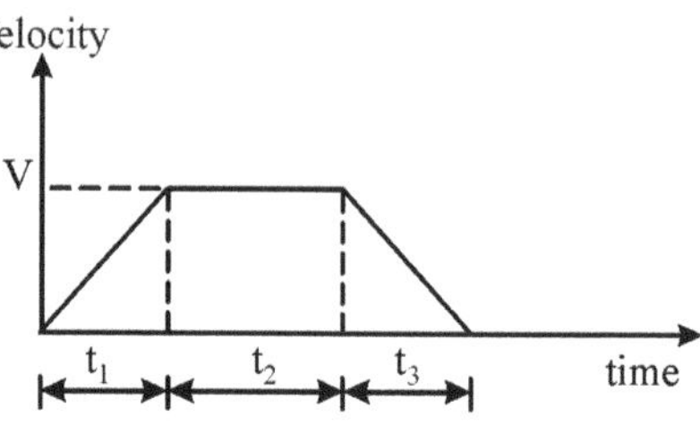

Solution :

Let t_1 be the time taken by the cage to reach its maximum velocity 'v'

Then, S_1 = Average velocity × time

or

$$750 / 4 = (v/2) \times t_1 \Rightarrow v \times t_1 = 375 \text{ m} \quad .....(1)$$

Let t_2 be the time taken by the cage to travel the central half of the shaft by uniform velocity

Then, S_2 = velocity × time or $750 / 2 = v \times t_2$

$$\Rightarrow \quad v \times t_2 = 375 \text{ m} \quad .....(2)$$

From equations (1) and (2), $t_1 = t_2$

Let the last quarter distance be covered in time 't_3'.

Since acceleration and retardation are equal, $t_1 = t_3$

Total time of travel, $t = t_1 + t_2 + t_3$ or $45 = 3 \times t_1 \Rightarrow t_1 = 15$ sec

Max. speed of the cage, from eq. (1), $v = 375 / 15 = 25$ m/sec

Example 10

Two trains P and Q start from stations A and B, facing each other, with accelerations of 0.5 m/sec^2 and 2/3 m/sec^2, reaching their maximum speeds of 90 kmph and 72 kmph respectively. If they cross each other midway between the stations, find the distance 'S' between the stations and the times taken by the two trains

Solution :

Time taken by train P to reach its maximum velocity,

$$t_1 = v_1/a_1 = (90/3.6)/0.5 = 50 \text{ sec}$$

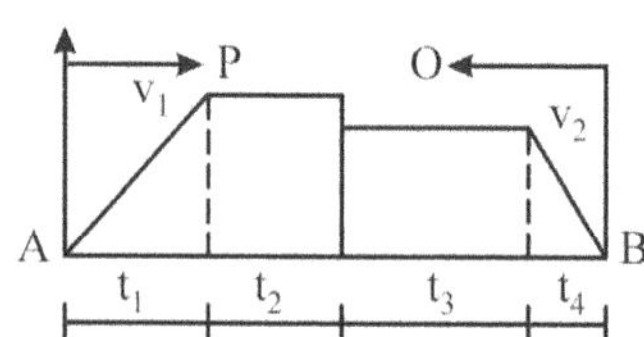

Similarly, time taken by train Q to reach its maximum velocity,

$t_4 = v_2/a_2 = (72/3.6) / (2/3) = 30$ sec

Let S be the distance covered by each train

Then, $S = (v_1 \times t_1)/2 + v_1 \times t_2$ for train P

$= (90/3.6) \times [(50/2) + t_2] = 625 + 25\, t_2$ .....(1)

Similarly,

$S = (v_2 \times t_4)/2 + v_2 \times t_3$ for train Q

$= (72/3.6) \times [(30/2) + t_3] = 300 + 20\, t_3$ .....(2)

Also, total time taken by each train is equal

i.e., $t_1 + t_2 = t_3 + t_4$ or $t_2 = t_3 - 20$ sec .....(3)

Solving the 3 equations, we get $t_2 = 15$ sec, $t_3 = 35$ sec and $S = 1000$m

Then, $AB = 2 \times S = 2000$m

Time taken by train P $= t_1 + t_2 = 50 + 15 = 65$ sec

Time taken by train Q $= t_3 + t_4 = 35 + 30 = 65$ sec

11.2.3 RELATIVE VELOCITY

If two particles A and B are moving with velocities v_A and v_B respectively, then velocity of B relative to A in vector form, as shown in Fig 11.1, is given by

$$v_B = v_A + v_{BA} \quad \text{or} \quad v_{BA} = v_B - v_A \qquad .....(11.6)$$

Similarly velocity of A relative to B is given by $v_{AB} = v_A - v_B$

It can also be considered as the vector sum of v_B and $(-v_A)$

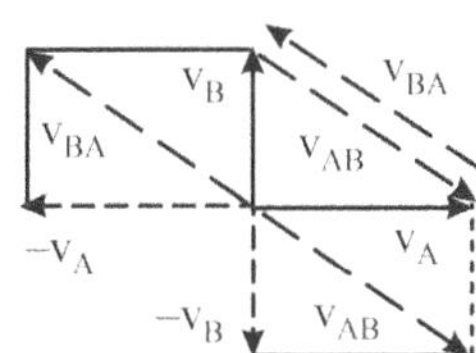

FIGURE 11.1 Relative velocity

If v_A and v_B are parallel, then

$|v_{BA}| = |v_B| - |v_A|$ if v_B and v_A are in the same direction

and $|v_{BA}| = |v_B| + |v_A|$ if v_B and v_A are in opposite directions

If v_B and v_A are inclined, then the relative velocity v_{BA} is obtained as vectorial sum of v_B and $(-v_A)$ or as the diagonal of the parallelogram, constructed with the vectors v_B and $(-v_A)$ as its sides, similar to the resultant of force vectors.

Such problems can also be solved by resolution of velocity vectors v_B and $(-v_A)$, say along X and Y directions. Then,

$$V_{BA} = V_R = \sqrt{(V_R)_X^2 + (V_R)_Y^2} \qquad .....(11.7)$$

where $(V_R)_X = \sum V_X$ and $(V_R)_Y = \sum V_Y$

These two approaches are explained through the following two examples.

Example 11

Two ships leave a port at the same time. The first steams North-West at 32kmph and the second 40^0 South of West at 24kmph. What is the velocity of the second ship relative to the first? After what time will they be 160 km apart.

Solution :

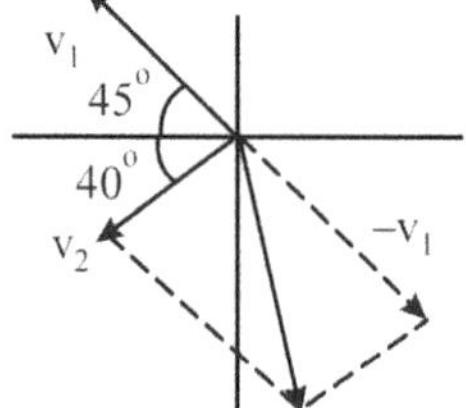

The vector diagram of velocities v_1 and v_2 is plotted in the figure. The angle between the two sides of the parallelogram, $\alpha = [180 - (40 + 45)]^0$

Relative velocity is the larger diagonal of the parallelogram and its magnitude is given by

$$R = \sqrt{(v_1 + v_2 \times \cos\alpha)^2 + (v_2 \times \sin\alpha)^2}$$

$$R = \sqrt{(32 + 24 \times \cos 95)^2 + (24 \times \sin 95)^2} = 38.29 \text{ kmph}$$

The two ships will be 160 km apart after

$t = 160 / 38.29 = 4.18$ hrs.

Alternative Method

$$R = \sqrt{\left[(v_1 \times \cos 45 - v_2 \times \cos 40)^2 + (-v_1 \times \sin 45 - v_2 \times \sin 40)^2\right]}$$

$$= \sqrt{\left[(22.624 - 18.385)^2 + (22.624 + 15.427)^2\right]}$$

$= 38.29$kmph

Example 12

An enemy ship is located at a distance of 25 km in N-W direction by a warship. If the enemy ship is moving with a velocity of 18 kmph N 30^0 E, in which direction warship must move with a velocity of 36 kmph, to strike at its earliest? When is the shell to be fired, if the fire range of warship is 5 km.

Solution :

Let A be the location of the enemy ship and B the location of warship.

Given, $V_A = 18$ kmph ; $V_B = 36$ kmph ; AB = 25 km

Then, $(V_A)_X = 18 \sin 30 = 9$ kmph ; $(V_A)_Y = 18 \cos 30 = 15.588$ kmph

and $(V_B)_X = -36 \sin\theta$; $(V_B)_Y = 36 \cos\theta$

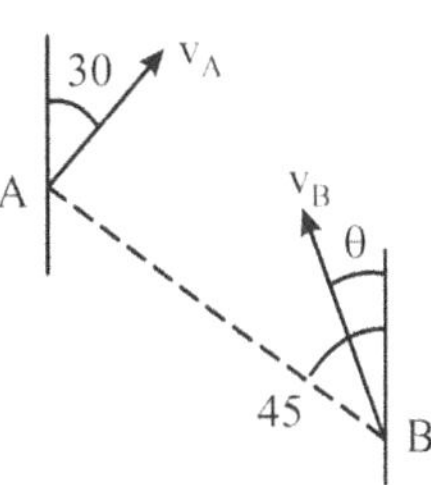

Components of relative velocity,

$(V_R)_X = -36 \sin\theta - 9$ and $(V_R)_Y = 36 \cos\theta - 15.588$

In order that the shell fired by B hits A, Relative velocity should be along BA.

Therefore, $\tan\alpha = (V_R)_X / (V_R)_Y = -1$

or $(V_R)_X = -(V_R)_Y$

since angle of BA with the vertical, $\alpha = 450$

$\Rightarrow$ $-36 \sin\theta - 9 = -(36 \cos\theta - 15.588)$

or $\cos\theta - \sin\theta = (9 + 15.588) / 36 = 0.683$

Let $\sin\theta = x$ Then, $\cos\theta = \sqrt{(1-x^2)}$

and $\sqrt{(1-x^2)} - x = 0.683$ or $1 - x^2 = (x + 0.683)^2$

$\Rightarrow$ $x = 0.277$ or $\theta = \sin^{-1}(0.277) = 16.120$

Then, $(V_R)_X = -36 \sin\theta - 9 = -18.972$kmph

and $(V_R)_Y = 36 \cos\theta - 15.588 = 19.003$kmph

$$V_R = \sqrt{(V_R)_X^2 + (V_R)_Y^2} = 26.87 \text{ kmph}$$

Distance to be traveled before shell is fired, S = 25 – 5 = 20 km

Therefore, Minimum time at which shell needs to be fired,

$t = S / V_R = 0.744$ hr

Example 13

Two trains A and B are moving on parallel tracks with velocities 72 kmph and 36 kmph, in the same direction. The driver of train A finds that it took 42 sec to overtake train B while driver of train B finds that train A took 30 sec to overtake him.

(a) Determine the length of each train
(b) If the two trains are moving in opposite directions, how much time is taken for complete crossing?

Solution:

Length of train B 'S_B' is the distance covered by train A while overtaking train B, with relative velocity, V_R, as observed by D_A (driver of train A). Similarly, length of train A 'S_A' is the distance covered by train B while overtaking train A, with relative velocity, V_R, as observed by D_B (driver of train B).

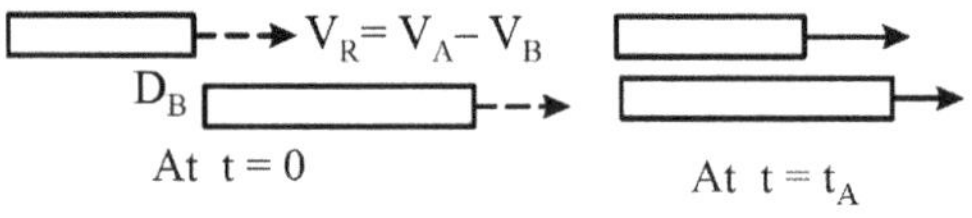

$V_R = V_A - V_B$

D_B

At $t = 0$ At $t = t_A$

(a) When the two trains are going in the same direction,

$$V_R = V_A - V_B = (72 - 36) / 3.6 = 10 \text{ m/sec}$$

$$S_B = V_R \times t_A = 10 \times 42 = 420 \text{ m}$$

Similarly, $S_A = V_R \times t_B = 10 \times 30 = 300$ m

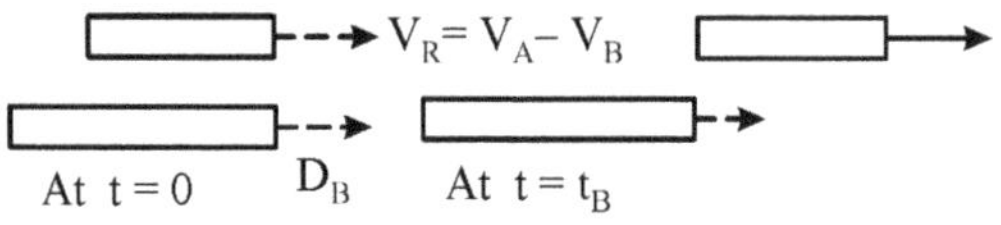

$V_R = V_A - V_B$

At $t = 0$ D_B At $t = t_B$

(b) When the two trains are going in the opposite directions,

$$V_R = V_A + V_B = (72 + 36) / 3.6 = 30 \text{ m/sec}$$

Time taken for complete crossing,

$$t' = (S_A + S_B) / V_R = (300 + 420) / 30 = 24 \text{ sec}$$

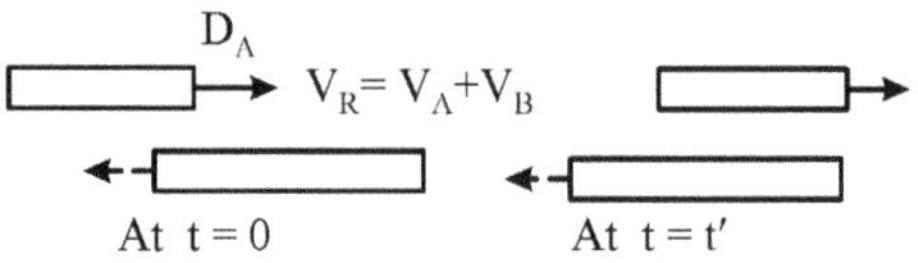

D_A

$V_R = V_A + V_B$

At $t = 0$ At $t = t'$

11.2.4 RECTILINEAR MOTION WITH VARIABLE ACCELERATION

In some cases, displacement or path of motion of a particle is expressed by an equation whose 2nd derivative is not a constant. Prominent example of such a motion is that of a simple harmonic motion, whose displacement is given by

$$s = k \times \sin \omega t.$$

Then, $v = ds/dt = \omega k \times \cos \omega t$ .....(11.8)

and $a = d^2s/dt^2 = -\omega^2 k \times \sin \omega t$

These three functions are graphically represented in Fig 11.2, with time on X-axis

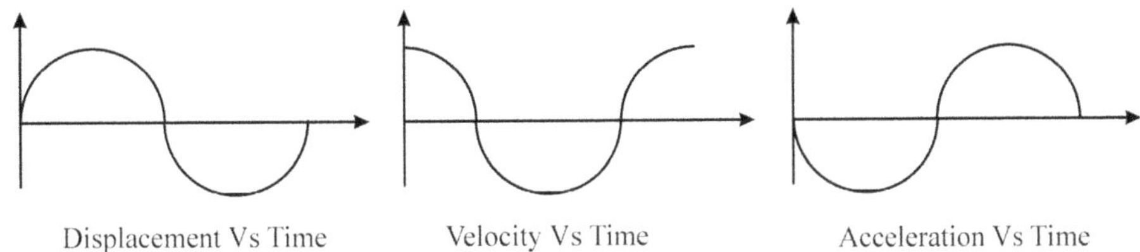

FIGURE 11.2 Displacement, velocity and acceleration as functions of time

Example 14

Velocity of a particle moving from a point 'A' in a straight line is given by $v = 2t^3 - t^2 - 2t + 4$. The particle is found to be at a distance of 10m from position A after 2 seconds. Determine acceleration and displacement after 6 seconds.

Solution :

$$v = dx / dt = 2t^3 - t^2 - 2t + 4$$

Integrating, $x = \int v\, dt = 2 \times (t^4/4) - (t^3/3) - 2 \times (t^2/2) + 4 \times t + C$,

where C is a constant

After t = 2 seconds, $10 = 2 \times (2^4/4) - (2^3/3) - 2 \times (2^2/2) + 4 \times 2 + C$

$\Rightarrow$ $C = 2/3$

After t = 6 sec, $x = 2 \times (6^4/4) - (6^3/3) - 2 \times (6^2/2) + 4 \times 6 + 2/3 = 564.67$ m

and $a = dv / dt = 6 \times t^2 - 2 \times t - 2 = 6 \times 6^2 - 2 \times 6 - 2 = 202$ m/sec^2

Example 15

Velocity of a particle moving along x-axis is defined by $v = kx^3 - 4x^2 + 6x$ where k is a constant. Calculate

(i) acceleration when x = 2 m, assuming k = 2

and (ii) value of k that will make acceleration = 16 m/sec^2 at x = 3 m.

Solution :

(i) $a = dv/dt = (dv/dx) \times (dx/dt) = (dv/dx) \times v$

$= (6x^2 - 8x + 6) \times (2x^3 - 4x^2 + 6x)$ with $k = 2$

$= (6 \times 2^2 - 8 \times 2 + 6) \times (2 \times 2^3 - 4 \times 2^2 + 6 \times 2)$ when $x = 2$ m

$= 14 \times 12 = 168$ m

(ii) $a = dv/dt = (dv/dx) \times (dx/dt) = (dv/dx) \times v$

$= (3kx^2 - 8x + 6) \times (kx^3 - 4x^2 + 6x)$

When $x = 3$ m, $16 = (3k \times 3^2 - 8 \times 3 + 6) \times (k \times 3^3 - 4 \times 3^2 + 6 \times 3)$

$= (27k - 18) \times (27k - 18)$

$= 9^2 \times (3k - 2)^2$

$\Rightarrow (3k - 2)^2 = 16/9^2$ or $3k - 2 = 4/9$

$\Rightarrow k = (4 + 9 \times 2) / (3 \times 9) = 22 / 27 = 0.815$

Example 16

A particle is constrained to move upwards along the path $2y^2 = x^3 + 26$ where x and y are in cm. The x-coordinate of the particle at any time is $x = t^2 - t + 4$. Determine y component of (a) velocity and (b) acceleration, both at x = 6 m

Solution :

From $x = t^2 - t + 4 = 6$, $t^2 - t - 2 = 0$ or $t = 2, -1$ Negative value is not feasible

Corresponding to x = 6 m, from $2y^2 = x^3 + 26$, $y = 11$ m

(a) At x = 6m, $v = dy/dt = (dy/dx) \times (dx/dt) = [(3x^2) / (4y)] \times (2t - 1)$

$= 7.36$ m/sec

(b) At x = 6m, differentiating $2y^2 = x^3 + 26$ twice w.r.t. t,

$4y \times (d^2y/dt^2) + 4 \times (dy/dt)^2 = 3x^2 \times (d^2x/dt^2) + 6x \times (dx/dt)^2$

$\Rightarrow$ $a_Y = d^2y/dt^2 = 7.35$ m/sec^2

Example 17

A particle moves along a straight line. Its motion is represented by the equation $S = 16t + 4t^2 - 3t^3$. Determine

(a) displacement, velocity and acceleration after 2 sec

(b) displacement and acceleration when velocity is zero

and

(c) displacement and velocity when acceleration is zero.

Solution :

$S = 16\,t + 4\,t^2 - 3\,t^3$ m

Therefore, $v = ds/dt = 16 + 8\,t - 9\,t^2$ and $a = dv/dt = 8 - 18\,t$

(a) After $t = 2$ sec, $S = 24$ m ; $v = -4$ m/sec and $a = -28$ m/sec^2

(b) When $v = 16 + 8\,t - 9\,t^2 = 0$, $t = 1.85$ sec.

Then, $S = 24.3$ m and $a = -25.3$ m/sec^2

(c) When $a = 8 - 18\,t = 0$, $t = 0.444$ sec.

Then, $S = 7.638$ m and $v = 17.78$ m/sec

Example 18

A particle starting from rest moves in a straight line with $a = 12 - 0.01\ S^2$. Determine (a) velocity of the particle after traveling 25 m and (b) max. distance traveled.

Solution :

$$a = dv/dt = (dv/ds) \times (ds/dt) = (dv/ds) \times v = 12 - 0.01\ S^2$$

$$\Rightarrow \qquad v\,dv = (12 - 0.01\ S^2)\ ds$$

Integrating both sides, $v^2/2 = 12\ S - 0.01\ S^3/3 + C_1$ where C_1 is a constant

Applying initial condition, $S = 0$ when $v = 0 \Rightarrow C_1 = 0$

Therefore, $v^2 = 24\ S - 0.02\ S^3/3$

(a) After traveling 25 m distance,

$$v^2 = 24 \times 25 - 0.02 \times 25^3 / 3$$

$$= 600 - 104.17 = 495.83$$

Therefore, $v = \sqrt{495.83} = 22.27$ m/sec

(b) Max. distance traveled corresponds to V=0.

Then, from eq (1), $0 = 24\ S - 0.02\ S^3/3$ or $0.02S^2 = 3 \times 24$

$$S = \sqrt{(3 \times 24 / 0.02)} = \sqrt{3600} = 60\text{m}$$

11.3 CURVILINEAR MOTION

Problems involving motion of a particle or body along a curved path are solved by resolving displacement, velocity or acceleration along horizontal and vertical directions.

Then, $x = f(t)$; $y = f(t)$; $u = dx/dt$; $v = dy/dt$

$a_X = du/dt$; $a_Y = dv/dt$

Resultant velocity, $v_R = \sqrt{(u^2 + v^2)}$

Resultant acceleration $a = \sqrt{(a_x^2 + a_y^2)}$ .....(11.9)

The **velocity and acceleration** of a point can also be expressed in intrinsic or path coordinates – **along tangential and normal directions** to the path of the particle.

Thus, in *vector form*, $V = V_t \times e_t + V_n \times e_n$

where e_t and e_n are unit vectors along tangential and normal directions

and V_t and V_n are the corresponding magnitudes

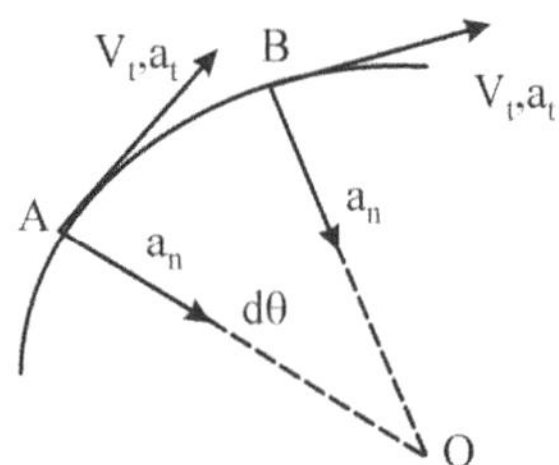

FIGURE 11.3 Components of acceleration

Velocity at any point of the path is tangential and, hence,

$V_n = 0$ and $V = V_t.e_t$

Similarly, $a = dV/dt = (dV_t/dt) \times e_t + (dV_n/dt) \times e_n$

$= a_t \times e_t + a_n \times e_n$ (Ref Fig 11.3)

For very small intervals of time 'dt', $dV \approx dV_n \approx V_t\, d\theta$

Therefore, $a_t = dV_t/dt$ acts along the forward motion of the particle

and $a_n = dV_n/dt \approx (V_t\, d\theta)/dt = V_t \times (d\theta/dt) = V_t \times (dS/Rdt)$

$= V_t \times (V_t/R) = V_t^2/R$

where, R is the radius of curvature of the path (OA, OB) at that point

a_n acts along the radius towards the center of curvature

Alternative approach

For very small interval of time 'dt', $dv_n \approx v \times d\theta = v \times (ds/R)$

$a_n = dv_n/dt = (v/R) \times (ds/dt) = v^2/R$ or $v \times \omega = R \times \omega^2$ since, $v = R \times \omega$

Special cases :

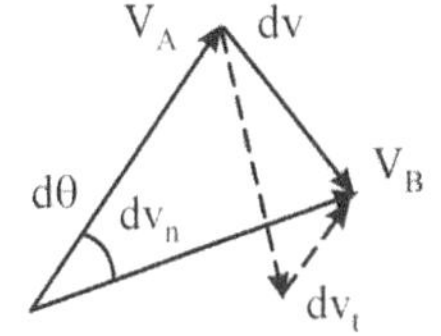

(a) Motion along a straight line : $1/R = 0$

Therefore, $an = 0$ and so, $a = dV/dt$

(b) Motion with uniform tangential velocity along a circular path:

$a_t = dV_t/dt = 0$; Therefore, $a = V^2/R$, is the centripetal acceleration.

Example 19

A stone is thrown with an initial velocity of 30 m/sec at 60^0 to the horizontal. Compute the radius of curvature of its path when it is at 20 m along the horizontal from its starting position ?

Solution :

Let U be the velocity at point A and V the velocity at point B of the path, inclined at 60^0 and θ respectively to the horizontal.

$$U_X = U \cos 60^0 = 30 \cos 60^0 = 15 \text{ m/sec}$$

$$U_Y = U \sin 60^0 = 30 \sin 60^0 = 26 \text{ m/sec}$$

Horizontal distance traveled, $S_X = U_X \times t + a_X \times t^2/2$

Assuming $a_X = 0$, $t = S_X/U_X = 20/15 = 4/3$ sec and $V_X = U_X$

After time t, $V_Y = U_Y - g \times t = 26 - 9.81 \times 4/3 = 12.92$ m/sec

Then, $V = [V_X^2 + V_Y^2]^{0.5} = 19.8$ m/sec

$$\tan \theta = V_Y/V_X = 12.92/15 = 0.8613 \Rightarrow \theta = 40.73^0$$

$$a_n = a_X \times \sin \theta + a_Y \times \cos \theta = 0 - 9.81 \times \cos 40.73^0$$

$$= -7.44 \text{ m/sec}$$

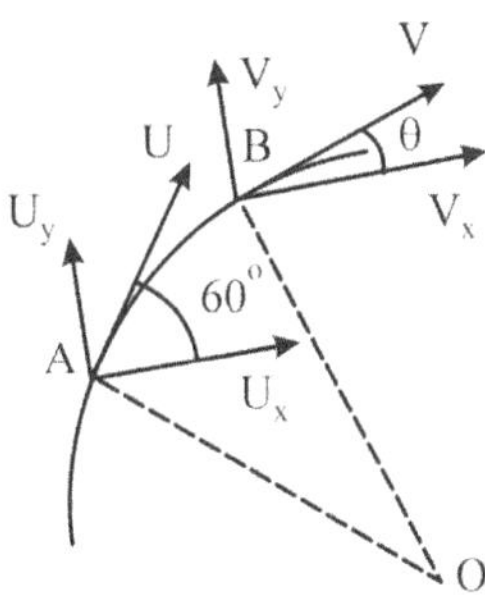

Radius of curvature, $R = V^2/a_n = 19.8^2 / 7.44 = 52.69$ m

11.3.1 PATH OF A PROJECTILE

It is one of the most prominent examples of curvilinear motion

If a particle is thrown up with initial velocity 'u', at an angle α with the horizontal, then the forces acting on the particle (Ref Fig 11.4) are given by

$$F_X = m \times a_X = 0 \text{ and } F_Y = m \times a_Y = -W = -m \times g$$

i.e., $a_X = 0$ and $a_Y = -g$

Integrating, $u = C_1$ and $v = -g \times t + C_2$

Applying initial conditions at $t = 0$, $u = u_0$ and $v = v_0$,

we get $C_1 = u_0$ and $C_2 = v_0$

Therefore, $u = u_0$ and $v = v_0 - g \times t$

Integrating again, $x = u_0 \times t + C_3$

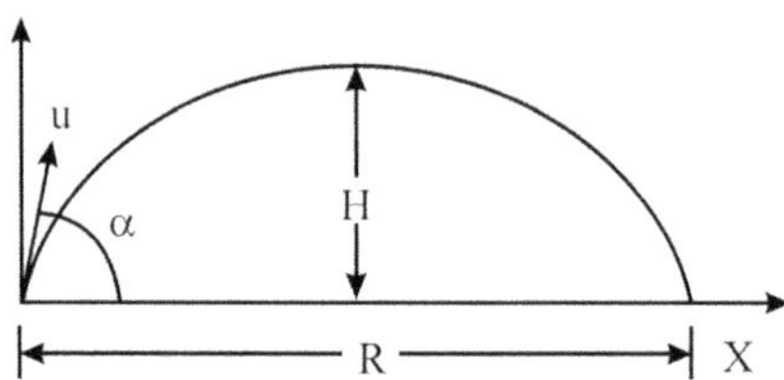

FIGURE 11.4 Path of a Projectile

and $$y = v_0 \times t - g \times t^2/2 + C_4$$

where, x and y are the horizontal and vertical components of the displacement s

Applying initial conditions at t=0, $x = y = 0$, we get $C_3 = 0$ and $C_4 = 0$

Thus, $x = u_0 \times t$ and $y = v_0 \times t - g \times t^2/2$

Eliminating t from these two equations, we get

$$\mathbf{y = v_0 \times x / u_0 - g \times x^2 / (2 \times u_0^2)} \quad .....(11.10)$$

Substituting, $u_0 = u \times \cos\alpha$ and $v_0 = u \times \sin\alpha$,

$$\mathbf{y = x \times \tan\alpha - (g \times \sec^2\alpha / 2\, v_0^2) \times x^2} \quad .....(11.11)$$

This is the equation of trajectory and is parabolic

At the point of max height (H) of the trajectory,

$$dy/dx = v_0 / u_0 - g \times x / (u_0^2) = 0$$

Corresponding horizontal distance, $x_1 = u_0 \times v_0 / g$

or Range, $\mathbf{R} = 2 \times x_1 = 2u_0 \times v_0 / g = \mathbf{u^2 \times \sin 2\alpha / g}$ .....(11.12)

Then, $\mathbf{H = v_0^2 / 2g}$ or $\mathbf{v_0 = \sqrt{(2gH)}}$ .....(11.13)

Time of travel, $\mathbf{t = 2 \times t_1 = 2 \times (x_1 / u_0) = 2v_0 / g}$ .....(11.14)

Max. range of a projectile, $\mathbf{R_{max} = u^2 / g}$

and corresponds to $\sin 2\alpha = 1$ or $\alpha = 45^0$

Example 20

A motor cyclist wants to jump over a ditch A-B of 6m long and 2m deep. Find the minimum velocity at A and resultant velocity at B

Solution :

Assuming uniform horizontal velocity of motor cycle, $H = g \times t^2/2$

or $t = \sqrt{2H / g} = \sqrt{(2 \times 2 / 9.81)} = 0.63855$ sec

$R = u \times t \Rightarrow u = R/t = 6/0.63855 = 9.4$ m/sec = 33.84 kmph

Vertical velocity at B,

$v = g \times t = 9.81 \times 0.063855$

$= 6.264$ m/sec

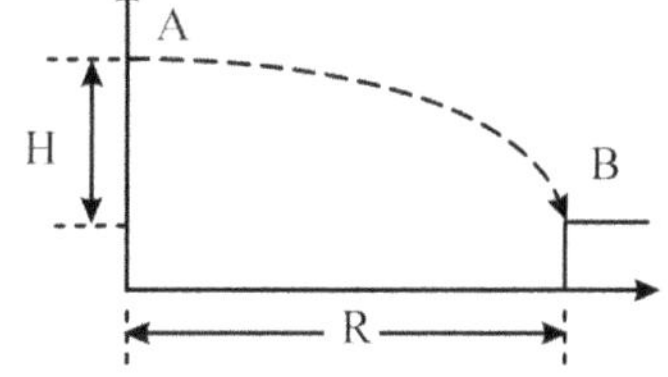

Resultant vel at B,

$v_R = \sqrt{(u^2 + v^2)}$

$= 11.3$m/sec or 40.68kmph

Angle made by the resultant velocity with the horizontal,

$\theta = \tan^{-1} (v/u) = 33.68^0$

Example 21

An airplane is moving with a horizontal velocity 'u' at a height H above the ground. If a projectile is fired from a gun, when the airplane is just above it, what should be its minimum velocity and angle so that the projectile hits the airplane ?

Solution :

Let u_0 and v_0 be the horizontal and vertical components of initial velocity U of the projectile. Let v be the vertical velocity of the projectile at height H.

Horizontal velocities of airplane and projectile should be equal, so that at some time they may meet.

i.e., $R = u \times t = u_0 \times t = (U \cos\alpha) \times t$ or $\alpha = \cos^{-1}(u/U)$

Assuming that H is the maximum height of the projectile,

$$H = (v^2 - v_0^2)/2(-g) = (v_0)^2 / 2g \quad \text{since } v = 0$$

$$\Rightarrow \quad v_0 = U \sin\alpha = \sqrt{2gH} \quad \text{and} \quad U \cos\alpha = u$$

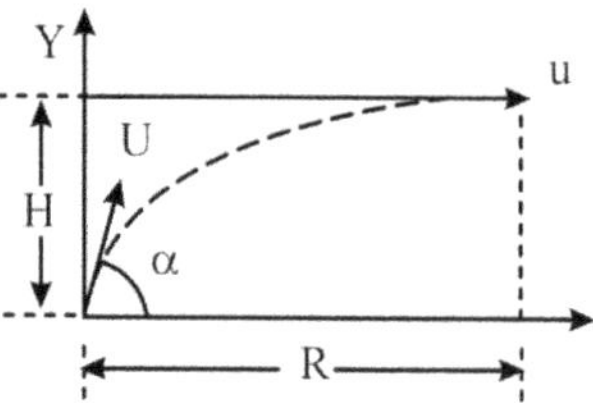

Then, $U = \sqrt{\left[(U\sin\alpha)^2 + (U\cos\alpha)^2\right]} = \sqrt{(2gH + u^2)}$

Example 22

A shot is fired with a velocity of 30 m/sec from a point 15 m in front of a vertical wall 6 m high. Find the angle of projection, to the horizontal to enable the shot just to clear the top of the wall.

Solution :

Let u_1 and u_2 be the initial velocities, inclined at α_1 and α_2, of the two possible trajectories so that the shot just clears top of the wall.

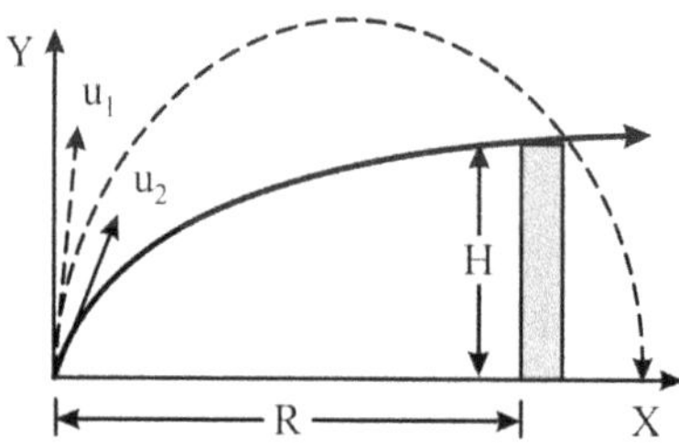

Given, $u_1 = u_2 = u$

Vertical component of velocity = $u \times \sin\alpha$

Time taken to cover the wall height of 6m is given by

$$s = u \times t - g \times t^2/2 \quad \text{or} \quad 6 = (30 \sin\alpha) \times t - 9.8 \times t^2 / 2$$

Similarly, horizontal component of velocity = u × cos α

Time taken to cover horizontal distance of 15 m, assuming uniform velocity, is given by

$$s = u \times t \quad \text{or} \quad 15 = (30 \cos \alpha) \times t \Rightarrow t = 15 / (30 \cos \alpha)$$

Since the horizontal distance and vertical height are covered during the same time 't',

$$6 = (30 \sin \alpha) \times [\ 15 / (30 \cos \alpha)\ ] - 9.8 \times [\ 15 / (30 \cos \alpha)\ ]^2 / 2$$

Simplifying, $1.225 \times \tan^2 \alpha - 15 \times \tan \alpha + 7.225 = 0$

or $\tan \alpha = 11.74 , 0.5$

$\Rightarrow$ $\alpha = 85^0 8'$ or $26^0 34'$

Example 23

A cricket ball thrown from a height of 1.8m at an angle of 30^0 with the horizontal with a speed of 18m/sec is caught by a person at a height of 0.6m from the ground. How far apart are the two players located ?

Solution :

$$y = (u \sin \alpha) \times t - g \times t^2 / 2$$

$$-(1.8 - 0.6) = 18 \sin 30^0 \times t - 9.81 \times t^2 / 2$$

$\Rightarrow$ $t = 1.96$ sec (neglecting 2nd answer, $t = -0.125$ sec)

$$R = (u \cos \alpha) \times t = 18 \times \cos 30^0 \times 1.96 = 30.55 \text{ m}$$

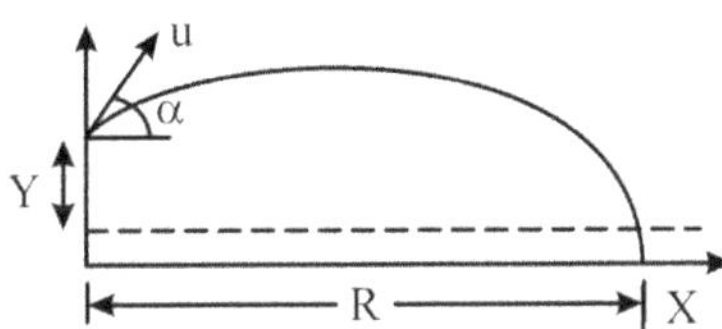

Example 24

Two guns are pointed at each other, one upward at an angle of 30^0 and the other at the same angle downwards. If the two guns, 30 m apart, are shot with velocities of 350 m/sec upwards and 300 m/sec downwards, find where and when they meet.

Solution:

Let u_1 (=350 m/sec) and u_2 (=300 m/sec) be inclined at 30^0 with the horizontal.

Distance between A and B = 30 m

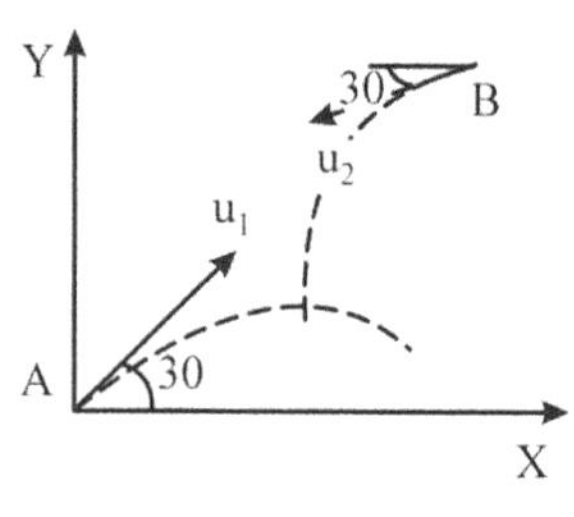

Horizontal distance between A and B

$= 30 \cos 30^0 = 15\sqrt{3}$ m

Vertical distance between A and B

$= 30 \sin 30^0 = 15$ m

Let 't' be the time when the two shots meet

Then,

$15\sqrt{3} = (u_1 \times \cos 30^0) \times t + (u_2 \times \cos 30^0) \times t$

$\Rightarrow t = 0.046$ sec

Horizontal and vertical distances of point of intersection of the two shots from A are given by $(u_1 \times \cos 30^0) \times 0.046$ and $(u_1 \times \sin 30^0) \times 0.046$

i.e., 13.943m and 8.05m

They can also be obtained, from B by

$(u_2 \times \cos 30^0) \times 0.046$ and $(u_2 \times \sin 30^0) \times 0.046$

i.e., 11.951m and 6.9m

Check : $(AB)_{Horizontal}$ or $AB \cos 30^0 = 13.943 + 11.951$ or $25.98 \approx 25.894$

and $(AB)_{Vertical}$ or $30 \sin 30^0 = 8.05 + 6.9$ or $15 \approx 14.95$

Example 25

A shot is fired with a velocity of 'u' at an angle 'θ' from the horizontal, aiming at a ball. If at the same time, the ball starts falling down, show that the shot hits the ball, irrespective of the velocity and the angle.

Solution :

The shot will hit the ball if the shot travels distance 'Z' vertically upwards while the ball drops by a height of 'S' such that H = S + Z, during the same time 't'

The time taken by the shot to travel distance 'R',

$t = R / (u \cos \alpha)$

$= [H / \tan \alpha] / [u \cos \alpha] = H / (u \sin \alpha)$

Vertical distance traveled by the shot during this time,

$Z = (u \times \sin \alpha) \times t - g \times t^2/2$

$= (u \sin \alpha) \times [H / (u \sin \alpha)] - g \times [H / (u \sin \alpha)]^2/2$

$= H - g \times [H / (u \sin \alpha)]^2/2$

Distance covered by the ball during this time,

$$S = u \times t + g \times t^2/2 = 0 + g \times [H/(u \sin \alpha)]^2/2 = H - Z$$

Hence, the shot hits the falling ball, irrespective of velocity 'u' and angle 'α'

Example 26

Maximum range of a field gun is 2000 m. If a target at a distance of 1200 m is to be hit, what should be the angle of projection ?

Solution :

Max. range of a projectile corresponds to angle of projection of 45^0 and is given by

$$R_{max} = 2000 = u^2/g$$

Angle of projection to meet the target range of 1200m is given by

$$R = 1200 = u^2 \times \sin 2\alpha / g = 2000 \times \sin 2\alpha$$

Therefore, $\sin 2\alpha = 0.6$ or $\alpha = 18.4^0$

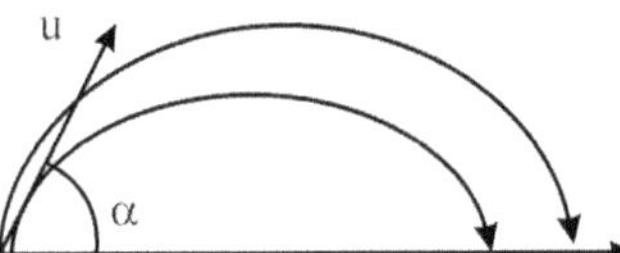

Example 27

A projectile is launched from a point A with an initial velocity of 120 m/sec at an angle of 40^0 with the horizontal. Calculate distance 'S' on the incline and the total time of flight.

Solution :

Let us consider the trajectory as made up of two parts A-C and C-B, with C as the top most point. Let S_1 be the horizontal distance between the points A and C.

For the part of travel A-C, Vertical component of velocity at C,

$$V = u \times \sin(40) - g \times t_1 = 0 \Rightarrow t_1 = 7.863 \text{ sec}$$

Corresponding horizontal and vertical distances are

$$h = -(120 \times \sin 40)^2/(-2\ g) = 733.945\text{m}$$

and $\quad S_1 = (120 \times \cos 40) \times t_1 = 722.81$ m

For the part of travel C-B, Vertical component of velocity at B,

$$V_B = 0 + g \times t_2 \qquad \text{.....(1)}$$

Corresponding horizontal and vertical components of distance traveled are

$$h + S \times \sin 20 = (V_B^2 - 0)/(2g) \qquad \text{.....(2)}$$

$$800 + S \times \cos 20 - S_! = (u \cos 40) \times t_2 \qquad \text{.....(3)}$$

Solving these three equations, we get, $S = 1057$ m

and $\quad t = t_1 + t_2 = 19.5$ sec

11.3.2 PROJECTILE ON AN INCLINED PLANE

If a particle is thrown upwards at an angle α on a plane inclined at an angle β (Ref Fig 11.5), let the horizontal distance covered by the projectile,

$$x = (u \cos \alpha) \times t = R \times \cos \beta$$

Similarly, $y = (u \sin \alpha) \times t - g \times t^2/2 = R \times \sin \beta$

Eliminating t, we get,

$$R = [2u^2 \times \cos^2 \alpha / (g \times \cos \beta)] \times (\tan \alpha - \tan \beta)$$

$$= [2u^2 \times \cos \alpha / (g \times \cos^2 \beta)] \times \sin (\alpha - \beta)$$

$$= [2u^2 / (g \times \cos^2 \beta)] \times [\sin (2\alpha - \beta) - \sin \beta] \qquad \text{.....(11.15)}$$

Obviously, for a horizontal plane, $\beta = 0$

and $\quad R = u^2 \times \sin 2\alpha / g$

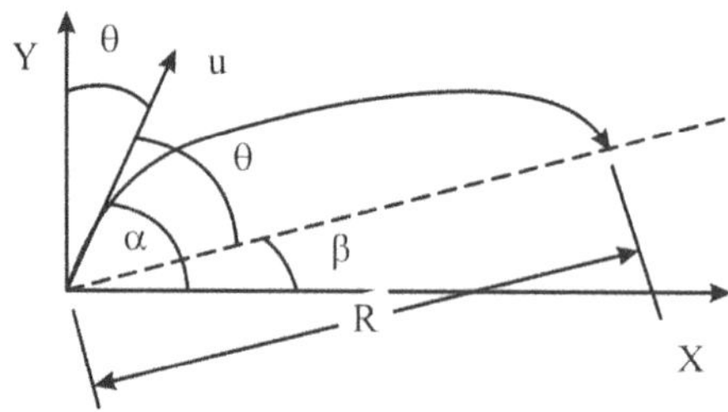

FIGURE 11.5 Path of a projectile on inclined plane

For Maximum range, $dR/d\alpha = 0 \Rightarrow \tan 2\alpha = -\cot\beta$ or $\alpha = 45^0 + \beta/2$

Then, $R_{MAX} = u^2 (1 - \sin\beta) / [\, g \cos^2\beta \,]$ .....(11.16)

Also, Angle, $\theta = \alpha - \beta = 45^0 - \beta/2$

i.e, The velocity vector bisects the angle between the vertical line and the inclined plane.

Example 28

A ball is thrown down an incline of 2:1 ratio between its horizontal and vertical distances and strikes it at a distance of 75m. If the ball rises to a max. height of 18m above the point of release, compute its initial velocity and inclination α with the horizontal

Solution :

Resolving distances and velocities, along X and Y axes,

$u_X = u \times \cos(\alpha - \theta)$; $u_Y = u \times \sin(\alpha - \theta)$

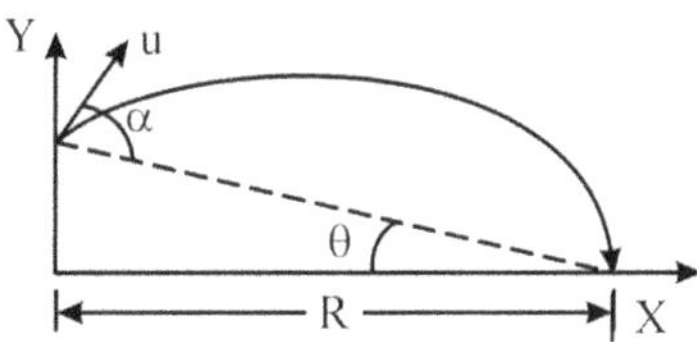

Assuming uniform velocity along X-axis,

$S_X = u_X \times t = 75 \times 2 / \sqrt{5} = 67.08$

When the ball rises to its maximum height,

$0 - u_Y^2 = -2 \times g \times H$ or $u_Y = 18.79$ m/sec

Also, $S_Y = u_Y \times t - g \times t^2/2 = -75 / \sqrt{5} = -33.54$

or $4.905 \times t^2 - 18.79 \times t - 33.54 = 0 \Rightarrow t = 5.1568$ sec

Then, $u_X = 67.08 / 5.1568 = 13.008$ m/sec

and $u_Y = [-33.54 + 4.905 \times 5.1568^2\,] / 5.1568 = 18.79$ m/sec

Therefore, $u = \sqrt{\left(u_x^2 + u_y^2\right)} = 22.85$ m/sec and inclination,

Inclination of u with the horizontal,

$$\alpha - \theta = \tan^{-1}(u_Y / u_X) = \tan^{-1}(1.4445) = 55.3^0$$

Example 29

A bullet is fired from a height of 120m at a velocity of 360kmph at an angle of 30^0 upwards. Find total time of flight.

Solution :

$$u = 360 \text{ kmph} = 360/3.6 \text{ m/sec} = 100 \text{ m/sec}$$

Method-1

Considering the travel from A to B along an imaginary inclined plane,

$$y = -120 = (100 \times \sin 30) \times t - g \times t^2 / 2$$

Taking only positive value of t as the realistic solution,

$$t = 12.2 \text{ sec}$$

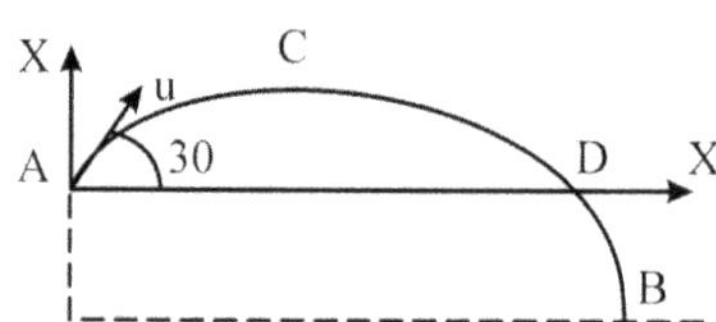

Method-2

Considering travel time from A to B in two parts from A to C (t_1) and from C to B (t_2),

$$0 = 100 \times \sin 30 - g \times t_1$$

$\Rightarrow$ $t_1 = 5.10$ sec, since vertical component of velocity at C = 0

Then, vertical distance from A to C, $h = (100^2 \times \sin^2 30) / (2g) = 127.42$ m

Also, $120 + 127.42 = 0 + g \times t_2^2 / 2$

$\Rightarrow$ $t_2 = 7.10$ sec (the other root is negative, not valid)

Therefore total time of flight, $t = t_1 + t_2 = 12.2$ sec

Method-3

Considering travel time from A to B in two parts from A to D (t_1) and from D to B (t_2),

$$t_1 = (2 \times 100 \times \sin 30) / g = 10.19 \text{ sec}$$

And, $120 = (100 \times \sin 30) \times t_2 + g \times t_2^2 / 2 \Rightarrow t_2 = 2.01 \text{ sec}$

Therefore total time of flight, $t = t_1 + t_2 = 12.2 \text{ sec}$

Example 30

A person can throw a ball at a max. velocity of 30 m/sec. If he wants to get max. range on the plane inclined at 20^0 to horizontal, at what angle should the ball be projected and what would be the max. range (a) up the plane and (b) down the plane

Solution :

(a) When the ball is thrown up the plane $(\beta = 20^0)$

For max. range, $\theta = 45^0 - \beta/2 = 45 - (20/2) = 35^0$

and $\alpha = \theta + \beta = 35 + 20 = 55^0$

Then, $R_{MAX} = u^2 (1 - \sin\beta) / [\, g \cos^2\beta \,]$

$= 68.362$ m

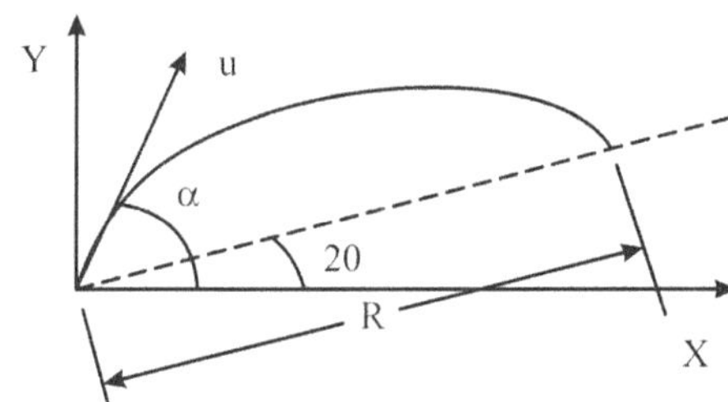

(b) When the ball is thrown down the plane $(\beta = -20^0)$

For max. range, $\theta = 45^0 + \beta/2 = 45 + (20/2) = 55^0$

and $\alpha = \theta - \beta = 55 - 20 = 35^0$

Then, $R_{MAX} = u^2 \times (1 - \sin\beta) / [\, g \times \cos^2\beta \,] = 139.432$ m

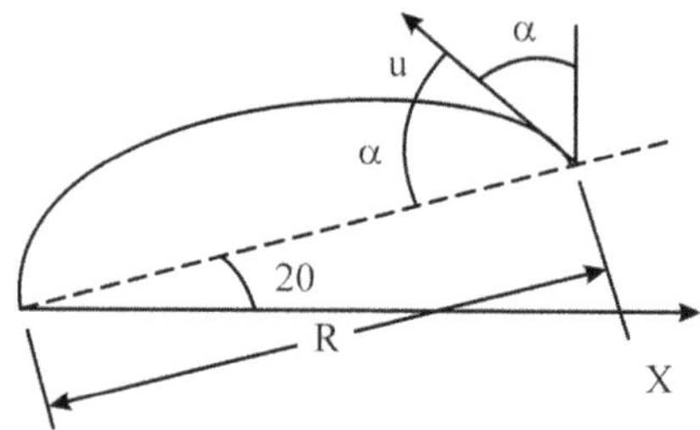

Example 31

A ball rebounds at A and strikes the inclined plane at B, 76m from A. If the ball rises to a maximum height of 19m above A, compute initial velocity and angle of projection at A. Inclination of the plane with the horizontal is given by $\tan\alpha = 1/3$

Solution :

AB = 76m

At C, vertical component of velocity, $V_C = 0$

i.e., $0^2 - V_A{}^2 = 2 \times (-g) \times 19 \quad \Rightarrow \quad V_A = u \times \sin\theta = 19.308$

For travel from A to B, vertical distance traveled

$$S = -76 \sin\alpha = (u \times \sin\theta) \times t - g \times t^2/2 \Rightarrow t = 4.93 \text{ sec}$$

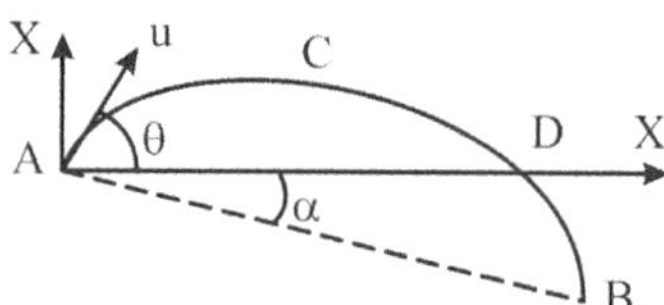

Horizontal distance traveled,

$$76 \cos\alpha = (u \times \cos\theta) \times t \Rightarrow u \times \cos\theta = 14.625$$

From the values of $u \times \sin\theta$ and $u \times \cos\theta$, we can obtain

$$\theta = 52.86^0 \text{ and } u = 24.222 \text{ m/sec}$$

11.4 ROTATION OF RIGID BODIES

If a particle or a rigid body is moving along a circular path about a fixed axis, the motion is more conveniently described in terms of angular displacement (θ), angular velocity (ω) and angular acceleration (α). These quantities are related to each other in the same way as the linear components. Thus,

$\omega = \omega_0 + \alpha \times t$ similar to $v = u + a \times t$

$\omega^2 - \omega_0{}^2 = 2\alpha \times \theta$ similar to $v^2 - u^2 = 2a \times s$

and $\theta = \omega_0 \times t + \alpha \times t^2/2$ similar to $s = u \times t + a \times t^2/2$ .....(11.17)

Expressions relating the angular and linear components can be obtained as follows.

$$v = ds / dt = (r \times d\theta) / dt = r \times \omega$$

$$a = dv / dt = d(r \times \omega) / dt = r \times \alpha$$

Radial or normal acceleration, $a_n = v^2/r = v \times \omega = r \times \omega^2$

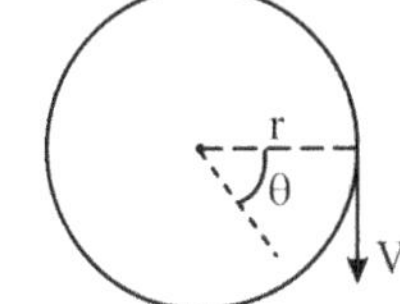

Example 32

A flywheel rotates with a constant retardation due to braking. At t=7.5 sec, its angular velocity was 40 π rad/sec. From t=0 to t=10 sec, it made 300 revolutions. Determine

(a) value of retardation
(b) total time taken to come to rest

and (c) number of revolutions made before it comes to rest.

Solution :

(a) $\theta = \omega_0 \times t - \alpha \times t^2 / 2$ $\Rightarrow>$ $300 \times 2\pi = \omega_0 \times 10 - \alpha \times 10^2 / 2$

$\omega = \omega_0 - \alpha\, t$ $\Rightarrow$ $40\,\pi = \omega_0 - \alpha \times 7.5$

From the above two relations, $\alpha = (\,600\,\pi - 400\,\pi\,) / (75\text{–}50) = 8\,\pi$ rad/sec^2

Therefore, $\omega_0 = 40\,\pi + 8\,\pi \times 7.5 = 100\,\pi$ rad/sec

(b) $0 = 100\,\pi - 8\,\pi \times t_1$ $\Rightarrow$ $t_1 = 12.5$ sec

(c) $\theta = \omega_0\, t_1 - \alpha\, t_1^2 / 2$ $\Rightarrow$ $\theta = 100\,\pi \times 12.5 - 8\,\pi \times 12.5^2 / 2$

$= 625\,\pi$ rad

$= (625\,\pi\,) / (\,2\,\pi\,) = 312.5$ revolutions

Example 33

A body rotates with $\alpha = 3\,t^2 - 3$ rad/sec^2. If its initial angular velocity is 2 rad/sec and initial angular acceleration is 0, what are the values of angular velocity and total angle covered at t = 5 sec.

Solution :

$\int d\omega = \int \alpha\, dt \;\Rightarrow\; \omega - 2 = t^3 - 3t$

At t = 5 sec, $\omega = t^3 - 3t + 2 = 112$ rad/sec

$\alpha = 3t^2 - 3 = 72$ rad/sec

and $\int d\theta = \int \omega\, dt \;\Rightarrow\; \theta = t^4/4 - 3t^2/2 + 2t = 128.75$ rad

Example 34

Two pulleys C and D of 1.5m and 1.0m radius respectively and two loads A and B are connected by strings, as shown. The load A has an initial velocity of 4.5 m/sec and acceleration of 3 m/sec^2, both acting downwards. Find

(a) number of revolutions executed by the pulley in 4 sec

(b) velocity and position of load B after 4 sec and

(c) acceleration of point E on pulley C at t = 0.

Solution :

Linear velocity of A is converted into angular velocity of pulley C. Since pulleys C and D are connected to the same shaft, angular velocities of both the pulleys will be equal. Finally, angular velocity of pulley D is converted into linear velocity of load B.

Thus, Initial angular velocity of pulleys, $\omega_0 = V_A/R_C = 3$ rad/sec

Angular acceleration of pulleys, $\alpha = a / R_C = 2$ rad/sec^2

After 4 sec,

(a) Angular velocity, $\omega = \omega_0 - \alpha\, t = 11$ rad/sec

Angular displacement, $\theta = \omega_0 \times t + \alpha \times t^2/2 = 28$ rad $= 4.456$ rev

(b) Linear velocity of B, $V_B = \omega \times R_D = 11$ m/sec

Distance moved by load B, $S_B = \theta \times R_D = 28$ m

(c) Linear acceleration of E, $a = [\, a_t^2 + a_n^2\,]^{0.5} = 13.829$ m/sec^2

where $a_t = R_C \times \alpha = 1.5 \times 2 = 3$ m/sec^2

and $a_n = \omega_0^2 \times R_C = 3^2 \times 1.5 = 13.5$ m/sec^2

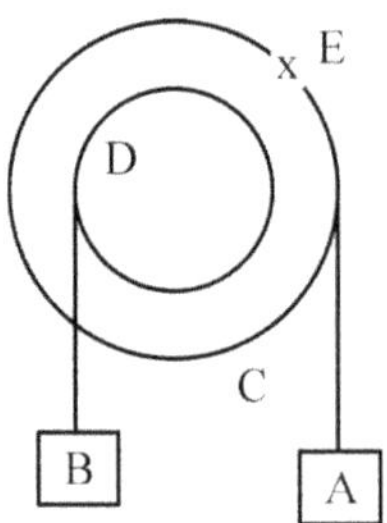

Example 35

A swing bridge turns through 90^0 in 120 sec. It is uniformly accelerated from rest in the first 40 sec. It then moves with uniform angular velocity in the next 60 sec and is uniformly retarded in the last 20 sec till it comes to rest. Find the acceleration, max. angular velocity and retardation.

Solution :

Let the max angular velocity be ω, acceleration α_1 and retardation α_2

In the accelerating phase, $\omega = 0 + \alpha_1 \times t_1 \quad \Rightarrow \alpha_1 = \omega / 40$

$\theta_1 = 0 \times t_1 + \alpha_1 \times t_1^2 / 2 = 20\ \omega$

In the uniform velocity phase, $\theta_2 = \omega \times t_2 = 60\ \omega$

In the retarding phase,

$0 = \omega + \alpha_2\ t_3 \quad \Rightarrow \quad \alpha_2 = -\omega / 20\ ;\ \theta_3 = \omega\ t_3 + \alpha_2\ t_3^2 / 2 = 10\ \omega$

Total angular displacement, $\theta = \theta_1 + \theta_2 + \theta_3 = 90\ \omega = \pi / 2$

Solving the above equations, we get

$\omega = 17.45 \times 10^{-3}$ rad/sec ; $\alpha_1 = 4.363 \times 10^{-4}$ rad/sec^2

and $\alpha_2 = -8.726 \times 10^{-4}$ rad/sec^2

Example 36

A horizontal bar of 1.5 m length and of small cross section rotates about a fixed axis through one end. It accelerates uniformly from 1200 rpm to 1500 rpm in an interval of 5 sec. Then,

(a) What is the linear velocity at the beginning and end of the interval and

(b) What are the normal and tangential components of acceleration of the mid point of the bar after 4 sec from the start of acceleration

Solution :

$\omega_0 = 1200 \times 2\ \pi / 60 = 40\ \pi$ rad/sec

After t = 5 sec, $\omega = 1500 \times 2\ \pi / 60 = 50\ \pi$ rad/sec

$\alpha = (\omega - \omega_0) / t = 2\ \pi$ rad/sec

(a) $V_0 = \pi\ d\ N_0 / 60 = \pi \times 2 \times 1.5 \times 1200 / 60 = 60\ \pi$ m/sec

$V = \pi \times 2 \times 1.5 \times 1500 / 60 = 75\ \pi$ m/sec

(b) After t = 4 sec,

$\omega = \omega_0 + \alpha\ t = 40\ \pi + 2\ \pi \times 4 = 48\ \pi$ rad/sec

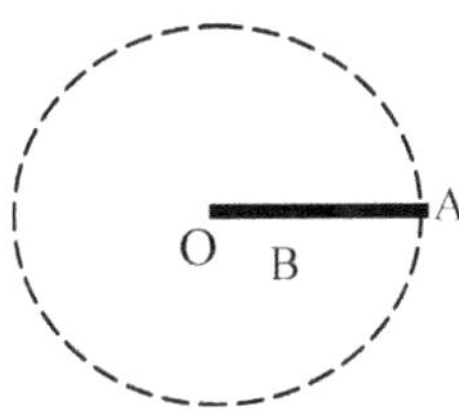

Tangential acceleration of mid point B, $a_t = \alpha \times R_B = 15\,\pi$ m/sec^2

Radial acceleration of mid point B, $a_n = \omega^2 R_B = 17052$ m/sec^2

SUMMARY

- ***Kinematics*** is a part of ***Dynamics***, which deals with motion of a body as a function of time (space-time relationship)
- Motion of a body can be divided into ***rectilinear*** (along a straight line), ***curvilinear*** (along a curved path) and ***rotational*** (about an axis)
- Dynamic motion is identified by rate of distance covered or ***velocity*** and rate of change of velocity or ***acceleration*** of the rigid body
- Velocity, $v = ds/dt$ and Acceleration, $a = dv/dt = d^2s/dt^2$
- The distance covered 's', velocity 'v' and acceleration 'a' are all vectors, along positive or negative axis. Negative acceleration is sometimes called '***deceleration***' or '***retardation***'. For a body moving with uniform velocity, acceleration is zero.
- If the motion takes place along a line inclined to the Cartesian coordinate system, then velocity and acceleration can be defined along X and Y axes, by taking components of distance covered 's' and velocity 'v' along the respective axes.
- **Rectilinear motion with *constant acceleration*** is governed by $a = (v-u)/t$ or $\mathbf{v = u + a \times t}$ where a body moves with initial velocity 'u' and covers distance 's' in time 't' with uniform acceleration 'a', reaching final velocity 'v'

$$\mathbf{s = u \times t + a \times t^2 / 2}$$

$$\mathbf{v^2 - u^2 = 2 \times a \times s}$$

If velocity 'v' is defined as a function of distance 's',

$$a = dv / dt = (ds/dt) \times (dv/ds) = v \times (dv/ds)$$

Distance moved in n^{th} second

$$s_n = \text{s in n sec} - \text{s in (n-1) sec} = u + (2n-1) \times a / 2$$

- If velocity is plotted as a function of time, acceleration ($a = dv/dt$) is represented by the slope of the velocity curve distance traveled ($s = \int v \times dt$) is represented by the area under the velocity curve during that time interval.
- **Relative velocity** is given by $v_{BA} = v_B - v_A$ or $v_B = v_A + v_{BA}$

- **Rectilinear motion with variable acceleration,** such as in a simple harmonic motion,

  $s = k \times \sin \omega t$; $v = ds/dt = \omega k \times \cos \omega t$

  and $a = d^2s/dt^2 = -\omega^2 k \times \sin \omega t$

- In **Curvilinear motion,** if components of distance travelled are x(t) and y(t), velocity and acceleration components are given by $u = dx/dt$; $v = dy/dt$; $a_X = du/dt$ and $a_Y = dv/dt$

  Resultant velocity, $v_R = \sqrt{(u^2 + v^2)}$ and Resultant acceleration $a = \sqrt{(a_X^2 + a_Y^2)}$

  The velocity and acceleration can also be expressed in intrinsic or path coordinates – **along tangential and normal directions** – to the path of the particle in *vector Form as,*

  $V = V_t \times e_t + V_n \times e_n$

  Velocity at any point of the path is tangential and, hence, $V_n = 0$ and $V = V_t \times e_t$

  Similarly, $a = dV/dt = (dV_t/dt) \times e_t + (dV_n/dt) \times e_n = a_t \times e_t + a_n \times e_n$

  For very small intervals of time 'dt', $dV \approx dV_n \approx V_t\, d\theta$

  Therefore, $a_t = dV_t/dt$ acts along the forward motion of the particle

  and $a_n = dV_n/dt \approx (V_t\, d\theta)/dt = V_t\, (d\theta/dt) = V_t\, (dS/Rdt) = V_t\, (V_t/R) = V_t^2/R$

- For motion with uniform tangential velocity along a circular path,

  $a_t = dV_t/dt = 0$ and $a = a_n = V^2/R$, is called **centripetal acceleration**

- For a projectile, $a_X = 0$ and $a_Y = -g$
- Integrating, $u = u_0 = u \times \cos \alpha$; $v = v_0 - g \times t = u \times \sin \alpha - g \times t$
- Distance travelled, $x = u_0 \times t$ and $y = v_0 \times t - g \times t^2/2$

  Equation of trajectory, $\mathbf{y = x \times \tan \alpha - (g \times \sec^2 \alpha / 2\, v_0^2) \times x^2}$

  Range, $\mathbf{R} = 2 \times u_0 \times v_0 / g = \mathbf{u^2 \times \sin 2\alpha / g}$

  Max height of travel, $\mathbf{H = v_0^2 / 2g}$

  or $\mathbf{v_0 = \sqrt{(2 \times g \times H)}}$

Time of travel, $\mathbf{t = 2 \times v_0 / g}$

Max. range of a projectile, $\mathbf{R_{max} = u^2 / g}$ and corresponds to $\sin 2\alpha = 1$ or $\alpha = 45^0$

- For a projectile on a plane inclined at β with the horizontal,

  $x = (u \cos \alpha) \times t = R \times \cos \beta$

  $y = (u \sin \alpha) \times t - g \times t^2/2 = R \times \sin \beta$ R

  $= [2u^2 \cos^2 \alpha / (g.\cos \beta)] \times (\tan \alpha - \tan \beta)$

  $= [2u^2 / (g.\cos^2 \beta)] \times [\sin (2\alpha - \beta) - \sin \beta]$

  For Maximum range, $dR/d\alpha = 0 \Rightarrow \tan 2\alpha = -\cot \beta$ or $\alpha = 45^0 + \beta/2$

  $R_{MAX} = u^2 (1 - \sin \beta) / [g \cos^2 \beta]$ and Angle, $\theta = \alpha - \beta = 45^0 - \beta/2$

  i.e, The velocity vector bisects the angle between the vertical line and the inclined plane

- If a particle or a rigid body moves along a circular path about a fixed axis, the motion is more conveniently described in terms of angular displacement (θ), angular velocity (ω) and angular acceleration (α) given by

  $\omega = \omega_0 + \alpha \times t$; $\omega^2 - \omega_0^2 = 2\alpha \times \theta$ and $\theta = \omega_0 \times t + \alpha \times t^2/2$

  The angular and linear components are related by

  $v = ds / dt = (r \times d\theta)/dt = r \times \omega$; $a = dv / dt = d(r \times \omega) / dt = r \times \alpha$

  Radial or normal acceleration, $a_n = v^2/r = v \times \omega = r \times \omega^2$

PROBLEMS FOR PRACTICE

1. A bomb is released from an aeroplane flying at a speed of 1000km/h on a straight level course 2000m above the ground. Find the time required for the bomb to reach the ground and the horizontal distance traveled by the bomb after its release. Assume $g = 9.81 m/s^2$.

   (*Ans:* 20.193 sec ; 5609m)

2. From the bottom of a tower 120m high a stone is thrown vertically upward with a velocity that would carry it to the top of the tower. After 1 second another stone is dropped from the top. When and where will the two stones meet?

   (*Ans :* at 100.9m from ground, after 2.973sec)

3. A cage descends in a mine shaft with an acceleration of 0.5 m/sec^2. After the cage has traveled 25 m, a stone is dropped from the top of the shaft. Determine

(i) the time taken by the stone to hit the cage

and (ii) the distance traveled by the cage before impact

(*Ans :* 2.92 sec, 41.7316 m)

4. The greatest possible acceleration and deceleration that a train may have is 'a' and its max. speed is 'v'. Find the minimum time in which the train can cover distance 'S' between two stations.

(*Ans :* S/v + v/a)

5. A police starts on his motor cycle from rest at A, 2 sec after a car goes past A with constant velocity of 120 kmph, and reaches max velocity of 150 kmph with an acceleration of 6 m/sec^2. Calculate the distance from A when the police overtakes the car.

(*Ans :* 911.4 m)

6. When a cyclist is riding west at 20 kmph, it is raining at an angle of 45^0 with the vertical. When he rides at 15 kmph, he feels it is raining at an angle of 30^0 with the vertical. What is the velocity of rain ?

(*Ans :* 3.994 m/sec or 14.377 kmph)

7. A projectile is aimed at a target on the horizontal plane and falls 12 m short when angle of projection is 15^0 while it overshoots by 24 m when the angle of projection is 45^0. Find the angle of projection to hit the target.

(*Ans :* 21.9^0)

8. A projectile is required to reach a point whose coordinates are x = 50m and y = 20m. If $\alpha = 45^0$, what should be the velocity of projection ?

(*Ans :* 28.59 m/sec)

CHAPTER **12**

KINETICS

While kinematics deals with motion of a body as a function of time (space-time relationship), kinetics deals with forces that need to be applied to produce a prescribed motion such as translation, rotation about a fixed axis, rotation of a rigid body or a combination of bodies on a surface.

The subject is essentially based on Newton's laws of motion -

(i) Every body continues to be in its state of rest or of uniform motion, until it is acted upon by an external force

(ii) Force acting on a body is proportional to the rate of change of momentum of the body. (Momentum of a body is defined as the product of its mass and velocity)

(iii) To every action, there will be an equal and opposite reaction

A body is considered to be in static equilibrium, if it satisfies all the equations of equilibrium. For a general system of non-concurrent forces in space, these equations are –

$$\sum F_X = 0\ ;\ \ \sum F_Y = 0\ ;\ \ \sum F_Z = 0\ ;\ \ \sum M_X = 0\ ;\ \ \sum M_Y = 0 \ \text{ and } \ \sum M_Z = 0$$

12.1 D'ALEMBERT PRINCIPLE

This principle states that the equations of static equilibrium are applicable for a body in motion, provided (an imaginary) inertia force of the body is included in the system of forces, in addition to the forces acting on the body (external applied forces and internal/inherent forces such as self weight, normal reaction, frictional resistance).

Inertia force, F_i is obtained from Newton's second law of motion as

F_i = d(mv)/dt and acts in a direction opposite to the resultant of external applied forces.

If mass of the body remains constant, as is the case in most practical situations,

$F_i \approx m \times (dv/dt) = k \times m \times (dv/dt) = m \times a$, if proportionality constant, k=1

Thus, $\sum F = \sum F_{Applied} - F_{Inertia} = 0$. This is also called a state of 'fictitious equilibrium'.

12.1.1 VELOCITY, ACCELERATION AND DISTANCE AS A FUNCTION OF EXTERNAL FORCE

External force can be a constant or a function of time, velocity or displacement. Depending on the type of force function, acceleration, velocity and displacement are calculated using the following relations.

(i) Force is a constant: $a = d^2s/dt^2 = F/m = dv/dt$

Then, $v = ds/dt = \int a\, dt = \int (F/m)\, dt = (F/m) \times t + C_1$

and $s = \int v\, dt = (F/m) \times t^2/2 + C_1 \times t + C_2$

(ii) Force is a function of time : $a = d^2s/dt^2 = F(t)/m$

Then, $v = ds/dt = \int a\, dt = \int [F(t)/m]\, dt + C_1$

and $s = \int v\, dt = \int [\int \{F(t)/m\}\, dt + C_1]\, dt + C_2$

(iii) Force is a function of velocity : $a = dv/dt = F(v)/m$

or $dv/F(v) = dt/m$

Then, $v = \int dv/F(v) = (1/m) \times t + C_1$

and $s = \int ds = \int v(t)\, dt$

(iv) Force is a function of displacement : $a = dv/dt = F(s)/m$

or $(dv/ds) \times (ds/dt) = v \times (dv/ds) = F(s)/m$

Then, $\int v\, dv = v^2/2 = (1/m) \int F(s)\, ds + C_1$

Here, C_1 and C_2 are integration constants, to be evaluated using the given conditions of motion. If conditions of motion are not specified, $C_1 = 0$ and $C_2 = 0$ are used

Example 1

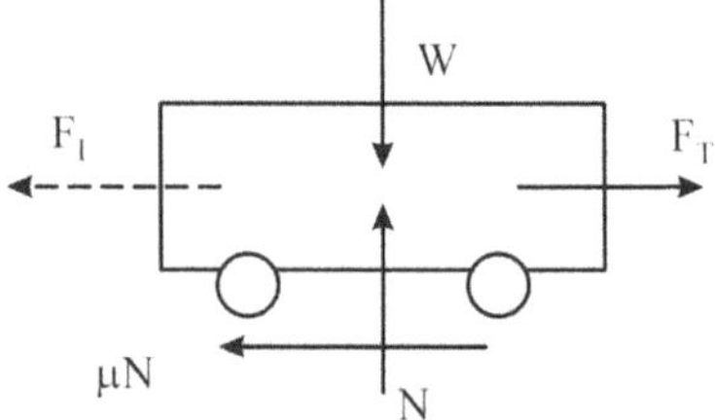

A train of 3000kN weight starts from rest and reaches a velocity of 54kmph after 1 min. If

the frictional resistance of the track is 5N per kN of train's weight, find the traction force developed by the engine.

Solution:

Weight of train W (3000kN) is balanced by the vertical reactions at the wheels, whose sum is N. Corresponding to this reaction, frictional resistance μN is produced at the wheels in the direction opposite to the motion.

Initial velocity of train, U = 0 (starting from rest)

Velocity of train after 1 min (or 60 sec), V = 54 kmph = 54/3.6 or 15 m/sec

and Acceleration, a = (V–U) / t = 15/60 = 0.25 m/sec2

Frictional resistance, μN = 5 × 3000 = 15000 N = 15 kN

Part of the traction force developed by the engine is used up to overcome frictional resistance and the balance force is responsible for the acceleration of the train.

Then, Equation of fictitious equilibrium,

$$\Sigma F_X = F_T - (F_I + \mu N) = 0$$

Thus, Traction force, $F_T = F_I + \mu N = (W/g) \times a + \mu N$

$$= (3000/9.81) \times 0.25 + 15 = 91.45 \text{ kN}$$

Example 2

A 500 N body is initially stationary on a plane inclined at 450 to the horizontal. The coefficient of kinetic friction between the body and the inclined plane is 0.5. What distance along the incline must the body slide before it reaches a speed of 12m/sec.

Solution :

Under the influence of the three forces W, N and μN, the body tends to move down. The corresponding inertia force (F_I) is 'm × a'.

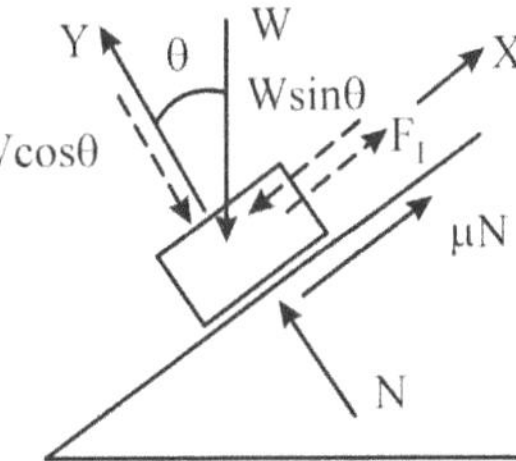

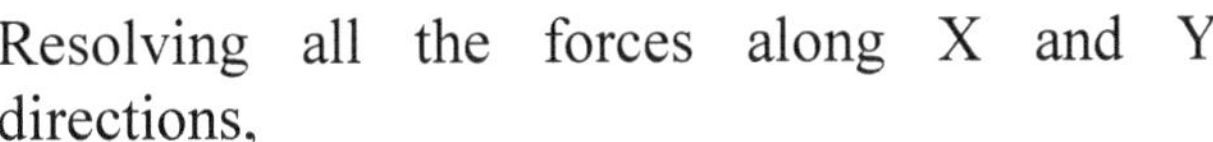
Resolving all the forces along X and Y directions,

$$\sum F_Y = N - W\cos\theta = 0$$

$$\Rightarrow \quad N = W\cos\theta$$

$$\sum F_X = F_I + \mu N - W\sin\theta = 0$$

$\Rightarrow$ $F_I = W \sin\theta - \mu N = (W/g) \times a$

$\Rightarrow$ $a = (W \sin\theta - \mu\, W \cos\theta) \times (g/W) = (\sin\theta - \mu \cos\theta)\, g$

$= (\sin 45 - 0.5 \times \cos 45) \times 9.81$

$= 0.5 \times 9.81 \times 0.707 = 3.4648\ \text{m/sec}^2$

Then, distance travelled 's' to gain velocity 'v' can be obtained from

$v^2 - u^2 = 2\, a \times s$

Since u = 0 and v = 12m/sec,

$s = (v^2 - u^2) / (2 \times a) = (12^2 - 0)/(2 \times 3.4648) = 20.78\text{m}$

Example 3

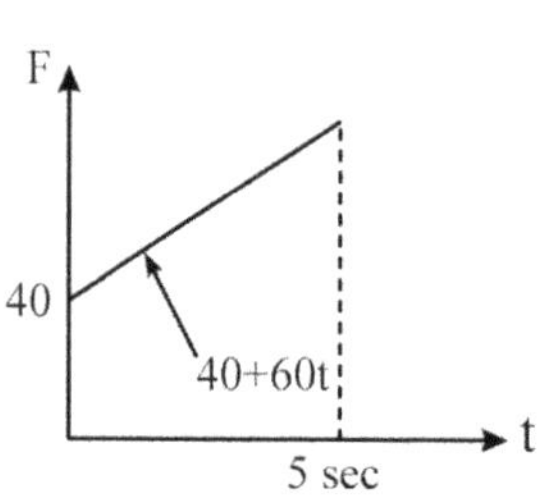

A particle of mass 1 kg moves in a straight line under the influence of a force which increases linearly with time at the rate of 60 N/sec, its initial value being 40 N. Determine the position, velocity and acceleration after 5 sec, if it started from rest.

Solution :

Force can be expressed as $F(t) = 40 + 60 \times t$

Acceleration, $a(t) = d^2s/dt^2 = F(t) / m = (40 + 60 \times t) / 1$

$= 40 + 60 \times t$

Then, velocity, $v(t) = ds/dt = \int (F(t)/m)\, dt + C_1$

$= 40 \times t + 60 \times (t^2/2) + C_1$

and displacement, $s(t) = \int v(t)\, dt = 40 \times (t^2/2) + 60 \times (t^3/6) + C_1 \times t + C_2$

From the initial condition that the body starts from rest,

$v(0) = 0$ and $s(0) = 0$ at $t = 0$

$\Rightarrow$ $C_1 = 0$ and $C_2 = 0$

Therefore, at t = 5 sec,

$s = 40 \times (5^2/2) + 60 \times (5^3/6) + 0 \times 5 + 0 = 1750\ \text{m}$

$v = 40 \times 5 + 60 \times (5^2/2) + 0 = 950\ \text{m/sec}$

and $a = 40 + 60 \times 5 = 340\ \text{m/sec}^2$

Example 4

Weights W and 2W are supported in a vertical plane by a string and pulleys as shown. Find the magnitude of Q applied along with W which will give a downward acceleration of 0.1g to the weight W. Neglect friction and inertia of pulleys

Solution :

Considering free body diagrams of the two pulleys,

$$\sum(F_Y)_W = T - W - Q + (W + Q)/g \times a = 0 \qquad .....(1)$$

Since downward movement of string by distance 'S' at the weight W is shared by the two parts of the string over pulley B, it causes weight 2W to move up by 'S/2'. Then, velocity and acceleration of weight 2W will also be half of those at weight W.

Therefore, $\sum(F_Y)_{2W} = 2T - 2W - (2W/g) \times (a/2) = 0 \qquad .....(2)$

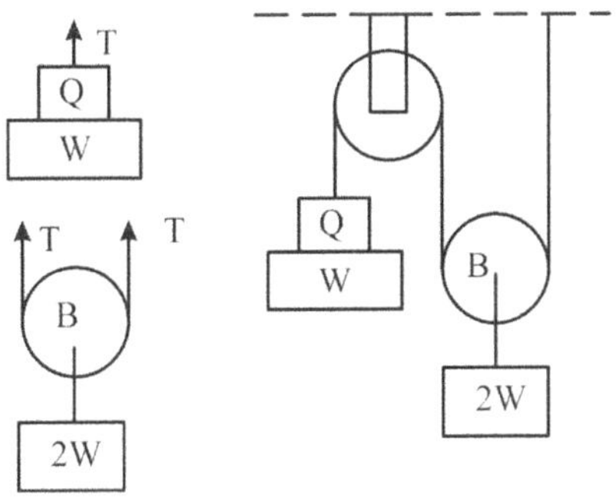

Eliminating T from eq (1) and (2),

we get $Q = (W+Q)/g \times a + (W/g) \times (a/2)$

$= 0.1\ Q + 0.15\ W \qquad$ (since $a = 0.1\ g$)

or $Q = (0.15/0.9)\ W = W/6$

Example 5

Determine the acceleration of the bodies A and B of weights 1000 N and 1500 N respectively and connected by a string passing over a pulley, if the kinetic friction coefficient is 0.2 at all contact surfaces.

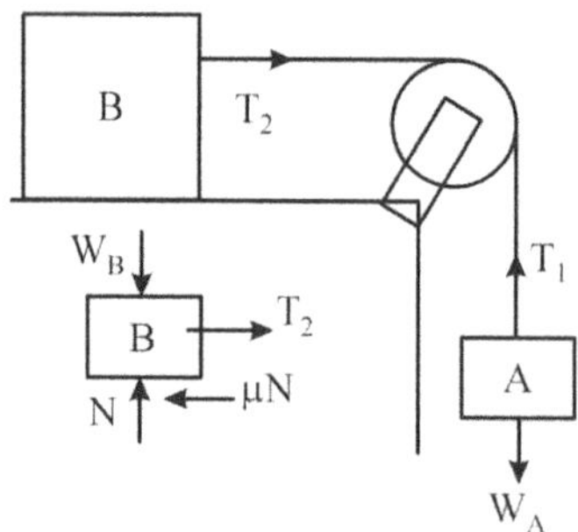

Solution:

Force balance along horizontal and vertical directions at A and B, since acceleration of A and B is the same in a single string, will give

$$\sum(F_Y)_A = T_1 - W_A + (W_A/g) \times a = 0$$

or $$T_1 = 1000 \times (1 - a/9.81) \quad .....(1)$$

$$\sum(F_Y)_B = N - W_B = 0 \quad \text{or} \quad N = 1500 \quad .....(2)$$

$$\sum(F_X)_B = T_2 - \mu N - (W_B/g) \times a = 0$$

or $$T_2 = 1500 \times (0.2 + a/9.81) \quad .....(3)$$

String tensions T_1 and T_2 are related by the friction between the string and pulley over its wrap angle of 90^0 or $\pi/2$ radians.

Then, $$T_1/T_2 = e^{\mu\pi/2} = 1.3691 \quad .....(4)$$

Solving these four equations, we get $a = 1.89$ m/sec^2

Example 6

Two weights A and B of weights 800 N and 200 N are connected by a string and move along a rough horizontal plane ($\mu = 0.3$). When A is pulled by 400 N, calculate acceleration and tension in the thread.

Solution :

Force balance along horizontal and vertical directions at A and B, since acceleration of A and B is the same, will give

$$\sum(F_Y)_A = N_A - W_A = 0 \Rightarrow N_A = 800 \text{ N} \quad .....(1)$$

$$\sum(F_X)_A = F - T - \mu N_A - (W_A/g)\times a = 0 \quad .....(2)$$

$$\sum(F_Y)_B = N_B - W_B = 0 \Rightarrow N_A = 200 \text{ N} \quad .....(3)$$

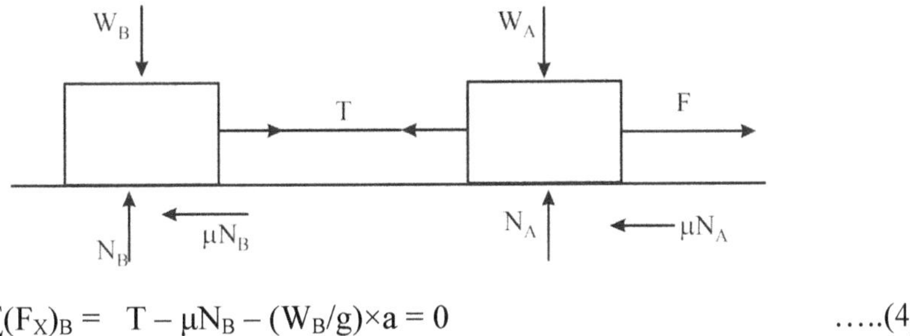

$\sum(F_X)_B = T - \mu N_B - (W_B/g)\times a = 0$ .....(4)

Solving these four equations, we get $T = 80$ N and $a = 0.981$ m/sec^2

Example 7

Two masses 14 kg and 3 kg, connected by a string rest on a plane inclined at 45^0 to the horizontal. What will be the tension in the string, when the masses are released, if coefficient of kinetic friction is 0.25

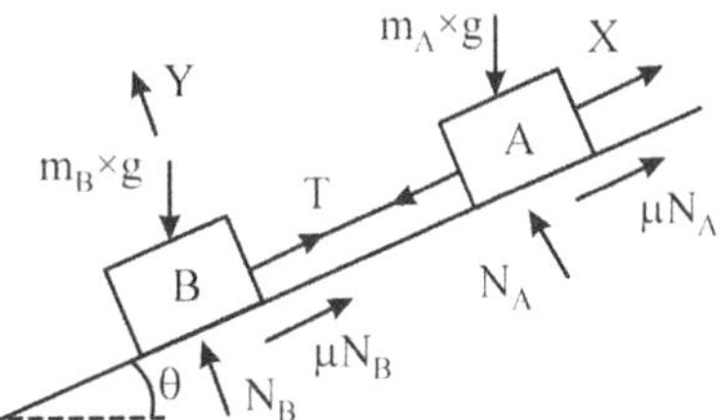

Solution :

Force balance along and perpendicular to the inclined plane at A and B, since acceleration of A and B is the same, will give

$\sum(F_Y)_A = N_A - (m_A \times g) \times \cos 45 = 0$

or $N_A = (14 \times 9.81) \times \cos 45 = 97$ N .....(1)

$\sum(F_X)_A = \mu \times N_A + m_A \times a - m_A \times g \sin 45 - T = 0$

or $14 \times a - T = (14 \times 9.81) \times \sin 45 - 0.25 \times 14$ .....(2)

$\sum(F_Y)_B = N_B - m_B \times g \cos 45 = 0$

or $N_B = (3 \times 9.81) \times \cos 45 = 20.81$ N .....(3)

$\sum(F_X)_B = T + m_B \times a + \mu \times N_B - m_B \times g \sin 45 = 0$

or $T + 3 \times a = (3 \times 9.81) \times \sin 45 - 0.25 \times 3$ .....(4)

From these 4 equations, we get

$(14 + 3) \times a = \{(14 + 3) \times 9.81/\sqrt{2} - 0.25 \times (97 + 20.81)\}$

or $a = 5.205$ m/sec^2 and $T = 0.005$ N

Example 8

A 750N crate rests on a 500N cart. The coefficient of friction between the crate and the cart is 0.3 and between cart and road is 0.2. If the cart is to be pulled by a force P such that the crate does not slip, determine

(a) the maximum allowable magnitude of P and

(b) the corresponding acceleration of the cart.

Solution:

Considering free body diagrams of crate and resolving forces along horizontal and vertical directions, $\sum(F_Y)_A = N_A - W_A = 0$ or $N_A = 750$ N

and $\sum(F_X)_A = \mu_A \times N_A - (W_A/g) \times a = 0$

$\Rightarrow$ $a = 2.943$ m/sec^2

Similarly, considering equilibrium of cart + crate,

$$\sum(F_Y)_B = N_B - (W_A+W_B) = 0 \Rightarrow N_B = 1250 \text{ N}$$

$$\sum(F_X)_B = P - \mu_B \times N_B - [(W_A+W_B)/g] \times a = 0$$

$\Rightarrow$ $P = 0.2 \times 1250 + [\,(750 + 500) / 9.81\,] \times 2.943 = 625$N

Example 9

A man of mass 65kg dives vertically downwards into a swimming pool from a tower of height 10m. He was found to go down in water by 2m and then start rising. Find the average resistance in water. Neglect air resistance.

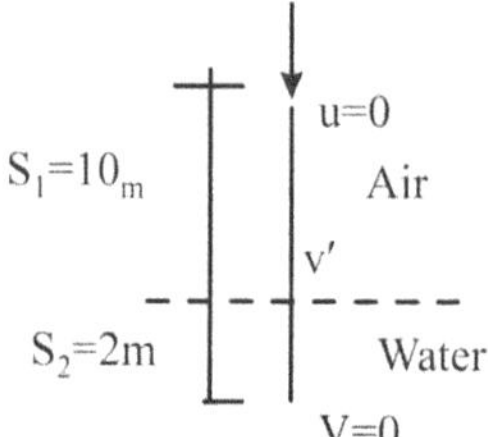

Solution :

Velocity of man while entering water,

$$v' = \sqrt{u^2 + 2 \times g \times S_1} = \sqrt{0^2 + 2 \times 9.81 \times 10} = 14\text{m / sec}$$

Assuming uniform retardation in water with initial velocity v′ of 14m/sec and final velocity V of 0m/sec (coming to rest) in 2m,

$$a = [V^2 - (v')^2] / (2S) = (0^2 - 14^2)/(2\times2) = 49 \text{ m/sec}^2$$

Average resistance in water = m × a = 65 × 49 = 3185 N

Example 10

A solid homogeneous cylinder of mass 3.6kg and radius 150mm is released from rest on a ramp. If the inclination of ramp with horizontal is 30^0 and coefficient of kinetic friction is 0.1, determine acceleration of the mass center of the cylinder and the friction force exerted by the ramp.

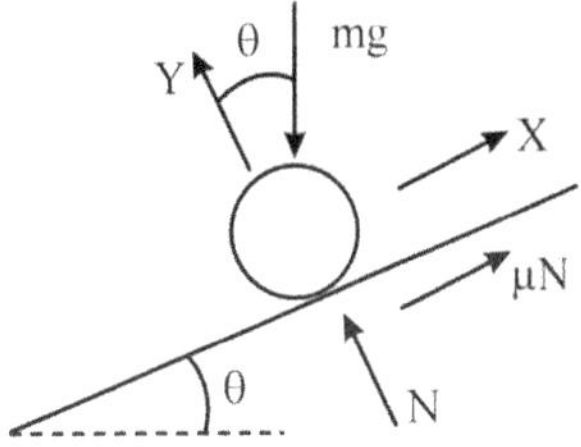

Solution:

Force balance along and perpendicular to the inclined plane gives

$$\sum F_Y = m\,g \cos 30 - N = 0 \Rightarrow N = 3.6 \times 9.81 \times \cos 30 = 30.58 \text{ N}$$

Friction force exerted by the ramp, $\mu \times N = 0.1 \times 30.58 = 3.058$ N

$$\sum F_X = \mu \times N - (m \times g) \times \sin 30 + m \times a = 0$$

or $\quad 3.058 - (3.6 \times 9.81) \times 0.5 + 3.6 \times a = 0 \Rightarrow a = 4.0555 \text{ m/sec}^2$

Example 11

Three blocks A, B and C are connected through a system of pulleys as shown. If the weights of blocks A, B and C are in the ratio 4:2:1, find the acceleration of the blocks. Also calculate tension in the strings. Assume that the pulleys are weightless.

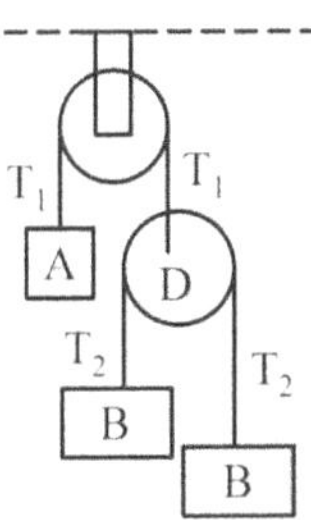

Solution:

Assuming acceleration of block A and pulley D as a_1 and acceleration of blocks B and C w.r.t. the pulley D as a_2, and considering equilibrium of block A, pulley D, block B and block C respectively,

$$\sum(F_Y)_A = T_1 - W_A + (W_A/g) \times a_1 = 0 \text{ or } T_1 = W_A \times (1 - a_1/g) \quad \text{.....(1)}$$

$$\sum(F_Y)_D = T_1 - (W_B + W_C) - [(W_B + W_C)/g] \times a_1 = 0 \quad \text{.....(2)}$$

$$\sum(F_Y)_B = T_2 - W_B - (W_B/g) \times a_2 = 0 \quad \text{.....(3)}$$

$$\sum(F_Y)_C = T_2 - W_C - (W_C/g) \times a_2 = 0 \quad \text{.....(4)}$$

Substituting, $W_A = 4\,W_C$ and $W_B = 2\,W_C$, eq (1) and (2) simplify to

$$T_1 = 4W_C\,(1 - a_1/g) = (2W_C + W_C) \times (1 + a_1/g) \Rightarrow a_1 = g/7$$

and eq (3) and (4) simplify to $T_2 = 2W_C\,[1-(a_2/g)] = W_C\,[1+(a_2/g)]$

$$\Rightarrow a_2 = g/3$$

Then, $a_A = a_1 = g / 17$; $a_B = a_2 - a_1 = 4g / 21$ and $a_C = a_2 + a_1 = 10g / 21$

Therefore, $T_1 = (24/7)\,W_C$ and $T_2 = (4/3)\,W_C$

Example 12

A jet aircraft has a total mass of 22000 kg acting at its mass center G. At take off, the engine provides a thrust T_1 of 4 kN at A and T_2 of 1.5 kN at B. Determine acceleration of the plane and normal reactions at the nose wheel C and two wing wheels at D. Neglect friction between wheels and runway. Relevant dimensions of the aircraft are : p-q = 1.2; p-r = 2.3; p-s = 2.5; e-g = 9; f-g = 6

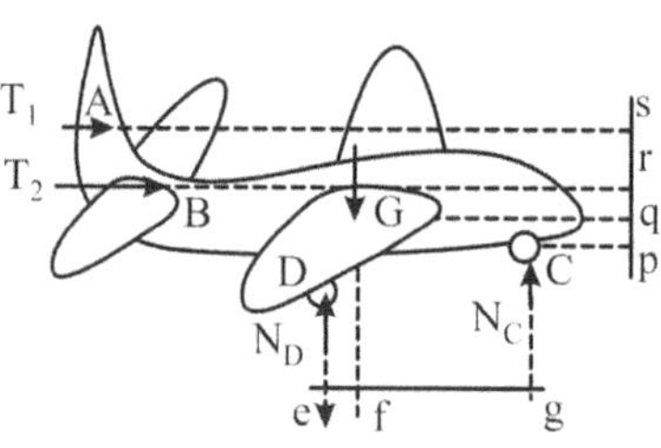

Solution :

Considering equilibrium of forces along and perpendicular to runway,

$$T_1 + T_2 = m \times a \Rightarrow a = (4 + 1.5)\text{kN} / 22000 \text{ kg} \qquad \text{.....(1)}$$

$$N_C + N_D = m \times g \qquad \text{.....(2)}$$

Taking moments about nose wheel C,

$$T_1 \times (p - s) + T_2 \times (p - r) + N_D \times (e - g)$$
$$= m \times g \times (f - g) + m \times a \times (p - q)$$
$$T_1 \times 2.5 + T_2 \times 2.3 + N_D \times 9 = m \times g \times 6 + m \times a \times 1.2 \qquad \text{.....(3)}$$

Then, reaction at each wing wheel, from (1) & (3)

$$N_D / 2 = 71.56 \text{ kN}$$

and reaction at the nose wheel, from eq. (2), is

$$N_C = m \times g - N_D = 72.767 \text{ kN}$$

Example 13

A string with weights of 1000 N and 500 N attached to its ends passes over a fixed drum. Calculate acceleration of weights

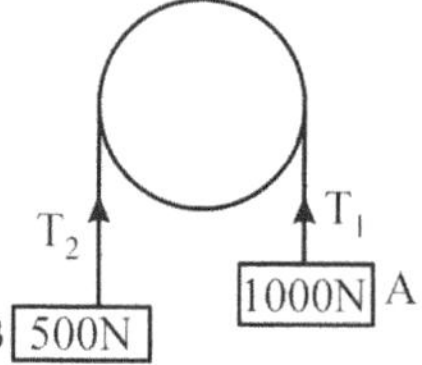

(a) if coefficient of friction between string and drum is 0.1

and (b) neglecting friction between string and drum.

Solution:

1000 N Weight moves down with acceleration 'a' while 500 N weight moves up with the same acceleration (assuming the string is inextensible. Then, Considering equilibrium of the two blocks,

$\sum(F_Y)_A = T_1 - 1000 + (1000/g) \times a = 0$ .....(1)

and $\sum(F_Y)_B = T_2 - 500 - (500/g) \times a = 0$ .....(2)

(a) If $\mu \neq 0$, $T_1 > T_2$ and $T_1/T_2 = e^{\mu\pi} = 1.3691$

Therefore, $a = 1.835$ m/sec^2

(b) If $\mu = 0$, $T_1 = T_2$ Then, adding eq.(1) and (2), we get $a = 3.27$ m/sec^2

Example 14

A lift weighing 7kN moves up with an acceleration of 2.5 m/sec^2. Determine tension 'T' in the cable of the lift

Solution :

$$\sum F_Y = T - W - m \times a = 0$$

$$\Rightarrow \quad T = W + m \times a = 7000 + (7000/9.81) \times 2.5 = 8783.9\text{N} \text{ or } 8.784\text{kN}$$

Example 15

A lift weighing 5kN starts to move upwards with a constant acceleration and acquires a velocity of 2 m/sec after traveling a distance of 3m. Determine the tension in the cable of the lift during accelerated motion. If the elevator while stopping moves with a constant deceleration from a constant velocity of 2 m/sec and comes to rest in 2 sec, calculate the pressure exerted by a man weighing 750N to the floor during stopping.

Solution:

During accelerated motion of the lift from rest (u=0),

$$2\,a \times S = V^2 - u^2$$

$$\Rightarrow \quad a = V^2/(2 \times S) = (2)^2/(2 \times 3)$$

$$= 2/3 \text{ m/sec}^2$$

$$\sum F_Y = T - W_L - m_L \times a = 0$$

$$T = W_L + m_L \times a = 5000 + (5000/9.81) \times (2/3)$$

$$= 5339.8\text{N} \quad \text{or} \quad 5.34\text{kN}$$

During decelerated motion of the lift,

$$a = (V - u) / t = (0 - 2)/2 = -1 \text{ m/sec}^2$$

Pressure exerted by a man on the floor of lift,

$$P = W_m - m \times a = [\,750 - (750 / 9.81) \times (-1)\,] = 826.45 \text{ N}$$

12.2 CENTRIFUGAL FORCE

When a body moves around a curve of radius 'R', a normal or centripetal acceleration of magnitude V^2/R acts radially inwards. The corresponding force ($m \times a_N$) is balanced by the inertia force acting radially outwards. This is called centrifugal force and is significant in the motion of automobiles along a curved path.

Thus, $F_C = (W/g) \times V^2/R$

12.2.1 MOTION OF AN AUTOMOBILE ON LEVEL GROUND

(a) ***Skidding*** takes place when frictional force in the radially inward direction reaches its limiting value and is less than the centrifugal force in the radially outward direction. Since centrifugal force is a function of velocity, limiting value of velocity of an automobile to prevent skidding while negotiating a curve is given by

$$\sum F_X = \text{Centrifugal force} - \text{Frictional force} = 0$$

or $(W/g) \times V^2/R - \mu \times W = 0$

since Frictional force at the four wheels

$$= 2\,(F_1 + F_2) = 2\,\mu \times (N_1 + N_2) = \mu \times W$$

Thus, $V = \sqrt{\mu \times g \times R}$

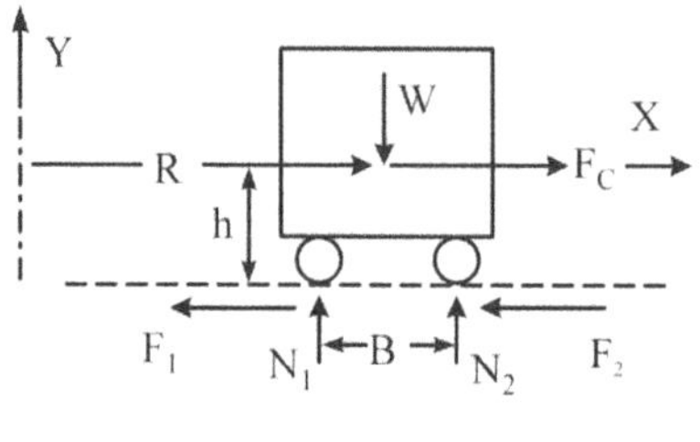

or safe velocity of vehicle without skidding on a curved road 'V' is more with larger radius of curvature of road and larger coefficient of friction

(b) ***Overturning*** takes place when the centrifugal force acting through the mass center of the automobile creates a moment more than the sum of the moments of the weight of the automobile and normal reaction at the inner wheels, about its outer wheels. The unbalanced moment tends to topple the automobile about the outer wheel. Taking moments of all the forces about the outer wheels, limiting velocity of an automobile to prevent overturning is given by

$$\sum M = N_1 \times B + (W/g) \times (V^2/R) \times h - W \times B/2 = 0$$

$$\Rightarrow \quad N_1 = (W/2) - (W/g) \times (V^2/R) \times (h/B)$$

and $\sum F_Y = N_1 + N_2 - W = 0 \quad \Rightarrow N_2 = (W/2) + (W/g) \times (V^2/R) \times (h/B)$

Thus, as velocity increases, N_1 reduces and N_2 increases. For a particular velocity, $N_1 = 0$ and the automobile tends to rotate in clockwise direction about the outer wheels. This velocity is called the critical velocity and is given by

$$N_1 = (W/2) - (W/g) \times (V^2/R) \times (h/B) = 0$$

$$\Rightarrow \quad V = \sqrt{g \times R \times B/(2h)}$$

or safe velocity of vehicle without overturning on a curved road 'V' is more with larger radius of curvature of road, larger wheel base and smaller height of center of mass of loaded vehicle from the road

12.2.2 BANKING (SLOPE) OF ROAD

It refers to the motion of an automobile on a road inclined to the horizontal. To minimize the risk of skidding as well as overturning while an automobile is negotiating a curve at a high speed, roads may be constructed with a slope towards the inner radius, at an angle 'θ' with the horizontal.

This is called ***banking***. Critical velocity of an automobile on such a road is obtained by resolving forces along and perpendicular to the road and considering equilibrium of the automobile

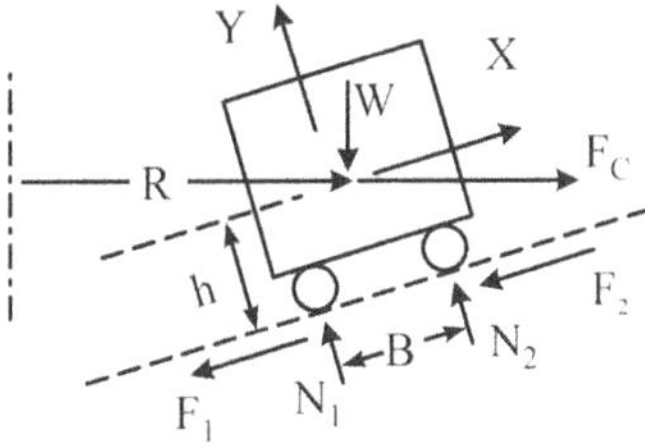

(a) *Skidding*:

$$\sum F_X = (W/g) \times (V^2/R) \times \cos\theta - W \sin\theta - \mu \times 2 \times (N_1 + N_2) = 0$$

$$\sum F_Y = (W/g) \times (V^2/R) \times \sin\theta + W \cos\theta - 2 \times (N_1 + N_2) = 0$$

Eliminating $(N_1 + N_2)$ from the above two equations, we get

$$V = \sqrt{g \times R\left[(\mu \cos\theta + \sin\theta)/(\cos\theta - \mu \sin\theta)\right]}$$

$$= \sqrt{g \times R\left[(\mu + \tan\theta)/(1 - \mu \tan\theta)\right]}$$

$$= \sqrt{g \times R \tan(\theta + \phi)} \qquad \text{by substituting} \quad \mu = \tan\varphi$$

Critical velocity 'V' can thus be increased with larger banking radius 'R' of the road and larger coefficient of friction 'μ'

(b) *Overturning*: Critical velocity is obtained by taking moments about the outer wheels and equating the reaction at inner wheels to zero. Thus,

$$\sum M = R_1 \times B - (W/g) \times (V^2/R) \times (\sin\theta) \times (B/2)$$
$$+ (W/g) \times (V^2/R) \times (\cos\theta) \times h$$
$$- W \times (\sin\theta) \times h - W \times (\cos\theta) \times B/2 = 0$$

Thus, $V = \sqrt{g \times R \times \left[(2g \tan\theta + B)/(2h - B \tan\theta)\right]}$

Example 14

A car weighing 15kN goes round a flat curve of 50m radius. The distance between inner and outer wheels is 1.5m and the center of mass is 0.75m above the road level. Assuming coefficient of kinetic friction as 0.4,

(a) What is the limiting speed of the car on this curve ?

(b) Determine the normal reactions developed at the inner and outer wheels, if the car negotiates the curve with a speed of 40kmph.

Solution :

(a) Limiting speed, considering skidding, is

$$V = \sqrt{\mu \times g \times R} = \sqrt{0.4 \times 9.81 \times 50} = 14.007 \text{ m/sec}$$

$$= 14.007 \times 3.6 \text{ or } 50.42\text{kmph}$$

Limiting speed, considering overturning, is

$$V = \sqrt{g \times R \times B/(2h)} = \sqrt{9.81 \times 50 \times 1.5/(2 \times 0.75)}$$

$$= 22.147 \text{ m/sec or } 79.73\text{kmph}$$

Therefore, limiting speed of the automobile = min (50.42, 79.73) = 50.42 kmph

(b) For $V = 40kmph = 40/3.6$ or 11.11 m/sec,

$R_1 = (W/2) - (W/g) \times (V^2/R) \times (h/B)$

$= (15/2) - (15/9.81) \times (11.11^2/50) \times (0.75/1.5) = 5.612$ kN

and $R_2 = (W/2) + (W/g)\times(V^2/R)\times(h/B)$

$= (15/2) + (15/9.81) \times (11.11^2/50) \times (0.75/1.5) = 9.388$ kN

Example 15

An automobile weighing 25kN moves with a uniform velocity of 50kmph on a hill road, as shown. What is the vertical reaction experienced at points A, B and C.

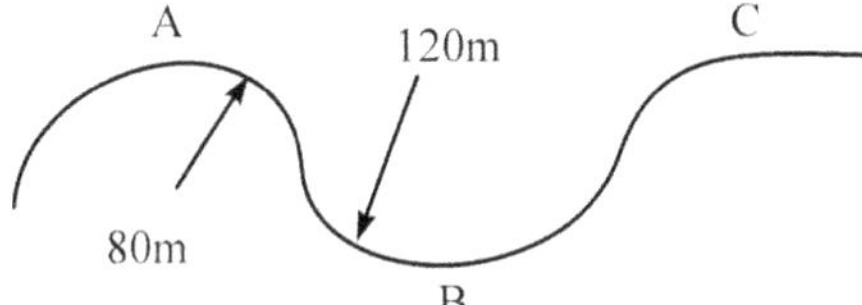

Solution:

Reaction experienced on the road at the three points depends on the net downward load (vector sum of self weight and the centrifugal force) acting on the road. Centrifugal force is upwards at A, downwards at B and zero on the level road at C.

Thus, $R_A = W - (W/g) \times (V^2/R) = 25 - (25/9.81) \times (50/3.6)^2/80$

$= 18.855$ kN

$R_B = W + (W/g) \times (V^2/R) = 25 + (25/9.81) \times (50/3.6)^2/120$

$= 29.097$ kN

and $R_C = W = 25$ kN

12.3 FIXED AXIS ROTATION

Just as any resultant force 'F' on a body produces linear motion with acceleration 'a' in the direction of the resultant force, given by $F = m \times a$, a resultant moment on a body produces rotary motion with angular acceleration 'α' given by $M = I \times \alpha$, where I is the area moment of inertia about the mass center.

Thus, if the axis of rotation passes through its mass center, $M = I \times \alpha$

However, if the rotation takes place through a point A, away from the mass center, then, $M = I_A \times \alpha = I \times \alpha + (W/g) \times (r \times \alpha) \times r$

where, r = distance of point A from mass center

Example 16

A bar AB of length 'L' and weight 'W' is rotating about a point C, at a distance 'b' from its mass center. Calculate X and Y components of reaction at C.

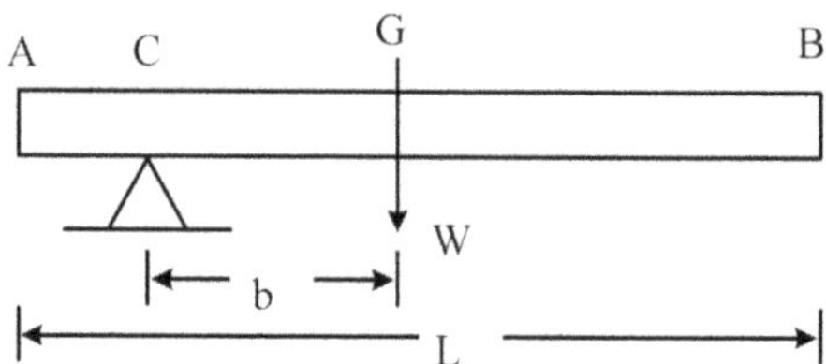

Solution :

$$M_A = I_A \times \alpha \text{ or } W \times b = (W/g) \times (L^2/12 + b^2)\times\alpha$$

Therefore, $\alpha = 12 \times b \times g / [L^2 + 12 \times b^2]$

Also, $F = m \times a$

Therefore, $(R_A)_X = (W/g) \times r \times \omega^2 = 0$

and $(R_A)_Y = W - (W/g) \times b \times \alpha$

$$= W \times [1 - 12 \times b^2 / (L^2 + 12 \times b^2)]$$

$$= W \times L^2 / [L^2 + 12 \times b^2]$$

Example 17

A bar AB of length 3m weighing 200N is rotating in the vertical plane and at a particular instant has an angular velocity of 5 rad/sec. Find its angular acceleration and X and Y components of reaction at A.

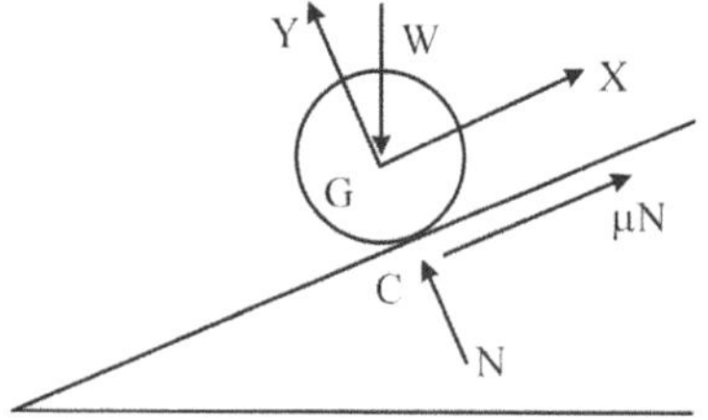

Solution:

$$M_A = I_A \times \alpha \quad \text{or} \quad 200 \times (3/2) = (200/g) \times (3^2/3) \times \alpha$$

Therefore, $\alpha = 4.905$ rad/sec^2

Also, $F = m \times a$

Therefore, $(R_A)_X = (W/g) \times r \times \omega^2 = 200 \times (3/2) \times 5^2 = 764.5$ N

and $(R_A)_Y = W - (W/g) \times r \times \alpha = 200 - (200/g) \times (3/2) \times 4.905$

$= 50$ N

12.4 ROLLING BODIES

In a body rolling along an inclined plane without slipping, if C is the point of contact between the cylinder and the inclined plane, then the cylinder is rotating about its instantaneous center of rotation C. Force and moments are given by the following equations -

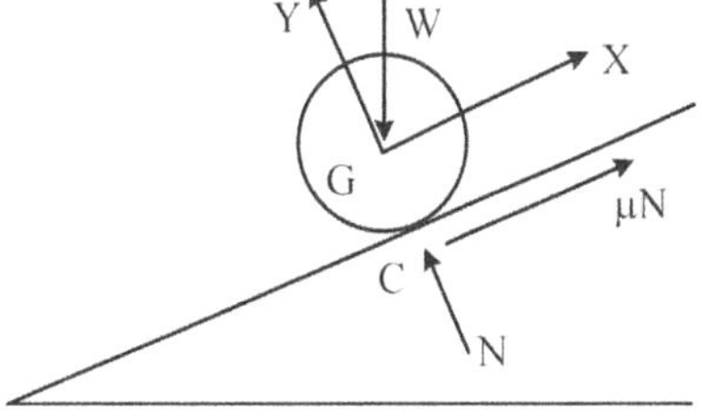

$\sum M_C = I_C \times \alpha = [I_G + (W/g) \times r^2] \times \alpha$;

$\sum F_X = (W/g) \times a$

and $\sum F_Y = 0$

Example 18

A solid cylinder of weight 'W' and radius 'r' rolls without slipping down a plane inclined at 'θ' with the horizontal. Determine minimum coefficient of friction μ to prevent slipping. Also calculate acceleration of mass center.

Solution :

$N = W \cos\theta$ and $F = \mu N = \mu W \cos\theta$

$\sum M_C = [I + (W/g) \times r^2] \times \alpha$

or $(W \sin\theta) \times r = [(W/g) \times r^2/2 + (W/g) \times r^2] \times \alpha$

Therefore, $\alpha = (2/3) \times (g/r) \times \sin\theta$

Linear acceleration of mass center, $a = r \times \alpha = (2/3) \times g \times \sin\theta$

About mass center, $M_G = I \times \alpha$

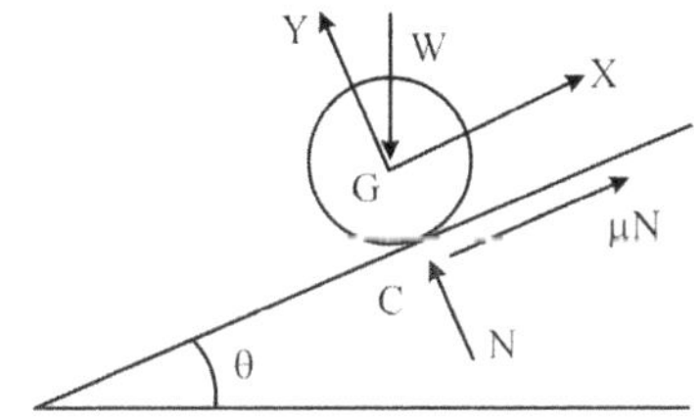

or $F \times r = [\ (W/g) \times r^2/2\ ] \times [\ (2/3) \times (g/r) \times \sin\theta\ ]$

⇒ $F = W \times \sin\theta / 3$

Thus, $\mu = F/N = [W\times\sin\theta / 3\ ] / [W\cos\theta\ ] = \tan\theta / 3$

Example 19

A solid sphere of weight 'W' and radius 'r' rolls without slipping down a plane inclined at 'θ' with the horizontal. Determine minimum coefficient of friction μ to prevent slipping. Also calculate acceleration of mass center.

Solution :

The solution for a cylinder as well as a sphere is the same, except for the change in its moment of inertia 'I' value.

Thus, $(W\sin\theta)\times r = [\ (W/g)\times 2r2/5 + (W/g)\times r2\ ] \times \alpha$

Therefore, $\alpha = (5/7)\times(g/r)\times\sin\theta$

and $a = r \times \alpha = (5/7) \times g \times \sin\theta$

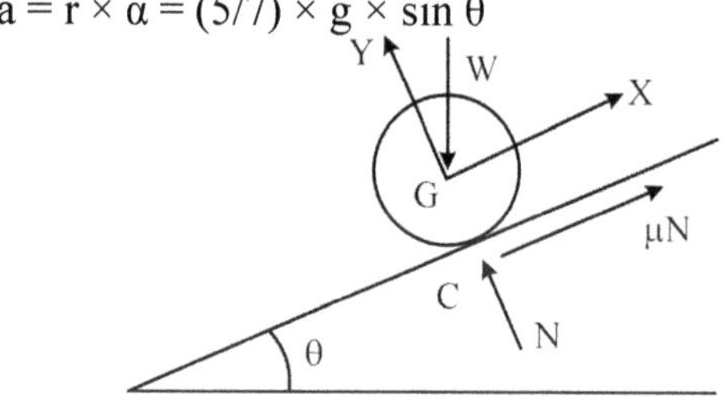

Also $F \times r = [\ (2/5) \times (W/g) \times r^2\ ] \times [\ (5/7) \times (g/r) \times \sin\theta\ ]$ and so,

$\mu = F/N = (2/7)\times\tan\theta$

Example 20

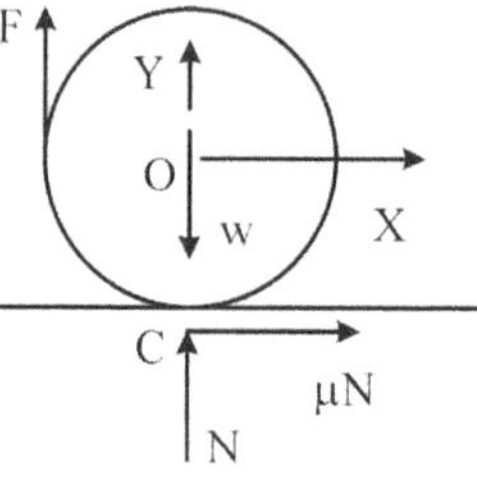

A solid cylinder of weight 3000 N and diameter 1.2 m is acted upon by an upward force of 500 N, applied by a cord wrapped around it. Determine minimum coefficient of friction μ to prevent slipping.

Solution:

$\sum M_C = I_C \times \alpha$ or $F \times r = [(W/g) \times (r^2/2 + r^2)\ ] \times \alpha$

⇒ $\alpha = 500 \times 0.6 / [(3000/g) \times (3/2) \times 0.6^2] = g / 5.4$

$\sum M_O = I_O \times \alpha$ or $F \times r - \mu N \times$

$r = [(W/g) \times (r^2/2)] \times \alpha$

$\Rightarrow \quad \mu N = 333.33$ N

But, $\sum F_Y = 0$ or $W = F + N \Rightarrow N = 3000 - 500 = 2500$N

Therefore, min. coefficient of friction, $\mu = 333.33/2500 = 0.1333$

12.5 MOMENTUM OF BODIES IN COLLISION

The product of mass and its velocity is called momentum and has units of kg×m/sec. This is related to the applied force, according to Newton's second law of motion which states

$F = d(mv)/dt = m \times (dv/dt) = m \times a$, assuming that mass of the body remains constant

Impulse is change of momentum produced in an infinitely short interval of time.

$$\text{Impulsive force, } F_I = m \times (V - U) / t$$

Thus, impulsive force has much larger effect than gradually applied force, since $t << 1.0$ sec

12.5.1 LAW OF CONSERVATION OF MOMENTUM

Total momentum of two ideally elastic bodies of masses m_1 and m_2 moving along a straight line is unaltered, due to impact between them, if the resultant force is zero. i.e., $m_1 \times u_1 + m_2 \times u_2 = m_1 \times v_1 + m_2 \times v_2$

If the two bodies move together after collision, then

$$m_1 \times u_1 + m_2 \times u_2 = (m_1 + m_2) \times v_2$$

If a moving body hits a stationary body, then

$$m_1 \times u_1 + m_2 \times 0 = m_1 \times v_1 + m_2 \times v_2$$

12.5.2 COEFFICIENT OF RESTITUTION (E)

It is defined as the ratio of change in final velocities of two colliding bodies to the change in initial velocities or ratio of final relative velocity to initial relative velocity with a negative sign and is a function of the material of colliding bodies.

i.e., $e = -(v_2 - v_1) / (u_2 - u_1)$

$e = 1$ for ideally elastic bodies

and $e = 0$ for perfectly plastic bodies

Example 21:

A hammer weighing 2450N falls down freely on a pile, from a height of 5m. If the hammer comes to rest in 1/100th of a second, find the impulsive force.

Solution :

Let velocity of hammer just before hitting the pile be 'V'.

$$V^2 - U^2 = 2 \times g \times S \quad \text{or} \quad V = \sqrt{(2 \times g \times S)} = \sqrt{(2 \times 9.81 \times 5)} = 9.9 \text{ m/sec}$$

and $F = m \times (V - U) / t = (2450/g) \times 9.9 / (1/100) = 247500 \text{ N}$

Example 22

Three balls A, B and C of masses 2kg, 4kg and 8kg respectively move along a straight line and in the same direction with velocities of 4m/sec, 1m/sec and ¾ m/sec. If A collides with B and subsequently B collides with C, show that balls A and B are brought to rest after collisions. Take coefficient of restitution as 1.0

Solution :

A → U_A B → U_B C → U_C

Let us first consider collision of balls A and B,

Conservation of momentum gives

$$M_A \times V_A + M_B \times V_B = M_A \times U_A + M_B \times U_B$$

or $2 \times V_A + 4 \times V_B = 2 \times 4 + 4 \times 1 = 12$ .....(1)

Also, $V_A - V_B = -e \times (U_A - U_B) = -(4-1) = -3$ .....(2)

Solving eq. (1) and (2), $V_A = 0$ and $V_B = 3\text{m/sec}$

Let us now consider collision of balls B and C.

Conservation of momentum gives

$$M_B \times V_B + M_C \times V_C = M_B \times U_B + M_C \times U_C$$

or $4 \times V_B + 8 \times V_C = 4 \times 3 + 8 \times (3/4) = 18$ .....(3)

Also, $V_B - V_C = -e \times (U_B - U_C) = -[3 - (3/4)] = -9/4$ .....(4)

Solving eq. (3) and (4), $V_B = 0$ and $V_C = 9/4 \text{ m/sec}$

Example 23

A 2N ball is bowled to a batsman horizontally with a velocity of 20m/sec. After hitting, it went away with a velocity of 48m/sec at 30^0 to the horizontal. If impact lasts for 0.02 sec, find the average force exerted by the batsman.

Solution :

Resolving the applied force along X and Y directions, the impulsive force components are

$$F_X \times t = (W/g) \times [\, V \times \cos\theta - (-U)] \Rightarrow F_X = 627.62 \text{ N}$$

and $$F_Y \times t = (W/g) \times [\, V \times \sin\theta - (0)] \Rightarrow F_Y = 244.65 \text{ N}$$

Therefore, actual force applied by the batsman,

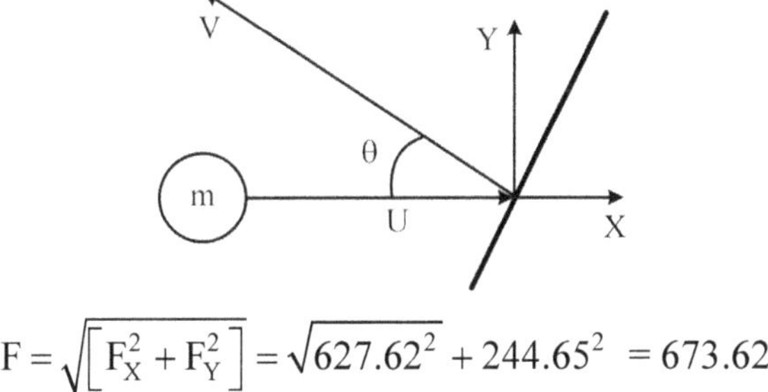

$$F = \sqrt{\left[F_X^2 + F_Y^2\right]} = \sqrt{627.62^2 + 244.65^2} = 673.62 \text{ N}$$

Inclination of force F with X-axis, $\varphi = \tan^{-1}(F_Y/F_X) = 21.3^0$

Example 24

A 100N ball is released from rest from its initial position, at 60^0 to the vertical, to strike a 75N ball freely suspended by a string of length 3m. After impact, the 75N ball is raised by an angle of 45^0. Determine coefficient of restitution.

Solution :

Velocity of ball A just before hitting B,

$$U_A = \sqrt{(2 \times g \times h_1)} = \sqrt{(2 \times 9.81 \times (3 - 3\cos 60))}$$

$$= 5.425 \text{m/sec}$$

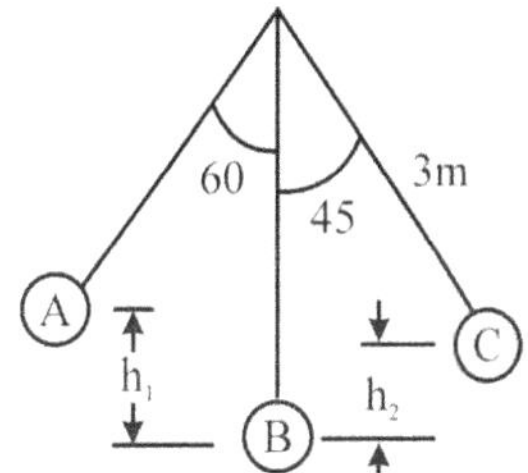

Velocity of ball B after collision by ball A, so that velocity is zero at its final position C,

$$V_B = \sqrt{(2 \times g \times h_2)} = \sqrt{(2 \times 9.81 \times (3 - 3\cos 45))}$$

$$= 4.151 \text{m/sec}$$

Coefficient of restitution,

$$e = -(V_B - V_A) / (U_B - U_A) = 4.151/5.425 = 0.765$$

Example 25

A ball of 10kg mass is dropped from a height of 5m, on a horizontal floor. Find the height upto which the ball will rebounce, if coefficient of restitution is 0.5

Solution : $h_2 = U_2^2/(2 \times g) = (-e \times V_1)^2 / (2 \times g)$

$$= \left[e \times \sqrt{(2 \times g \times h_1)}\right]^2 / (2 \times g)$$

$$= e^2 \times h_1 = 0.5^2 \times 5 = 1.25 \text{ m}$$

SUMMARY

- Kinetics deals with forces that need to be applied to produce a prescribed motion such as translation, rotation about a fixed axis, rotation of a rigid body or a combination of bodies on a surface
- It is based on D'ALEMBERT principle which states that the equations of static equilibrium are applicable for a body in motion, provided (an imaginary) inertia force of the body is included in the system of forces, in addition to the forces acting on the body (external applied forces and internal/inherent forces such as self weight, normal reaction, frictional resistance)
- Acceleration, $a = d^2s/dt^2 = F/m = dv/dt$; Velocity, $v = ds/dt = \int a \, dt$ and distance travelled, $s = \int v \, dt$
- When a body moves around a curve of radius 'R', a normal or centripetal acceleration of magnitude V^2/R acts radially inwards. The corresponding force ($m \times a_N$) is balanced by the inertia force (also called centrifugal force) acting radially outwards given by $F_C = (W/g) \times V^2/R$
- Max velocity to avoid skidding on a level road, $v = \sqrt{(\mu \times g \times R)}$
- Max velocity to avoid overturning on a level road,

  $$v = \sqrt{[(g \times R \times B)/(2h)]}$$
- Max velocity to avoid skidding on a road with banking (slope of θ with horizontal), $v = \sqrt{[g \times R\text{ran}(\theta + \phi)]}$ where $\mu = \tan \varphi$

- Max velocity to avoid overturning on a road with banking (slope of θ with horizontal), with height of CG from ground as h and wheel base or distance between wheels as B

$$v = \sqrt{g \times R \times \left[(2h \tan\theta + B) / (2h - B \tan\theta) \right]}$$

- A resultant moment on a body produces rotary motion with angular acceleration 'α' given by

  $M = I \times \alpha$, where, I is the area moment of inertia about the mass center

- In a body rolling along an inclined plane without slipping, if C is the point of contact between the cylinder and the inclined plane, then the cylinder is rotating about its instantaneous center of rotation C. Force and moments are given by the equations

  $\sum M_C = I_C \times \alpha = [I_G + (W/g) \times r^2] \times \alpha$; $\sum F_X = (W/g) \times a$ and $\sum F_Y = 0$

- **Impulse** is change of momentum produced in an infinitely short interval of time.

  Impulsive force, $F_I = m \times (V - U) / t$

  Impulsive force has much larger effect than gradually applied force, since $t \ll 1.0$ sec

- **Law of conservation of momentum** states that total momentum of two ideally elastic bodies of masses m_1 and m_2 moving along a straight line is unaltered, due to impact between them, if the resultant force is zero.

  i.e., $m_1 \times u_1 + m_2 \times u_2 = m_1 \times v_1 + m_2 \times v_2$

- **Coefficient of restitution (e)** is defined as the ratio of change in final velocities of two colliding bodies to the change in initial velocities or ratio of final relative velocity to initial relative velocity with a negative sign and is a function of the material of colliding bodies.

  i.e., $e = -(v_2 - v_1) / (u_2 - u_1)$

  $e = 1$ for ideally elastic bodies and $e = 0$ for perfectly plastic bodies

CHAPTER **13**

VIBRATIONS

Variation or displacement of a body with time over its undeformed position is, in general, defined as ***vibration***. It may result from application or release of a load suddenly (for example, in a spring-mass system) or from application of a dynamic load varying with time (for example, earthquake).

13.1 CLASSIFICATION OF VIBRATIONS

Vibrations can be broadly classified as ***periodic*** motion (which repeats itself after a ***period***, τ) and ***aperiodic*** (with no specific pattern of repetition). ***Frequency*** of vibrations is the inverse of period ($f = 1/\tau$ cycles/sec or Hertz, abbreviated as Hz). Vibrations can also be classified based on the direction of motion – ***longitudinal*** (motion of the component along its axis, such as a spring under axial load), ***transverse*** (motion of the component normal to its axis, such as a beam bending due to transverse load or moment) and ***torsional*** (motion of the component is rotation about its axis, such as a shaft connected to an oscillating pendulum). Each of these categories of vibration can be ***free vibrations*** (initiated by a load in the appropriate direction) or ***forced vibrations*** (caused by time dependent load, acting continuously on the component). These vibrations may be ***damped*** (amplitude of vibration reducing with every cycle till it is zero) or ***undamped***.

In this chapter, we limit ourselves to periodic, free, undamped vibrations. One of the most common forms of periodic vibrations is the ***simple harmonic motion*** (SHM), whose acceleration is proportional to the displacement and is in the opposite direction.

13.2 SIMPLE HARMONIC MOTION

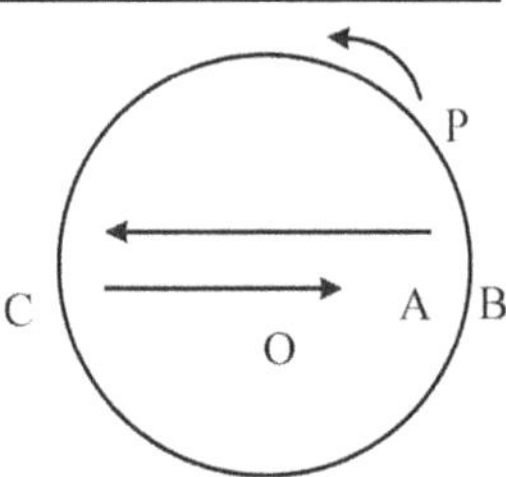

Simple harmonic motion can be visualized as the motion of projection (A) on a diameter (BC), when a particle (P) is rotating with uniform velocity on the periphery of an imaginary circle, about its center (O). As P moves from B to C on the circle, its projection A moves along the diameter from B to C. In the same way, while P moves from C to B on the circle, its projection A moves along the diameter from C to B.

Let 'ω' be the uniform angular velocity of the particle. Then, the angle made by the particle at time 't' with the diameter, representing position of the particle at time t = 0, is given by $\theta = \omega \times t$.

Distance of this projection 'A' from the center of the circle, $OA = x = R \times \cos\theta = R \times \cos\omega t$

Velocity of A along the diameter,

$$V_x = dx/dt = -\omega \times R \times \sin\omega t$$

Also, $$V_x^2 = \omega^2 \times R^2 \times (1 - \cos^2\theta) \quad \text{or} \quad V_x = \omega \times \sqrt{(R^2 - x^2)}$$

Acceleration of A along the diameter,

$$a = dV_x/dt = -\omega^2 \times R \times \cos\omega t = -\omega^2 \times x \quad \text{or} \quad a \propto x$$

Negative sign indicates that acceleration of A increases as the projection of the particle moves from B to O (decreasing the distance x) and decreases as the projection of the particle moves from O to B (increasing the distance x). Radius of the circle is the ***amplitude*** (maximum distance from the mean position O) of oscillation of A ***One rotation*** of P corresponds to ***one complete oscillation*** of A from B to C and from C to B. Therefore,

Frequency of oscillation of A,

$$f = \omega = \sqrt{(a/x)} \text{ rad/sec or } \omega/2\pi \text{ cycles/sec (or Hz)}$$

and time for one oscillation or period, $\tau = 1/f = 2\pi/\omega$ sec

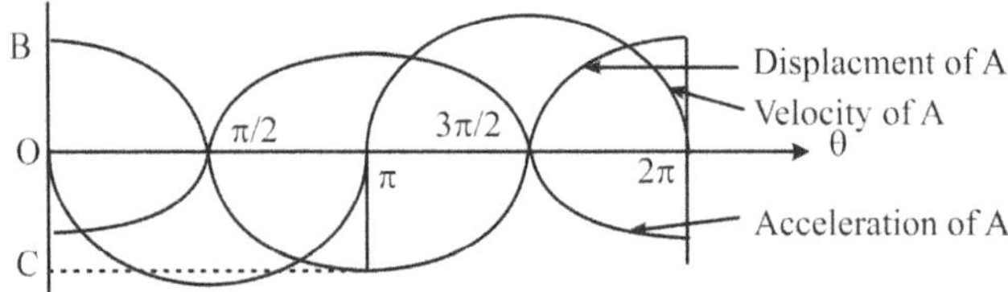

FIGURE 13.1 Motion of A for one rotation of P

Variation of displacement, velocity and acceleration of A for one rotation of P are plotted in Fig 13.1. As can be seen,

Corresponding to $\theta = 0$ or 2π,

$x = R \times \cos\theta = R$ or OB, max displacement (on +ve X-axis)

$v = -\omega \times R \times \sin\theta = 0$

$a = -\omega^2 \times R \times \cos\theta = -\omega^2 \times R$, min acceleration

Corresponding to $\theta = \pi/2$,

$x = R \times \cos\theta = 0$

$v = -\omega \times R \times \sin\theta = -\omega \times R$, min velocity

$a = -\omega^2 \times R \times \cos\theta = 0$

Corresponding to $\theta = \pi$,

$x = R \times \cos\theta = -R$ or OC, min displacement (max on -ve X-axis)

$v = -\omega \times R \times \sin\theta = 0$

$a = -\omega^2 \times R \times \cos\theta = \omega^2 \times R$, max acceleration

Corresponding to $\theta = 3\pi/2$,

$x = R \times \cos\theta = 0$

$v = -\omega \times R \times \sin\theta = \omega \times R$, max velocity

$a = -\omega^2 \times R \times \cos\theta = 0$

Example 1

A particle moving with simple harmonic motion is observed to have a velocity of 40cm/sec when it is 15cm from the center of its path and 30cm/sec when it is at 20cm. Determine the max. velocity, max. acceleration and frequency of oscillation.

Solution :

$$V_x = \omega \times \sqrt{(R^2 - x^2)}$$

or $\omega^2 = V_x^2 / (R^2 - x^2) = 40^2 / (R^2 - 15^2) = 30^2 / (R^2 - 20^2)$

$\Rightarrow$ $30^2 \times (R^2 - 15^2) = 40^2 \times (R^2 - 20^2)$

or $R^2 \times (40^2 - 30^2) = 40^2 \times 20^2 - 30^2 \times 15^2 = 640000 - 202500$

$\Rightarrow$ $R = \sqrt{(437500 / 700)} = \sqrt{625} = 25\text{cm}$

and $\omega = \sqrt{\left[40^2 / \left(25^2 - 15^2\right)\right]} = 2$ rad/sec

Therefore, $V_{max} = R\,\omega = 25 \times 2 = 50$ cm/sec

$a_{max} = \omega^2 R = 2^2 \times 25 = 100$ cm/sec^2

and $f = \omega / (2\pi) = 2 / (2\pi) = 1/\pi$ cycles/sec

Example 2

A weight 'W' is attached to one end of a vertical spring of stiffness 'k'. The weight is displaced vertically by 'h' from its equilibrium position and released. Show that the motion of the weight is SHM and calculate its period.

Solution :

Let 'δ' be the deflection of the mass in static equilibrium.

Then, $W = m \times g = k \times \delta$

If the mass is pulled by an additional displacement 'h', considering equilibrium of the weight,

$$\sum F_Y = -W + k \times (\delta + h) = m \times a$$

or $a = (k/m) \times h$ since, $W = k \times \delta$

⇒ $a \propto h$ Therefore, motion of weight is SHM with

$$a = \omega^2 h$$

and $\omega = \sqrt{(k/m)} = \sqrt{\left[(mg/\delta)/m\right]} = \sqrt{(g/\delta)}$

Frequency of oscillation,

$$f = \omega/2\pi = (1/2\pi) \times \sqrt{(m/k)} = (1/2\pi) \times \sqrt{(g/h)}$$

and Period of oscillation, $\tau = 1/f = 2\pi/\omega = 2\pi\sqrt{(m/k)}$ sec

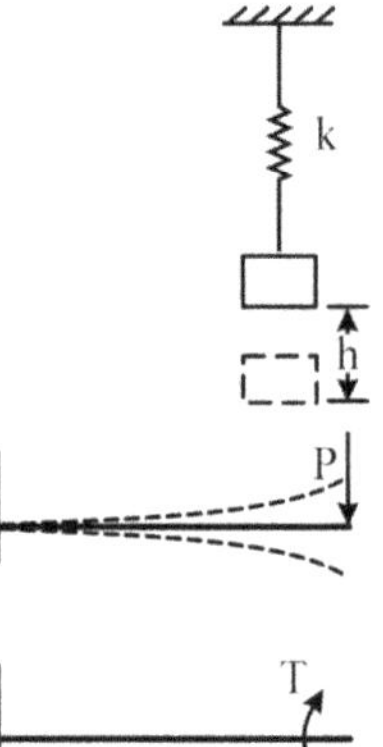

In a similar way, frequency of *transverse vibrations* of a cantilever with concentrated load 'P' at its free end,

$$f = (1/2\pi) \times \sqrt{(g/\delta)} \quad \text{where } \delta = P L^3/(3EI)$$

and frequency of *torsional vibrations* of a cantilever with torsional moment 'T' at its free end,

$$f = (1/2\pi) \times \sqrt{(q/I)}$$

where $q = T/\theta = GJ/L$; $I = M \times k^2$ and k is the radius of gyration

13.3 SIMPLE PENDULUM

A freely oscillating suspended weight (such as in a wall clock) is called a ***simple pendulum***. Time of travel from one extreme position of the weight to the other extreme position is called one ***beat.*** A simple pendulum with period of oscillation of 2 sec or 1 beat/sec is called ***seconds pendulum***. The frequency or period of oscillation of a simple pendulum is a function of only length and is independent of the weight.

For a seconds pendulum, $\tau = 2\pi\sqrt{(L/g)} = 2$ sec

$\Rightarrow$ $$L = g/\pi^2 = 99.316\text{cm} \quad \text{for } g = 981\text{cm/sec}^2$$

If frequency or number of beats/sec increases, pendulum gains time. Length of the pendulum should be increased to get the correct time.

If T is the time taken to perform 'n' beats, $T = n \times \text{time/beat} = n\,\pi\sqrt{(L/g)}$

No. of beats, $n = T/\,[\pi\sqrt{(L/g)}\,]$

Taking logarithms on both sides,

$$\text{Log } n = \text{Log } T - \text{Log } \pi + (1/2) \times [\,\text{Log } g - \text{Log } L\,]$$

Differentiating both sides,

$$dn/n = 0 - 0 + (1/2) \times [dg/g - dL/L]$$

since, T and π are constants

In a single location (constant g), $dn/n = -\,dL/2L$

Example 3

Determine the period of oscillation of a simple pendulum, consisting of a weight 'W' suspended from a mass-less string of length 'L', oscillating about the other (fixed) end of the string.

Solution :

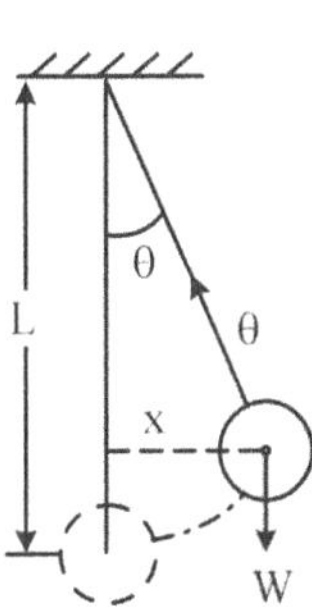

Considering equilibrium of the weight, and denoting tension in the string by 'T',

Tangential force along the path of oscillation

$$= T \times \sin\theta = W \times \tan\theta \approx W \times \theta = W\times(x/L)$$

for small angles of θ

$$= (W/g) \times a \qquad \Rightarrow \qquad a = (g/L) \times x = \omega^2 \times x$$

Therefore, motion of weight is SHM with $\omega = \sqrt{(g/L)}$

Frequency of oscillation, $f = \omega/2\pi$

and Period of oscillation, $\tau = 1/f = 2\pi/\omega = 2\pi\sqrt{(L/g)}$ sec

Example 4

A pendulum is having a length of 99.39cm. If it loses 5sec/day, by how much length of the pendulum should be shortened to ensure correct time.

Solution:

Since acceleration due to gravity 'g' remains constant at a particular place,

$$dn/n = -dL/(2L) \quad \text{or} \quad (-5)/(24 \times 60 \times 60) = -dL/(2 \times 99.39)$$

$$\Rightarrow \quad dL = 0.0115\text{cm}$$

Example 5

A body moving with SHM has an amplitude of 1m and the period of oscillation is 2sec. What will be the velocity and acceleration of the body after 0.4sec from the extreme position?

Solution :

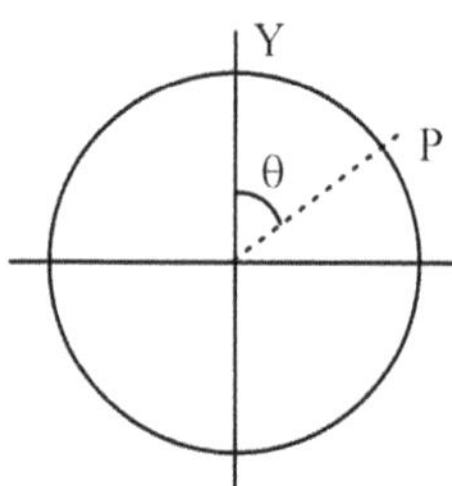

Considering Y as the extreme position, time taken by the body to travel to the position P, $t = 0.4$ sec

$$\omega = 2\pi/\tau = 2\pi/2 = \pi \text{ rad/sec}$$

Corresponding angle,

$$\theta = \omega t = \pi \times 0.4 = 1.257 \text{ rad} \quad \text{or} \quad 72^0$$

Velocity at P,

$$V_Y = \omega \times R \times \sin\theta = \pi \times 1 \times \sin 72 = 2.989 \text{ m/sec}$$

Acceleration, $a = \omega^2 \times R \times \cos\theta = \pi^2 \times 1 \times \cos 72 = 3.052 \text{ m/sec}^2$

Example 6

An elastic string of length 2L, tightly stretched between two rigid supports carries a small ball of weight W at its mid-point. Show that for small displacements, the ball will have a SHM. Compute the period of oscillation.

Solution :

$$\sum F_x = -2T\sin\theta + (W/g) \times a = 0$$

$$0 = -2T \times (x/L) + (W/g) \times a$$

since $\sin\theta \approx \tan\theta = X/L$ when θ is very small

$$\Rightarrow \quad a = [2T \times g / (W \times L)] \times x = \omega^2 \times x$$

Since $a \;\alpha\; x$, motion of the weight is SHM and $\omega = \sqrt{[2T \times g/(W \times L)]}$

Period, $\tau = 2\pi/\omega = 2\pi\sqrt{[W \times L/(2T \times g)]}$ sec

13.4 COMPOUND PENDULUM

An oscillating body of distributed mass suspended from any one point of the body (other than its center of gravity, G) is called a compound pendulum.

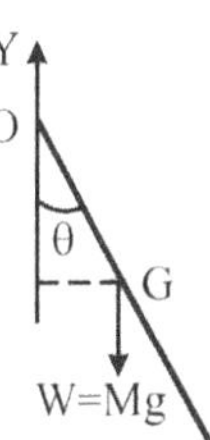

$$\sum(M_z)_O = I_O \times \alpha - W \times R \times \sin\theta = 0$$

where, $OG = L/2 = R$

or $\quad W \times R \times \sin\theta = (W/g) \times k_o^2 \times \alpha$

where, k is the radius of gyration

$$W \times R \times \theta = (W/g) \times k_o^2 \times \alpha, \text{ for small angles of } \theta$$

Therefore, $\alpha = (g \times R/k_o^2) \times \theta = -\omega^2 \times \theta \quad \Rightarrow \omega = \sqrt{g \times R/k_o^2}$

Period of oscillation, $\tau = 2\pi/\omega = 2\pi\sqrt{k_o^2/(g \times R)}$

Frequency of oscillation, $f = \omega/2\pi = (1/2\pi)\sqrt{g \times R/k_o^2}$

$$= (1/2\pi)\sqrt{(g/L_e)}$$

where, $L_e = k_o^2/R = (k^2 + R^2)/R = k^2/R + R$

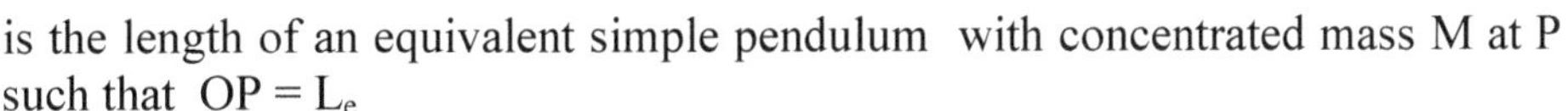

is the length of an equivalent simple pendulum with concentrated mass M at P such that $OP = L_e$

and $\quad I_O = M \times k_o^2 = I_G + M \times R^2 = M \times k^2 + M \times R^2$

Example 7

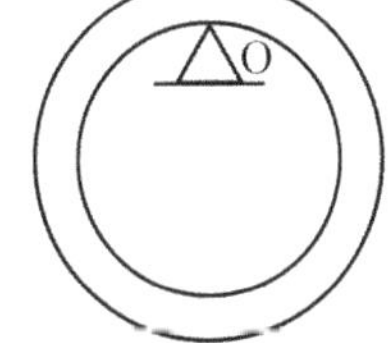

A circular ring of outer radius 'R', inner radius 'r' and weighing 320N is suspended from a knife edge. Determine the period of oscillation, if R = 0.9m and r = 0.6m.

Assume $\rho \times t = 1$

Solution:

This is considered as a circular plate of radius R from which a concentric circular area of radius r is removed. Moment of inertia about O, perpendicular to the plane of ring,

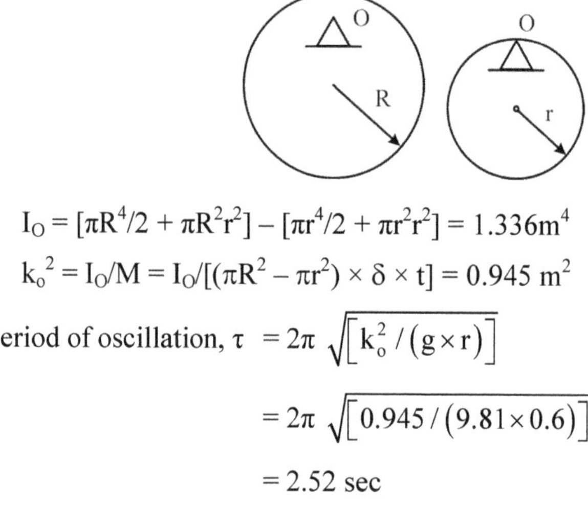

$$I_O = [\pi R^4/2 + \pi R^2 r^2] - [\pi r^4/2 + \pi r^2 r^2] = 1.336 m^4$$

$$k_o^2 = I_O/M = I_O/[(\pi R^2 - \pi r^2) \times \delta \times t] = 0.945\ m^2$$

$$\text{Period of oscillation, } \tau = 2\pi \sqrt{\left[k_o^2/(g \times r)\right]}$$

$$= 2\pi \sqrt{\left[0.945/(9.81 \times 0.6)\right]}$$

$$= 2.52 \text{ sec}$$

Example 8

Determine the period of oscillation of the combination of two identical slender bars, as shown.

Solution :

Given AB = OD = L and OG = L/2

Moment of inertia about O, perpendicular to the plane of bars,

$$I_O = [I_O]_{AB} + [I_O]_{OD} = [M \times L^2/12 + M \times L^2] + M \times L^2/3$$

$$= (17/12) \times (M \times L^2)$$

$$k_o^2 = I_O/2M = (17/24) \times L^2$$

Distance of centroid from O, $R = [M \times L + M \times (L/2)] / (M + M) = 3L/4$

$$\text{Period of oscillation, } \tau = 2\pi \sqrt{\left[k_o^2/(g \times R)\right]}$$

Example 9

A horizontal board performs a SHM, moving horizontally to and fro through a distance of 1m, making 15 complete oscillations per minute. Find the minimum value of coefficient of friction in order that a heavy body of mass 'm' placed on the board may not slide on the board.

Solution :

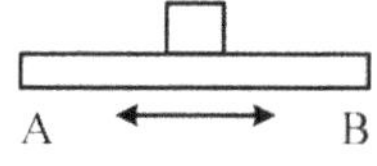

Amplitude of oscillation, $r = ½$ m

Period of oscillation, $\tau = 1/15$ min

Angular velocity, $\omega = 2\pi \times f = 2\pi / \tau = \pi/2$ rad/sec

Max. acceleration, $a = \omega^2 \times r = (\pi/2)^2 \times (1/2) = \pi^2 / 8$ m/sec^2

If the body does not slide on the board, in limiting equilibrium,

$m \times a = \mu_{min} \times mg$ or $\mu_{min} = a/g = \pi^2 / (8g) = 0.0125$

13.5 SPRINGS CONNECTED IN SERIES AND PARALLEL

A single spring-mass system is used to explain the type of motion and its attributes like frequency, amplitude etc. Many practical situations will have multiple springs (or bodies with different mass and stiffness values) in the system, connected in series or parallel or a combination of both. Calculation of relevant attributes in these cases are explained through the following example.

Example 10

A 50kg block supported by two springs connected in (a) series (b) parallel hangs from the ceiling. It can move between two smooth vertical guides. The spring constants are 4kN/m and 6kN/m. The block is pulled 40mm down from its position of equilibrium and then released. Determine period of vibrations, maximum velocity and maximum acceleration of the block.

Solution:

Amplitude of oscillation, r = 40mm or 0.04 m

(a) Springs in series –

Static extension, $\delta = \delta_1 + \delta_2 = m \times g/k_1 + m \times g/k_2$

$= 50 \times 9.81 \times (1/4000 + 1/6000) = 0.20435$m

Period of oscillation, $\tau = 2\pi \sqrt{(\delta / g)} = 0.9072$ sec

Alternatively, Equivalent spring stiffness, $k_{eq} = k_1 \times k_2 / (k_1 + k_2)$

$= 2.4$kN/m $= 2400 \times 1$ (kg/m) $\times$ (m/sec^2) $= 2400$ kg/sec^2

Period of oscillation, $\tau = 2\pi\sqrt{(m/k_{eq})}$

$= 2\pi\sqrt{(50/2400)}$ sec = 0.90726sec

Angular velocity, $\omega = 2\pi/\tau = 6.9286$ rad/sec

Max. velocity, $V_{max} = \omega \times r = 6.9286 \times 0.04 = 0.277$ m/sec

Max. acceleration, $a_{max} = \omega^2 \times r = 6.9286^2 \times 0.04 = 1.92$ m/sec^2

(b) Springs in parallel –

Load, $P = mg = P_1 + P_2 = k_1 \times \delta + k_2 \times \delta$

Equivalent spring stiffness, $k_{eq} = k_1 + k_2 = 10$kN/m

$= 10000$ N/m $= 10000$ kg/sec^2

Period of oscillation, $\tau = 2\pi\sqrt{(m/k_{eq})}$

$= 2\pi\sqrt{(50/10000)}$ sec = 0.44467sec

Angular velocity, $\omega = 2\pi \times f = 2\pi/\tau = 14.1421$ rad/sec

Max. velocity, $V_{max} = \omega \times r = 14.1421 \times 0.04$

$= 0.565685$ m/sec

Max. acceleration, $a = \omega^2 \times r = (14.1421)^2 \times (0.04) = 8$ m/sec^2

PROBLEMS FOR PRACTICE

1. A body performing SHM has a velocity of 12 m/sec when the displacement is 50 mm and 3 m/sec when the displacement is 100 mm, the displacement being measured from the mid-point. (a) Calculate the frequency and amplitude of the motion. (b) What is the acceleration when the displacement is 75 mm.

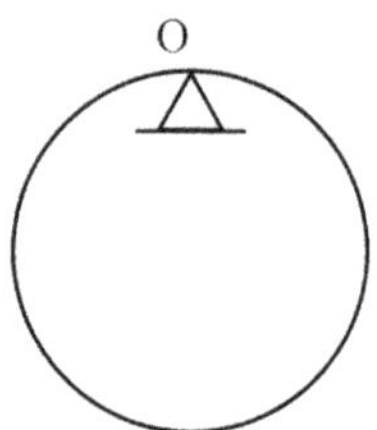

(***Ans:*** (a) 133.3rad/sec, 0.1025m ; (b) 1333 m/sec^2)

2. A thin circular ring of diameter 'd' hangs over a knife edge at 'O'. Find the period for small amplitudes of swing in its own vertical plane.

(***Ans :*** $2\pi\sqrt{(d/g)}$)

3. A body moving with a simple harmonic motion, has an amplitude of 0.7m and a period of oscillations of 3 sec. Find the velocity and acceleration of the body 0.3 sec after passing the mid position.

(***Ans:*** –0.8624 m/sec ; –2.4857 m/sec^2)

4. Determine the stiffness of a spring to which a weight of 100N is attached and set vibrating vertically. The weight makes 5 oscillations / sec.

(***Ans:*** 10069 N/m)

5. A body, moving with simple harmonic motion, has an amplitude of 0.8m and a period of oscillation of 2 sec. Find the velocity and acceleration of the body 0.2 sec after passing the mid position.

(***Ans:*** –1.47837 m/sec ; –6.3917 m/sec^2)

6. A simple harmonic motion is defined by the relation $\alpha = -36x$ where, α is the linear acceleration and x is the displacement from fixed point. Determine frequency and time period. Do they change with amplitude ?

(***Ans:*** 0.9545 cyc/sec ; 1.04762 sec)

7. A particle in SHM has an amplitude of 300mm and a period of $\pi/2$ sec. Find the velocity and acceleration of the particle when it has traveled 180mm to the right of the center of its path. What time is required for this displacement?

(***Ans:*** 0.2318 sec ; –0.96 m/sec ; –2.88m/sec^2)

8. A weight of W Newtons is suspended from a vertical spring, whose modulus is k N/m. The weight is pulled down by 'y' meters from its equilibrium position and then released. Determine its velocity when it returns to its equilibrium position.

(***Ans :*** $v = v_{max} = -y\sqrt{(k \times g / W)}$

CHAPTER 14

WORK - ENERGY

14.1 ENERGY

It is the capacity to do work and has the same units (Nm) as the work done. It is broadly classified into atomic energy, chemical energy, thermal energy, electrical energy, mechanical energy,.. ***Energy can neither be created nor destroyed***. It can only be converted from one form to another in various equipment such as – atomic to thermal in reactors, chemical to thermal in boilers and I.C. engines, thermal to mechanical in turbines, mechanical to electrical in generators, electrical to chemical in storage batteries, etc.. and may include some loss or partial conversion into other form which can not be utilized effectively.

In this subject, our discussion is limited to mechanical energy, which can be further divided into potential energy (depending on the altitude 'h' of a moving or stationary body, defined by "$m \times g \times h$") and kinetic energy (depending upon the velocity of a moving body, defined by "$m \times v^2/2$"). Neglecting loss of energy into some other form, the sum of potential energy and kinetic energy at different positions of a moving body remain the same. This is called ***law of conservation of energy***.

14.2 WORK DONE BY A MOVING BODY

If the resultant of a system of forces acting on a body is zero, the body remains stationary (static). However, if the resultant is a force, then the body moves in the direction of the force; and if the resultant is a moment, the body rotates.

The product of force applied on a body and the distance by which it moves in the direction of force is called the work done. $W = F \times S$

It has the units of 'Nm' or 'joules', similar to energy.

If the force F is a function of distance 'x' (Ref Fig 14.1),

$$W = \int F \times dx = \text{Area below the curve, on F - x plot}$$

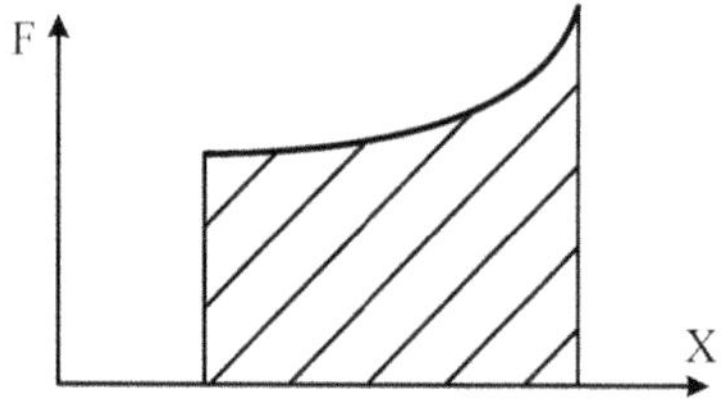

FIGURE 14.1 Work done by a force

14.3 RELATION BETWEEN ENERGY AND WORK

If a body moving with velocity 'v' on a level ground comes to rest,

Change in energy = Change in potential energy + Change in kinetic energy

or $$\delta E = \delta(PE) + \delta(KE) = 0 + m \times (v^2 - 0) / 2 = m \times v^2 / 2$$

According to law of conservation of energy, Total change in energy = 0

Therefore, δE = Loss of energy due to friction

or work done against friction

Example 1

Two railway wagons weighing 120kN and 100kN moving in the same direction at speeds of 3m/sec and 5m/sec respectively, collide and subsequently move together. Calculate their velocity after collision and loss of kinetic energy due to impact.

Solution :

From law of conservation of momentum,

$$(120{,}000/g) \times 3 + (100{,}000/g) \times 5 = (120{,}000+100{,}000) \times V / g$$

$$\Rightarrow \quad V = 3.91\text{m/sec}$$

Kinetic energy of the two wagons before collision

$$= (120{,}000/g) \times 3^2/2 + (100{,}000/g) \times 5^2/2 = 182{,}467 \text{ Nm}$$

Change in Kinetic energy $= 182{,}467 - (120{,}000 + 100{,}000) \times V^2 / (2g)$

$$= 11{,}052 \text{ Nm}$$

Percentage loss of K.E. $= (11{,}052 / 182{,}467) \times 100 = 6.057\ \%$

Example 2

A ball dropped from a height of 1.6m is observed to rebound to a height of 1.1m Determine coefficient of restitution and percentage energy lost during impact.

Solution :

Velocity just before impact, $V_1 = \sqrt{(2gh)} = \sqrt{(2g \times 1.6)} = 5.6 \text{ m / sec}$

Velocity just after bouncing, $U_2 = \sqrt{(2gh_2)} = \sqrt{(2g \times 1.1)} = 4.65 \text{ m / sec}$

But, $e = U_2/V_1 = 4.65 / 5.6 = 0.83$

Percentage loss of energy $= [(m \times g \times h_1 - m \times g \times h_2) / (m \times g \times h_1)] \times 100$

$= [(h_1 - h_2)/h_1] \times 100$

$= [\,(1.6 - 1.1) / 1.6\,] \times 100 = 31.25\ \%$

Example 3

Determine the constant force P that will give the system of bodies A, B and C of weights 250N, 1000N and 500N respectively, a velocity of 3m/sec after moving by 4.5m from rest. Assume coefficient of friction to be 0.2

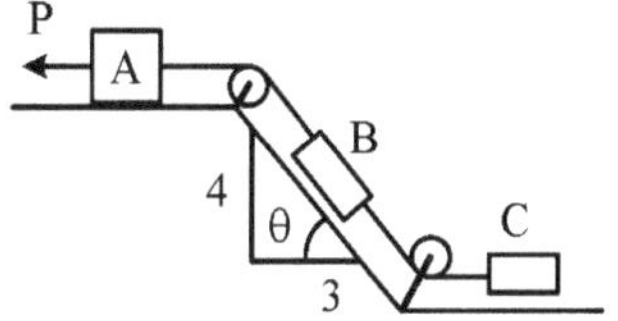

Solution :

Resolving forces of free bodies perpendicular to the direction of motion,

$N_A = W_A = 250\text{ N}\,;\quad F_A = \mu \times N_A = 50\text{ N}$

$N_B = W_B \times \cos\theta = 1000 \times (3/5) = 600\text{ N}$

$F_B = \mu \times N_B = 120\text{ N}$

and $N_C = W_C = 500\text{ N}\,;\quad F_C = \mu \times N_C = 100\text{ N}$

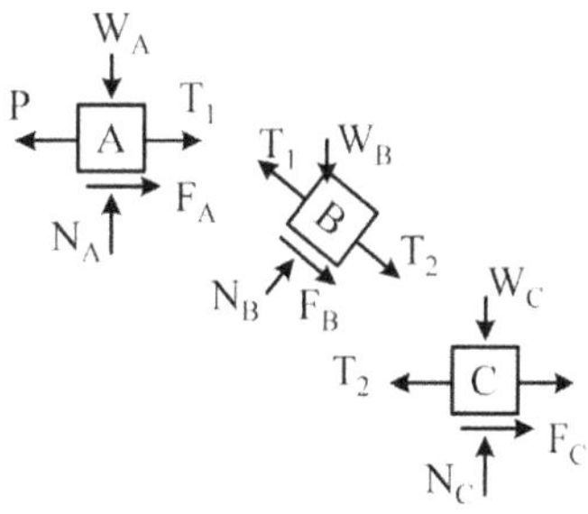

Resolving forces along the direction of motion,

$\sum F_X = P - F_A - T_1 = (P - F_A) - (F_B + W_B \times \sin\theta + T_2)$

$= (P - F_A) - (F_B + W_B \times \sin\theta) - (F_C)$

Method-1

While the three bodies move by 4.5m,

$$\sum \text{Work done} = \sum \text{Change in Kinetic energy}$$

$$(P - F_A - T_1) \times S_A + (T_1 - F_B + W_B \times \sin\theta - T_2) \times S_B + (T_2 - F_C) \times S_C$$
$$= (1/2) \times [(W_A + W_B + W_C)/g] \times (V_A^2 - U_A^2)$$

Assuming inextensible ropes, $S_A = S_B = S_C$

Therefore,

$$(P - 50 - T_1) \times 4.5 + [T_1 - 120 - 1000 \times (4/5) - T_2] \times 4.5 + (T_1 - 100) \times 4.5$$
$$= (1/2) \times [(250 + 1000 + 500) / 9.81] \times (3^2 - 0^2)$$

$\Rightarrow \quad P = 1248.5\text{ N}$

Method-2

The three connected bodies A, B and C move with the same displacement and, hence, with the same acceleration given by,

$$a = (V^2 - U^2) / 2S = (3^2 - 0^2) / (2 \times 4.5) = 1\text{ m/sec}^2$$

At C, $\quad \sum F_X = T_2 - F_C - m_C \times a = 0$

$\Rightarrow \quad T_2 = F_C + m_c \times a = 100 + (500/9.81) \times 1 = 150.97\text{N}$

At B, $\quad \sum F_X = T_1 - T_2 - F_B - W_B \sin\theta - m_B \times a = 0$ along slope

$\Rightarrow \quad T_1 = T_2 + F_B + W_B \sin\theta + m_B \times a$

$= 150.97 + 120 + 1000 \times (4/5) \times (1000/9.81) \times 1$

$= 1172.31\text{ N}$

At A, $\quad \sum F_X = P - T_1 - F_A - m_A \times a = 0$

$\Rightarrow \quad P = T_1 + F_A + m_A \times a = 1172.31 + 50 + (250/9.81) \times 1$

$= 1248.4\text{ N}$

Example 4

A 300N crate slides down the curved path from A to B. If the velocity of crate is 1.2m/sec at A and 8m/sec at B, find the work done against friction during its motion from A to B.

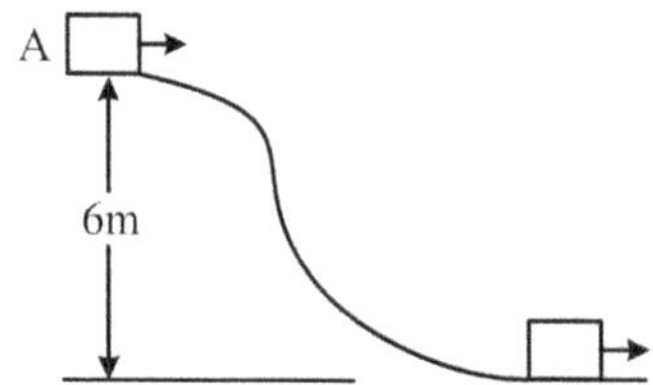

Solution :

Total energy at A,

$$E_A = (PE)_A + (KE)_A = W \times h + (W/g) \times V_A^2/2$$

$$= 1822 \text{ Nm}$$

Similarly, $E_B = 0 + (W/g) \times V_B^2/2 = 978.59$ Nm

Work done against friction = Loss of total energy

$$= E_A - E_B = 843.43 \text{ Nm}$$

Example 5

A wagon weighing 490 kN starts from rest, runs 30m down a 1% gradient and strikes a post. If the rolling resistance of the track is 5N per kN, find the velocity of the wagon when it strikes the post. If the impact is received by a spring which compresses by 1mm per 14.7kN, determine total compression of the spring.

Solution :

Let δ be the compression of the spring.

Slope θ is given by $\sin \theta = 1/100$

Frictional resistance, $f = 5 \times 490 = 2450$N

Method-1 :

Resolving forces along the inclined plane,

$$\sum F = W \times \sin \theta - f = (W/g) \times a$$

or $490{,}000 \times (1/100) - 2450 = (490{,}000/9.81) \times a$

$\Rightarrow$ $a = 0.049 \text{ m/sec}^2$

$$V^2 - U^2 = 2a \times S \Rightarrow V = 1.715 \text{ m/sec}$$

Work done by spring (or average force × compression)

= Change in KE

or $(15\,\delta / 2) \times \delta = m \times V^2/2 \Rightarrow \delta = 100\text{mm}$

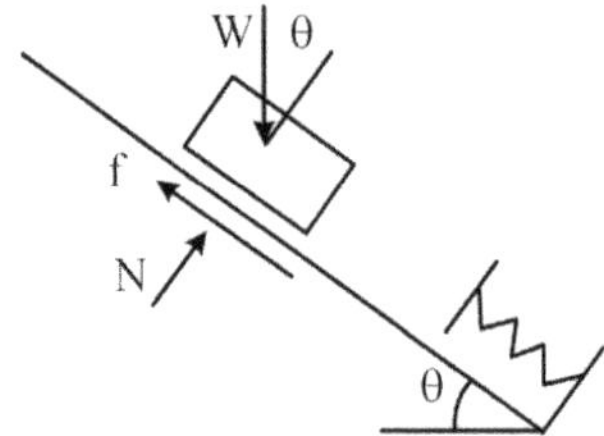

Method-2:

Total energy before rolling

= Total energy after rolling + Energy lost due to friction

$W \times (S.\sin\theta) + 0 = 0 + m \times V^2/2 + f$

or $490{,}000 \times 30 \times (1/100) + 0 = 0 + (490{,}000/9.81) \times V^2/2 + 2450$

$\Rightarrow$ $V = 1.715$ m/sec

Work done by spring (or average force x compression) = Change in KE

or $(15\,\delta / 2) \times \delta = m \times V^2/2 \Rightarrow \delta = 100\text{mm}$

Example 6

A bullet loses 1/25 of its velocity in passing through a wooden plank. Find how many such planks will bring the bullet to rest, assuming resistance of all planks to be uniform.

Solution :

Loss of kinetic energy of bullet in passing through one plank

$= m \times (U^2 - V^2)/2 = m \times [\,U^2 - \{(24/25) \times U\}^2\,] / 2$

$= (49/1250) \times m \times U^2$

If n is the number of planks, Total loss of energy = Initial kinetic energy

or $n \times (49/1250) \times m \times U^2 = m \times U^2/2$

$\Rightarrow$ $n = 625/49 = 12.755 \approx 13$

14.4 WORK AND ENERGY IN FIXED AXIS ROTATION

If the resultant of a system of forces acting on a body is a moment, it causes pure rotation of the body and angular displacement ‘θ’. Then, work done

$W = M \times \theta$ and $M = I_m \times \alpha$

If the moment is a function of angular displacement, then

$$W = \int M \times d\theta = \int (I_m \times \alpha) \times d\theta = \int I_m \times \omega \times d\omega$$
$$= I_m \times (\omega_2^2 - \omega_1^2) / 2$$

Example 7

Determine the distance moved by A in changing its velocity from 2 m/sec to 4 m/sec. Weights of A and B are 300 N and 400 N respectively, while the radii of the two pulleys connected to a common shaft are 3m and 2m. Assume $\mu = 0.2$; $\tan\theta = 3/4$ and combined weight of the two pulleys = 322 N

Solution :

Let T_A and T_B be the tensions in the rope connecting bodies A and B respectively.

Pulleys connected to a common shaft will have the same torque

i.e., $T_A \times R_A = T_B \times R_B$ or $T_A = (R_B/R_A) \times T_B = (2/3) \times T_B$

Also, for an angular displacement of α,

$$\alpha = S_A/R_A = S_B/R_B \quad \text{or} \quad S_A = (R_A/R_B) \times S_B = (3/2) \times S_B$$

Differentiating, $V_A = (3/2) \times V_B = R_A \times \omega = 2\,\omega$

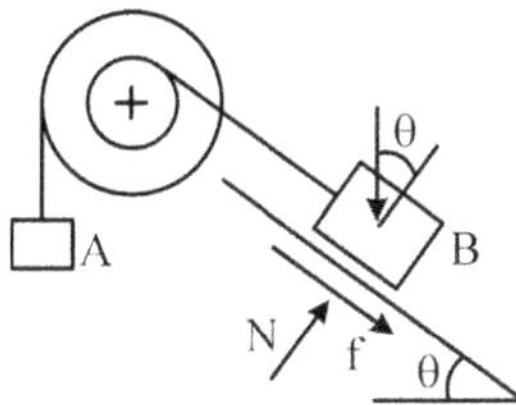

Work done by the two bodies = Change in KE

or $(W_A - T_A) \times S_A + (T_B - W_B \times \sin\theta - \mu \times W_B \times \cos\theta) \times S_B$

$$= (W_A/2g) \times (V_A^2 - U_A^2) + (W_B/2g) \times (V_B^2 - U_B^2) + I_m \times (\omega_2^2 - \omega_1^2)/2$$

Since, $T_A \times S_A = [(2/3) \times T_B] \times [(3/2) \times S_B] = T_B \times S_B \quad \Rightarrow \quad S_A = 4.53$ m

Example 8

A pulley of radius 2m, weighing 300N is mounted over an axle of diameter 0.25m. Friction force between pulley and axle is 50N. It supports a block of weight 200N. How many turns does the pulley make before it stops, if the block is dropped from rest from a height of 12m.

Solution :

The solution consists of two phases. In the first phase, the rope and the body A move down together, simultaneously rotating the pulley in the counterclockwise direction, until the block touches the ground. In the second phase, the block remains stationary while the pulley continues to rotate because of the initial momentum until the friction between the pulley and the axle retards it to zero speed.

Phase 1: Distance covered by block A, $S_A = R_B \times \theta$

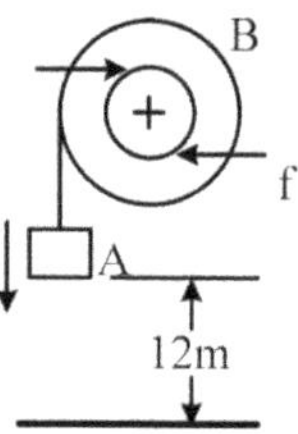

Velocity of block A, $V_A = dS_A/dt = R_B \times (d\theta/dt) = R_B \times \omega$

Angular displacement of pulley while block A moves by 12m to touch the ground,

$$\theta = S_A / R_B = 12/2 = 6 \text{ rad}$$

Work done by the block and pulley,

$$WD = W_A \times S_A - M_B \times \theta \quad \text{and} \quad M_B = f \times r_B$$

$$= 200 \times 12 - (50 \times 0.25/2) \times (6) = 2362.5 \text{ Nm}$$

$$WD = \text{Change in KE} = (200/g) \times V_A^2/2 + [(300/g) \times (2^2/2)] \times \omega^2/2$$

$$\Rightarrow \qquad \omega = 5.754 \text{ rad/sec}$$

Phase 2: Let n be the number of revolutions of pulley after block A touches the ground.

$$WD = \text{Change in KE} \quad \text{or} \quad -(F\times r) \times \theta' = (300/g) \times (2^2/2) \times (0^2 - \omega^2)/2$$

$$\Rightarrow \qquad \theta' = 162 \text{ rad}$$

Total rotation of the pulley, $(\theta + \theta')/2\pi = (6 + 162) / 2\pi = 26.73$ revolutions

14.5 WORK AND ENERGY IN PLANE MOTION

In general, a system of forces acting on a body may result in a force and a moment about an instantaneous center of rotation. Then, total work done

$$W_{TOTAL} = W_{TRANSLATION} + W_{ROTATION} = F \times S + M \times \theta$$

$$KE_{TOTAL} = KE_{TRANSLATION} + KE_{ROTATION}$$

$$= m \times (V^2 - U^2)/2 + I_m \times (\omega_2{}^2 - \omega_1{}^2)/2$$

Example 9

Find the distance traveled by A in changing its velocity from 1m/sec to 2m/sec. Assume weight of block A as 200N, weight of pulley as 100N and its radius of gyration about centroidal axis as √2. The disc rolls on horizontal surface without slipping. Radius of inner pulley of the disc is 1m while that of the outer pulley is 2m.

Solution :

The disc rolls about its instantaneous center of rotation C.

$$S_A = R_A \times \theta = \theta \;;\quad S_B = R_B \times \theta = 2 \times \theta \quad \text{Therefore,}\ S_A = S_B / 2$$

Differentiating, $V_A = V_B/2 = R_A \times \omega = \omega$

$$(I_m)_C = (I_m)_O + M \times R_B{}^2 = M \times k^2 + M \times R_B{}^2$$

$$= (100/g) \times [(\sqrt{2}\,)2 + 22] = 61.16$$

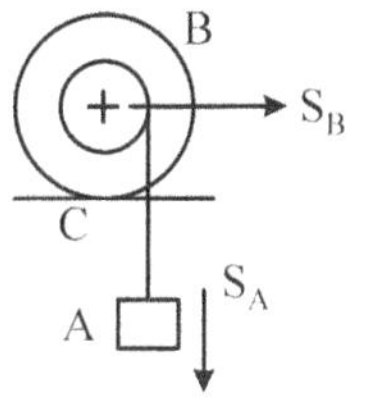

Applying work energy equation,

$$W_A \times S_A = m \times (V_A{}^2 - U_A{}^2)/2 + (I_m)_C \times (\omega_2{}^2 - \omega_1{}^2)/2$$

or $(200/g) \times S_A = (200/g) \times (2^2 - 1^2)/2 + 61.16 \times (2^2 - 1^2) \Rightarrow S_A = 0.61\text{m}$

Example 10

Slider A, weighing 1000N, moves in vertical frictionless guides. It is connected to the center of a solid cylinder B of 1.2m diameter and 1500N weight, by a uniform slender rod of 3m length and 900N weight. If the system starts from rest when B is vertically below A, compute velocity of A in the given position.

Solution :

Normal reactions at A and F do not contribute to work, since there is no displacement in each of their directions and hence can be neglected.

$$DG = \sqrt{GC^2 + DC^2} = \sqrt{1.2^2 + 0.9^2} = 1.5\text{m}$$

$$AD = \sqrt{AB^2 - DB^2} = \sqrt{3^2 - 1.8^2} = 2.4\text{m}$$

$$V_A = DB \times \omega_{AB} = 1.8\ \omega_{AB} \quad \text{and} \quad V_B = AD \times \omega_{AB} = 2.4\ \omega_{AB} = 0.6\ \omega_B$$

For the rod AB

$$(I_C)_{AB} = (W_{AB}/g) \times (L^2/12) + (W_{AB}/g) \times DG^2$$
$$= (900/g) \times (3^2/12) + (900/g) \times 1.5^2 = 2700/g \text{ kg m}^2$$

For the cylinder B

$$(I_C)_B = (W_B/g) \times (R^2/2) + (W_B/g) \times R^2$$
$$= (1500/g) \times (0.6^2/2) + (1500/g) \times 0.6^2 = 810/g \text{ kg m}^2$$

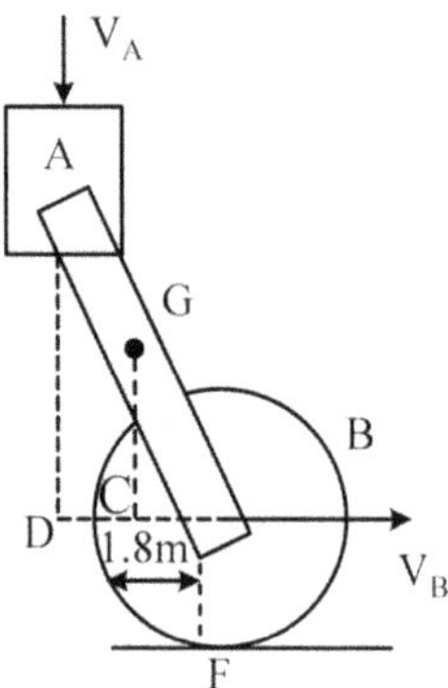

Using Work-energy relation,

$$W_A \times 0.6 + W_{AB} \times 300 = (W_A/g) \times V_A^2 / 2 + (I_C)_{AB} \times \omega_{AB}^2 + (I_C)_B \times \omega_B^2$$

Therefore, $\omega_{AB} = 0.94985$ rad/sec

since, $\omega_B = (2.4/2g) \times (3240 + 2700 + 12960)$

and $V_A = 1.8 \times 0.94985 = 1.7$ m/sec

Example 11

A wheel of 1m radius rolls freely with an angular velocity of 5rad/sec and angular acceleration of 4rad/sec^2, both clockwise. Determine velocity and acceleration of points B and D, where D is at 0.6m and line AD makes an angle of 60^0 with X-axis.

Solution :

Motion of B and D is split into two parts – translation and rotation.

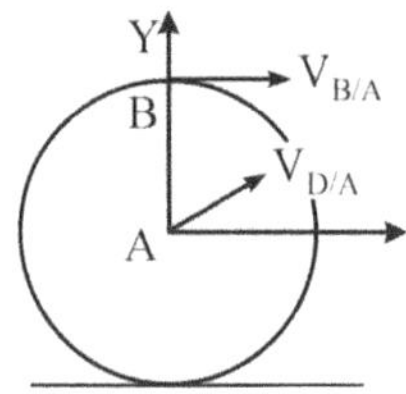

During translation, velocity $V_A = r \times \omega = 1 \times 5$ m/sec

and acceleration $a_A = r \times \alpha = 1 \times 4$ m/sec^2

(a) During rotation of point B about A,

Linear relative velocity, $V_{B/A} = AB \times \omega = 1 \times 5$ m/sec

Resultant linear velocity, $V_B = V_A + V_{B/A} = 5 + 5$ m/sec

Tangential acceleration, $(a_{B/A})_t = r \times \alpha = 1 \times 4$ m/sec^2

Normal acceleration, $(a_{B/A})_n = V^2_{B/A} / r = 5^2/1 = 25$ m/sec^2

Then, acceleration $a_B = \sqrt{\left[a_A + (a_{B/A})_t\right]^2 + (a_{B/A})_n^2}$

$$\sqrt{(4+4)^2 + 25^2} = 26.249 \text{m / sec}^2$$

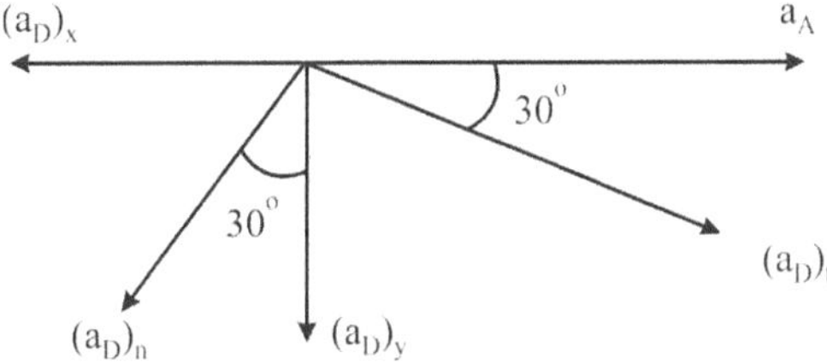

Its inclination with the horizontal, $\theta = \tan^{-1}(25/8) = 72.260$

(b) During rotation of point D about A,

Linear relative velocity $V_{D/A} = AD \times \omega = 0.6 \times 5 = 3$ m/sec

Therefore, linear velocity $(V_D)_X = V_A + V_{D/A} \times \sin 60$

$$= 5 + 3 \times \left(\sqrt{3/2}\right)$$

$$= 7.598 \text{ m/sec}$$

$$(V_D)_Y = V_{D/A} \times \cos 60 = 3 \times (1/2) = 1.5 \text{ m/sec}$$

Resultant linear velocity,

$$V_D = \sqrt{(V_D)_X^2 + (V_D)_Y^2} = \sqrt{7.598^2 + 1.5^2}$$

$$= 7.745 \text{ m/sec}$$

Its inclination with the horizontal, $\theta = \tan^{-1}(1.5/7.598) = 11.170$

At point D, we have three components of acceleration, simultaneously existing.

- Tangential acceleration of D, due to rotation about A,

$$(a_D)_t = r \times \alpha = 0.6 \times 4 \text{ m/sec}^2$$

- Radial inward acceleration, $(a_D)_n = V^2_{D/A} / r = 32/0.6 = 15 \text{ m/sec}^2$

and - Linear acceleration of A, $a_A = r \times \alpha = 1 \times 4 \text{ m/sec}^2$

Taking horizontal and vertical components of the three accelerations,

$$(a_D)_x = (a_D)_t \times \cos 30 + a_A - (a_D)_n \times \sin 30$$

$$= 2.4 \cos 30 + 4 - 15 \sin 30 = -1.422 \text{ m/sec}^2$$

and

$$(a_D)_y = -(a_D)_t \times \cos 60 - (a_D)_n \times \cos 30$$

$$= 2.4 \cos 60 + 4 - 15 \cos 30 = -14.19 \text{ m/sec}^2$$

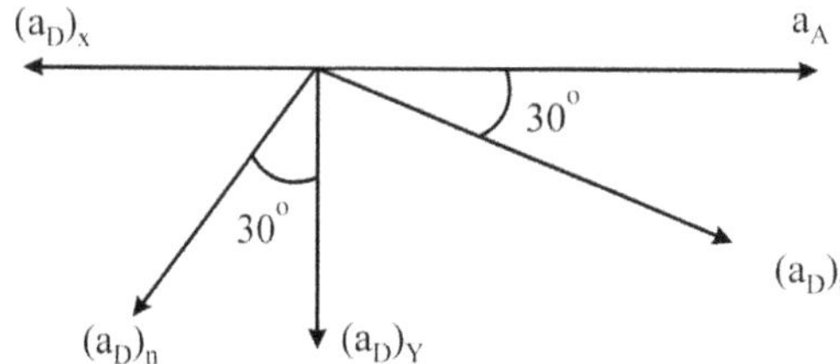

Then, acceleration

$$a_D = \sqrt{(a_D)_X^2 + (a_D)_Y^2} = \sqrt{(-1.422)^2 + (-14.19)^2} = 14.261 \text{m/sec}^2$$

Its inclination with the horizontal, $\theta = \tan^{-1}[(a_D)_X / (a_D)_Y] = 84.277°$

Example 12

A slender ladder AB of length 3 m has its ends A and B constrained to move in contact with the horizontal floor and the vertical wall, respectively as shown. Determine the velocity and acceleration of end B, if end A has a velocity of 2 m/sec and acceleration of 1.6 m/sec^2 towards left.

Solution :

Motion of B is split into two parts –

1. translation as rigid body

and 2. rotation w.r.t. A.

Tangential acceleration,

$$(a_B)_t = r \times \alpha = 3\,\alpha \text{ m/sec}^2$$

Normal acceleration,

$$(a_B)_n = V^2_{B/A}/r$$

$$= 2.31^2/3 = 1.778 \text{ m/sec}^2$$

Since B is moving down,

Horizontal component of acceleration,

$$(a_B)_X = -a_A + \text{Downward velocity}$$

$$V_B = V_A \times \cot 60 = 1.555 \text{ m/sec}$$

Relative velocity of B w.r.t. A, $V_{B/A} = V_A/\sin 60 = 2.31$ m/sec

$$(a_B)_t \times \cos 30 - (a_B)_n \times \sin 30 = 0$$

$$\Rightarrow \quad \alpha = 0.958 \text{ rad/sec}^2$$

$$V_{B/A} = r \times \omega = 3\omega \quad \Rightarrow \quad \omega = 0.77 \text{ rad/sec}$$

Vertical component of acceleration,

$$(a_B)_Y = -(a_B)_t \times \sin 30 - (a_B)_n \times \cos 30 = -2.9766 \text{ m/sec}^2$$

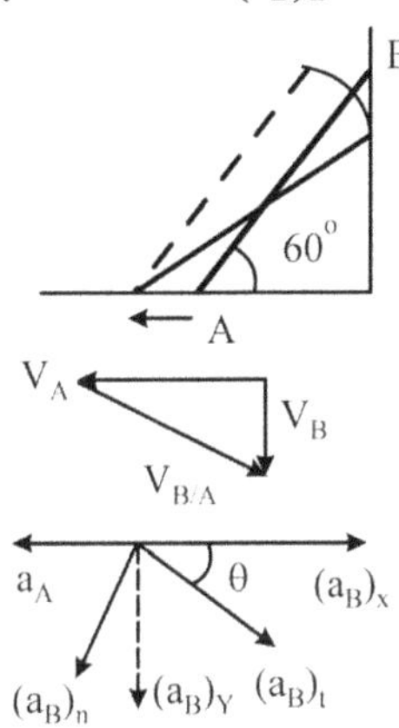

www.ingramcontent.com/pod-product-compliance
Ingram Content Group UK Ltd.
Pitfield, Milton Keynes, MK11 3LW, UK
UKHW052227270726
14060UKWH00004B/645

9 789391 910891